The Political Economy of the Middle East

Series Editor
Ashraf Mishrif
Centre for Middle East & Mediterranean Studies
King's College London
London, UK

This series explores the nature of Middle Eastern political regimes and their approaches to economic development. In light of the region's distinctive political, social and economic structures and the dramatic changes that took place in the wake of the Arab spring, this series puts forward a critical body of high-quality, research-based scholarship that reflects current political and economic transitions across the Middle East. It offers original research and new insights on the causes and consequences of the Arab uprisings; economic reforms and liberalization; political institutions and governance; regional and sub-regional integration arrangements; foreign trade and investment; political economy of energy, water and food security; finance and Islamic finance; and the politics of welfare, labor market and human development. Other themes of interest include the role of the private sector in economic development, economic diversification, entrepreneurship and innovation; state-business relationships; and the capacity of regimes and public institutions to lead the development process.

More information about this series at
http://www.palgrave.com/gp/series/14415

Giacomo Luciani • Tom Moerenhout
Editors

# When Can Oil Economies Be Deemed Sustainable?

palgrave
macmillan

*Editors*
Giacomo Luciani
Graduate Institute of International and Development Studies
Geneva, Switzerland

Paris School of International Affairs
Sciences Po
Paris, France

Tom Moerenhout
Columbia University
School of International and Public Affairs
New York, NY, USA

ISSN 2522-8854 ISSN 2522-8862 (electronic)
The Political Economy of the Middle East
ISBN 978-981-15-5727-9 ISBN 978-981-15-5728-6 (eBook)
https://doi.org/10.1007/978-981-15-5728-6



This Palgrave Macmillan imprint is published by the registered company Springer Nature Singapore Pte Ltd.
The registered company address is: 152 Beach Road, #21-01/04 Gateway East, Singapore 189721, Singapore

# Acknowledgements

This book originates from discussion in a workshop held in the context of the Gulf Research Meeting of August 2018 at Cambridge University (UK), supported by the Gulf Research Center Cambridge. We are grateful to GRC Cambridge and its Chairman, Dr. Abdulaziz Sager, for the generous support of the annual Gulf Research Meeting, that has allowed us to convene the seminar and, more broadly, has given extraordinary impulse to scholarship on the Gulf in multiple disciplines. Thanks also to the GRC team, notably Dr. Christian Koch, Dr. Oskar Ziemelis and Ms. Sanya Kapasi, who validly supported us as well as all other workshop conveners.

We are also grateful to Prof. Ashraf Mishrif for inviting us to publish in Palgrave's "Political Economy of the Middle East" series that he is the editor of.

We are indebted to Prof. Robert Springborg for many useful comments that allowed us to improve the manuscript and for his support to open access publication.

Finally, we are very grateful to the Swiss National Foundation for the generous grant that has made publication of this volume in Gold Open Access possible. Tom Moerenhout also expresses his gratitude to the Swiss National Foundation for funding his research

At a moment in history when the world is entering a global recession widely seen as the worst in a century, the issue of sustainability of the oil economies is more urgent than ever: we hope a wide public will find our reflections of use.

April 2020 Giacomo Luciani & Tom Moerenhout

# Contents

# Notes on Contributors

**Ali Al-Saffar** is the Middle East and North Africa programme manager at the International Energy Agency (IEA), leading the IEA's engagement with the region and acting as a key advisor to the executive management. Prior to this role, he was an energy analyst at the Directorate for Sustainability, Technology and Outlooks, where he co-authored seven editions of the IEA's flagship long-term forecast, the *World Energy Outlook*. He was the lead author of the *Outlook for Producer Economies* report, which focused on the issue of how the changing global energy landscape could impact oil-dependent economies. Prior to joining the IEA in 2012, he was a Middle East economist and chief automotive analyst at the Economist Intelligence Unit in London. He holds degrees in Economics and Development Studies.

**Joerg Beutel** is Professor of Economics and Environmental Sciences at Konstanz University of Applied Sciences in Konstanz, Germany. He served as Dean of the Department of Economics and Social Sciences from 1991 to 1998 and retired in 2008 from the university. He is still active in research and as economic advisor. He conducted major projects for the European Commission, Eurostat, United Nations Statistics Division (UNSD) of the United Nations, the Ministry of Economy and Planning (MOEP) of Saudi Arabia, the Prime Minister's Office of Brunei Darussalam (JPKE) and the Kuwait Institute for Scientific Research (KISR). He was the editor of the *Eurostat Manual of Supply, Use and Input-Output Tables* (2008) and became a member of the editorial board for the *Handbook of Supply, Use and Input-Output Tables* *with Extensions and Applications* (2018) of the United Nations. Main activities in research are development planning, macroeconomics, national accounts, and input-output analysis. He was engaged in applied economics throughout his professional career.

**Ishac Diwan** is Professor of Economics at Paris Sciences et Lettres (a consortium of Parisian universities) where he holds the chair of the Economy of the Arab World. He currently teaches at the Ecole Normale Superieure in Paris, and has held in recent years teaching positions at Columbia University, School for International Public Affairs, and at the Harvard Kennedy School. His research focuses on the political economy of the Middle East. His recent (co-authored) books include *A Political Economy of the Middle East* (2015); and *Crony Capitalism in the Middle East* (2019). He is also widely published on issues of global finance, macroeconomic and development strategies.

**Ibrahim A. Elbadawi** was the Minister of Finance and Economic Planning, Sudan, from September 2019 to July 2020; Managing Director, The Economic Research Forum for the Arab World, Iran and Turkey (since January 2017, currently on leave since September 2019); before that he was Director at the Economic Policy and Research Center, the Dubai Economic Council (2009–2016); Lead Economist at the Development Research Group of the World Bank, which he joined in 1989; and Professor of Economics at the University of Gezira in Sudan. He holds a Ph.D. in economics and statistics from North Carolina State and Northwestern universities in the USA. Elbadawi is the editor of *Understanding and Avoiding the Oil Curse in Resource-rich Arab Economies* (2016, with Hoda Selim) and *Democratic Transitions in the Arab World* (2016, with Samir Makdisi).

**Bassam Fattouh** is Director of the Oxford Institute for Energy Studies and a professor at the School of Oriental and African Studies, University of London. He specialises in international oil pricing systems, OPEC pricing power, security of Middle Eastern oil supplies, and the dynamics of oil prices and oil price differentials. He has published a variety of articles on oil and gas topics and his publications have appeared in academic and professional journals. Fattouh served as a member of an independent expert group established to provide recommendations to the 12th International Energy Forum (IEF) Ministerial Meeting for strengthening the architecture of the producer-consumer dialogue through the IEF. He also acts as an adviser to governments and industry, and is a regular speaker at international conferences. Fattouh focuses his research on aspects of the international oil pricing system such as the relationship between the futures market and spot market, the relationship between OPEC and the market, the causes of oil price volatility and the dynamics of oil price differentials. He also focuses his research on the IOC-NOC relationship and its implications for investment behaviour. He has a strong background in the economic environment of the Middle East.

**Martin Hvidt** is an associate professor at the Center for Contemporary Middle East Studies, University of Southern Denmark. He received his Ph.D. degree from Odense University and Colorado State University, Fort Collins, in 1995 and was in 2019 bestowed the prestigious Dr. phil degree by University of Southern Denmark based on a higher doctoral dissertation titled "Social and Economic Development in the Contemporary Arab Gulf States". For the past 25 years, his research has been related to these four broad topics/projects: (1) Irrigation Management in Egypt; (2) The Middle East and North Africa Region in a Globalized World; (3) Dubai: A Middle Eastern Developmental State; and (4) Gulf Rising. Developmental Patterns in the Resource Rich Economies of the Arab Gulf Region. Hvidt is the author of several books and a long range of articles and book chapters.

**Jim Krane** is the Wallace S. Wilson Fellow for Energy Studies at Rice University's Baker Institute for Public Policy in Houston. He specializes in energy geopolitics, with a focus on oil-exporting countries and the challenges they face from energy subsidies, internal demand, and climate change. His scholarly articles have been published in *Nature Energy*, *Energy Policy*, *Energy Journal*, *MRS Energy and Sustainability*, and the *Bulletin of the Atomic Scientists*. Krane spent nearly 20 years as a journalist, six of them in the Middle East. He is the author of two books. His acclaimed 2009 volume *City of Gold: Dubai and the Dream of Capitalism* is widely recognized as the seminal work on the iconic city-state, while his award-winning 2019 book *Energy Kingdoms: Oil and Political Survival in the Persian Gulf* is the definitive study of energy demand in the Gulf region. Krane was a long-time correspondent for the Associated Press and has written for myriad other publications including the *Washington Post*, *Wall Street Journal* and *Financial Times*. He is the winner of several journalism awards. He holds his Ph.D. from Cambridge University.

**Giacomo Luciani** teaches at the Paris School of International Affairs, Sciences Po, and at the Graduate Institute of International and Development Studies in Geneva. During 2010–2013 he was Princeton Global Scholar at the Woodrow Wilson School of Public and International Affairs and the Department of Near Eastern Studies.

His work has focused on the political economy of MENA. With Hazem Beblawi, he edited the book *The Rentier State* (1987), which is frequently cited as one of the origins of the concept. His latest edited book, *Combining Economic and Political Development* (2017), discusses economic policies to support democratic transitions.

**Samir Makdisi** is Professor Emeritus of Economics and Founding Director of the Institute of Financial Economics, 1984–2007 and Senior Fellow since 2008, the American University of Beirut (AUB). He is a former Minister of Economy and Trade, Republic of Lebanon, 1992; Deputy President of AUB, 1992–1998; and Chair of the Board of Trustees, Economic Research Forum, 1993–2001; and since 2012 an elected member of the Board. He has served as advisor to the Lebanese and other national governments and to various regional and international economic and financial organizations and has given lectures at major universities in the US, Europe, Asia and the Middle East. Recipient of several honors including the AUB medal (1998), he is widely published in academic journals and books in the areas of financial policies, civil conflicts and democratic transitions, among others, with special reference to the Arab World. He is the author of *The Lessons of Lebanon: The Economics of War and Development* (2004) and co-editor (with Ibrahim Elbadawi) of *Democracy in the Arab World: Explaining the Deficit* (2011) and *Democratic Transitions in the Arab World* (2017). He holds his Ph.D. in Economics from Columbia University, New York.

**Monica Malik** has over 25 years of experience as an economist specialising in the MENA region. Malik holds the position of Chief Economist at Abu Dhabi Commercial Bank, where she established and manages the Economic research team. Previously, she was the Chief Economist at EFG Hermes and was the lead person in the Economics team. Before this, she was the Senior Economist for the MENA region at Standard Chartered, Dubai, and at Dun & Bradstreet, London.

Malik has authored a number of academic books and articles on economic development in the GCC. She has presented at a number of high profile conferences and participated in policy round table discussions, as well as appearing regularly in the international media. She holds a Ph.D. in Economic Development in the Middle East focusing on private sector development in Saudi Arabia from the University of Durham.

**Tom Moerenhout** is Professor of International and Public Affairs at Columbia University and a resident scholar at the Center for Global Energy Policy. He also works with the World Bank's Energy and Extractives team and the International Institute for Sustainable Development. In recent years, Moerenhout has provided practical, in-country support to energy and development policy reforms in Lebanon, India, Nigeria, Egypt, Iraq and Iraqi Kurdistan. His research focuses on the sustainability dimension of economic globalization and energy sector reforms in developing countries. His work on energy subsidies and taxation and the social contract in oil and gas producing countries has been published extensively. He recently helped set up a COVID-19 stimulus tracker to verify energy policy decision in economic recovery packages.

**Thirumalai Nagesh** has been a member of the ADCB Economics team for the last three years. Prior to ADCB, he worked with research teams at RBS and Deutsche Bank in India. Nagesh has a total of ten years of experience in macro research, covering emerging market economies (India, GCC, Turkey) and oil.

**Anupama Sen** joined the Oxford Institute for Energy Studies in November 2009. She is Executive Director of the Electricity Programme and Senior Research Fellow. Her research interests lie in the applied economics of energy in developing countries and her research has spanned the oil, gas and electricity sectors, focusing largely on non-OECD Asia. In addition to OIES papers, her work has appeared in peer-reviewed academic journals and professional publications, as well as in several book chapters and Op-Eds. Sen is a Fellow of the Cambridge Commonwealth Society and has been a Visiting Fellow at Wolfson College, Cambridge. She holds a B.A. (Hons) from the University of Mumbai, M.Sc. from the London School of Economics and Ph.D. from the University of Cambridge.

**Manal Shehabi** is the OIES-KFAS Supernumerary Research Fellow at the Oxford Institute for Energy Studies. She is also an SCR Member of St. Antony's College, University of Oxford. An applied economist, she has expertise in economic modeling, energy, policy assessment and design, and resource and economic sustainability in resource-rich economies, with a focus on the Middle East and North Africa. Her research made important contributions to the analysis of economic adjustments and policy alternatives in resource economies facing volatility and energy transitions, specifically policy reform in economic diversification, labor reform, energy subsidy, alternative energy sources, and impacts of response measures to climate change. Combining evidence-based policy assessment with local political economy analysis, her research contributed to key policy debates at national levels and at the international level (such as the WTO and the UNFCCC). She also taught university-level courses in economics and conducted capacity building for policy makers in various countries. Shehabi is the author of various academic publications and policy reports, including being the co-author of *Kuwait Energy Outlook* and a contributing author to the *Intergovernmental Panel on Climate Change* (IPCC) *Special Report on the Ocean and Cryosphere in a Changing Climate*. A polyglot, her professional experience beyond academia spans economic research, consulting, energy multinationals, and public policy in the US, the Middle East, Europe, and Australia.

# List of Figures

# List of Tables

# List of Boxes

CHAPTER 1

# Introduction: Economic Sustainability of Oil Economies

*Giacomo Luciani and Tom Moerenhout*

## 1 What Is Sustainable?

When can oil economies be deemed sustainable? This complex question has been profusely debated in the last 5 years. The timing of this debate is no coincidence. On the one hand, the pace of environmental degradation has aggravated, in terms of air pollution, carbon emissions and the impacts of the planet's warming. On the other hand, volatile international oil markets and the acceleration of the energy transition have challenged the notion that oil revenues are sufficient to sustain oil economies in the near to medium term. With current challenges predicted to pursue in the near future, a more in-depth discussion about sustainability is warranted.

G. Luciani
Graduate Institute of International and Development Studies, Geneva, Switzerland

Paris School of International Affairs, Sciences Po, Paris, France
e-mail: giacomo.luciani@graduateinstitute.ch

T. Moerenhout (✉)
Columbia University, School of International and Public Affairs, New York, NY, USA
e-mail: tm2794@columbia.edu



G. Luciani, T. Moerenhout (eds.), *When Can Oil Economies Be Deemed Sustainable?*, The Political Economy of the Middle East, https://doi.org/10.1007/978-981-15-5728-6_1

The debate on sustainability in general, and on economic sustainability in particular, suffers from lack of conceptual clarity. What is the meaning of economic sustainability? As technology and the availability of goods and services constantly evolve, no economy is sustainable in a stationary state. Every economy, enterprise and individual needs to constantly evolve and adapt to changing circumstances. What is sustainable today may not be sustainable tomorrow: in the debate about sustainability, a concept of adaptability is implicit. As we cannot predict the future, and economies are constantly subjected to unforeseen shocks, we do not know what will be sustainable tomorrow, or for how long. Oil economies are confronted with this exact challenge. The current reform drive begs the question: What reforms should really be driven?

Adaptability is a function of many things. It is often considered as a function of diversification. Per se, economic logic rather favors specialization in one's comparative advantage than diversification. The increasing importance of global value chains for development emphasizes this economic logic, by moving competition from entire sectors to single stages of production, and even individual jobs. However, changing circumstances (movements in terms of trade) can be unfavorable for the given specialization. Economic history provides several examples of the rapid demise of specific industries due to technological progress or exhaustion of natural resources. In some cases, economic actors whose specialization has been challenged have successfully reinvented themselves, adapting more or less radically to a new specialization; in other cases, this has not succeeded, and economic actors have decayed and disappeared.

The pressure on many Gulf countries is high, but it is not equal as Gulf Cooperation Council (GCC) countries are heterogeneous. Some countries still hold a very strong comparative advantage in oil and gas production, so why should their governments not tailor their economies to this sector and approach their own diversification in the context of the value chain of petroleum products? One might argue that countries with large resource reserves and small populations could then simply accept, for the time being, that price shocks will happen periodically. Other GCC states, however, are faced with declining reserves and sizeable populations. It is in this context, with the additional drivers of technological advancements and environmental unsustainability, that it is more urgent to discuss diversification. As they accelerate diversification of value added and revenue, the pressure is on to invest in those sectors and stages of production that can yield results for some time to come.

## 2 What Is Unsustainable?

More than understanding what is sustainable, we may more easily know what is not sustainable. Adaptability, after all, is also a function of the ability to identify unsustainable trends early on, and correct them before very large costs are incurred. Current trends may be clearly unsustainable if extrapolated into the future: this is especially the case with respect to environmental impact and some socioeconomic trends such as rising inequalities and poverty. Thus, one way to discuss sustainability is to point to aspects, which are clearly unsustainable. All economies display such traits that are unsustainable; the question is whether they can be corrected on time before they cause excessive costs. It is very likely that some costs will need to be borne anyhow, before issues are corrected; the question is: what type of costs and how large can they be, before they become acutely "unsustainable"?

The lack of economic sustainability and the lack of diversification in oil economies in the Gulf have been discussed for decades and have often been equated. They have also featured in many previous development visions. In recent years, however, and in part due to the oil price collapse and a changing international energy landscape, diversification has become a more pronounced objective. But it appears that the debate on diversification suffers from path dependency, lacks specificity and still insufficiently takes into account the particular political economy constraints of GCC countries.

The first challenge, for example, lies in its definition: diversification of what? Of value added? Of exports? Of revenue? Quite clearly, it is the latter that lies at the basis of the post-2014 obsession with diversification, but a sole focus on revenue can undermine a fruitful discussion about what is possible in terms of policy-making on the medium term. A focus on what is acutely unsustainable would be a significant contribution to this debate on economic sustainability.

## 3 This Book

The chapters in this edited volume document discussions held in the context of the 2018 Gulf Research Meeting held by the Gulf Research Center at the University of Cambridge (UK). It starts from the acknowledgment that there is no widely accepted definition and measure of sustainability. All authors however try to capture not only political and economic factors, but also social and cultural influences on economic reform in GCC countries.

While authors have argued in total autonomy from each other and some divergent opinions remain, the thrust of the book is to conclude that some GCC economies have made significant progress toward diversification, reducing exclusive reliance on oil with respect to both composition of GDP and exports. This book also investigates a number of pronounced economic sustainability challenges in the Gulf's oil producers, in terms of not only the threats to fiscal balances, but also the nationalization and privatization of labor markets, environmental pressures on GCC countries and soaring income inequalities within Gulf countries.

This edited volume has four parts, discussing various facets of economic sustainability. In a first part, authors provide a holistic discussion of current trends in, and projections about, the sustainability of oil economies in the Gulf. The second part of this edited volume discusses trends in fiscal sustainability, given the quest, and need, of governments in the Gulf to diversify not just the economy, but especially their revenue base. These chapters also tie in fiscal reforms to the goal of Gulf economies to diversify and to adjust labor market structures. The third part addresses labor market policies and labor market reforms. The fourth part discusses strategies of oil economies toward international climate action. This fourth part misses an assessment of environmental consequences of climate change that can threaten Gulf economies. The editors consider it essential that the content of this volume is complemented by such assessments.

## 4 A Holistic, and Realistic, Picture

In Chap. 2, Giacomo Luciani frames the debate on economic sustainability of oil economies. He questions the popular notion that Gulf economies are not sustainable by discussing the progress on diversification as well as the continued relevance of hydrocarbons. Highlighting that the oil rent is likely to shrink over time, he points out the core challenge of developing taxation and fostering the development of an independent and competitive private sector. He also emphasizes the need for further labor sector reform and the challenges of income inequality.

In Chap. 3, Ishac Diwan argues that the current focus in Saudi Arabia of replacing a rentier economic model with economic diversification in tradables is unrealistically ambitious. Rather, he proposes a Norway-like model where public sector jobs are complemented with private sector employment in middle productivity and high-wage jobs. He discusses the challenges in labor market reform and the trade-off between slow adjustment and large-scale financing of new investment to support this transition.

In Chap. 4, Ali Al-Saffar explores the outlook for producer economies. He discusses current risks and emerging pressures to producer economies, arguing that uncertainties in oil demand and supply have strengthened the need for further diversification. He reviews the results of two scenarios toward 2040, one in which oil demand continues to grow and another where oil demand tapers off. He concludes by pressing the role of a more efficient and productive energy sector in the Gulf's economic reform initiatives.

In Chap. 5, Bassam Fattouh and Anupama Sen argue that the broader characteristics of the current energy transition toward low carbon energy sources will be more important than when oil demand will eventually peak. They expect that the diversification strategy of exporters will be conditioned by the speed of that transition, and that during the transition, the oil sector will continue to play a key strategic role. In reverse, also the pace of the global transition will be influenced by the speed and success of economic diversification in oil economies.

In Chap. 6, Joerg Beutel reveals the usefulness of an input-output approach to better understand the growing importance of intermediates when discussing economic diversification. He compares the performance on economic diversification and sustainable development of GCC economies with a reference case (Norway). He concludes that GCC countries have performed well on diversification, but less so on sustainable development, which includes, among others, the depletion of natural resources and environmental pollution.

In Chap. 7, Manal Shehabi redefines economic sustainability in resource-dependent states by offering a new definition that accounts for the economic challenges of resource dependence. She elaborates on the changing and elusive definition of economic sustainability, and then suggests an alternative definition that centers around volatility and depletion. She discusses the specific policies that successful growth-sustaining policies would imply, such as the regulation of oligopolies and human capital development.

## 5 Fiscal Sustainability

In Chap. 8, Tom Moerenhout discusses fuel and electricity pricing reform as core tenets of Gulf reform programs. He argues that pricing reforms were successfully used post-2014 to avoid an even more menacing collapse of state budgets. While future reforms will still be needed, he warns for a potential double-edged sword. If the extent and pace of pricing reforms is too ambitious, then it might erode the comparative advantage of industrial consumers and as such complicate efforts to achieve (within sector) diversification.

In Chap. 9, Monica Malik and Thirumalai Nagesh look at fiscal sustainability and the types of fiscal reforms (including the introduction of value-added tax) that have been implemented following the oil price collapse in 2014. They argue that those reforms have been deeper than earlier cycles, but that it remains difficult for many GCC governments to implement a multi-year fiscal reform process given the pressure that his would cause on the private sector and the population.

## 6 Labor Market Reform

In Chap. 10, Monica Malik and Thirumalai Nagesh discuss the relevance of population dynamics and labor markets to fiscal capacity. They elaborate on the opposition between the existing social contract and fiscal labor market needs in the area of reducing public sector employment. While noting that the greater burden of fiscal reforms has fallen on the expatriate population, they argue that an evolution in the social contract remains essential as governments seek to create private sector jobs for the youth population and downsize the public sector.

In Chap. 11, Martin Hvidt applies a value chain perspective to the question of economic diversification and job creation in oil-producing countries in the Gulf. He suggests that such a value chain perspective can provide more clarity on whether newly created jobs are in the lower or higher value-added segment. He suggests that Gulf countries should aim at creating jobs in the high-value segment, as this will create a virtuous process including the upskilling of the labor force. It will also require a readjustment of the current reliance on expatriate labor.

## 7 Environmental Sustainability

In Chap. 12, Jim Krane analyzes how oil export-dependent states have adopted three near-term strategies to protect oil rents and revenue. A "dig in" strategy aimed at insulating the oil industry from climate risk and undermining global action on climate change. A "join in" strategy aimed at rational domestic energy policies and with greenhouse gas reduction benefits. And a "throw in" strategy seeks to frame mitigation efforts as more expensive than climate change damages. He provides a case study on Saudi Arabia's response to global climate action.

## 8 Sustainability of GCC development

In Chap. 13, Ibrahim Elbadawi and Samir Makdisi review the development experiences of the six countries of the Gulf Cooperation Council, starting from the observation that they have been able to avoid the most explicit consequences of the resource curse. They highlight that the unusual high rents per capita have resulted in a developmental and political equilibrium. The authors conclude that the dependence of GCC countries on such high rents casts a shadow over the future sustainability of this equilibrium, especially under a new emerging global oil order, which will likely be characterized by lower oil prices.

In Chap. 14, Giacomo Luciani and Tom Moerenhout conclude. They underline that no unequivocal and final answer to what determines the sustainability of oil economies exists, but that, equally, not all oil economies should be stereotyped as unsustainable.

If one overall message can be distilled from this book, it is that we should move on from the simplistic branding of the Gulf economies as unsustainable, and tackle the details of which adaptations they might need to undertake. In this approach, developing fiscal tools to greatly reduce reliance of recurrent expenditure on the oil rent, and moving away from a model of low-cost, low-productivity expatriate labor to a model of higher wages and productivity emerge as the two decisive challenges. Both have direct implications for political institutions, which consequently will also need to adapt.

CHAPTER 2

# Framing the Economic Sustainability of Oil Economies

*Giacomo Luciani*

## 1 Introduction: My First Visit to Abu Dhabi

I first visited Abu Dhabi in the early 1980s, traveling with an economist colleague and friend. As we were driving into town, I was pleasantly surprised at the greenery along the streets (which was a priority for Sheikh Zayed, the Amir) and the well laid-out urban environment (which has become more chaotic in the intervening decades), and positively commented on it. Yet my friend objected: "But you see, Giacomo, this will all revert to dust as soon as oil is finished". That was also the time when *The Economist* was advising Kuwait not to invest in the country, as it would be futile anyhow, and the prevailing opinion was that Saudi Arabia would never succeed in establishing a petrochemical industry.

The stereotype according to which Gulf oil economies are not sustainable continues to be widely maintained. The original fear that "oil at some

G. Luciani (✉)
Graduate Institute of International and Development Studies, Geneva, Switzerland

Paris School of International Affairs, Sciences Po, Paris, France
e-mail: giacomo.luciani@graduateinstitute.ch



G. Luciani, T. Moerenhout (eds.), *When Can Oil Economies Be Deemed Sustainable?*, The Political Economy of the Middle East, https://doi.org/10.1007/978-981-15-5728-6_2

point will run out" has lately been replaced by the expectation that oil will lose all economic value because of global warming and "unburnable carbon", but the essence of the narrative has not changed.

The unsustainability mantra is rooted in the poor understanding of economics and the way markets work. There has never been in contemporary history a country that has reverted to dust. Economies are constantly adapting to changing circumstances, and no economy is sustainable in its steady state if no change is allowed. The same is true for political systems: some systems are more flexible and capable of adapting gradually, while others are rigid and may cause forfeiting potential growth for a period of time—but eventually change comes about. In some (too numerous) cases, resistance to change, or the inability to agree on the governance of it, has led to violent conflict entailing extensive losses of capital and human life. In some such cases, however, the economic recovery process has been swift and has opened the door to phases of relative prosperity.

This chapter elaborates on what sustainability actually entails. I devote the second section to the simple but frequently forgotten truth that no economy is sustainable without change, thus putting the case of the Gulf oil economies in a broader global perspective; the third section is devoted to the importance of identifying the needed change in order to be able to discuss the appropriate development strategies. While the standard line is that the Gulf oil economies have not diversified and "remain too dependent on oil", I argue in the fourth section that measurable and significant diversification has taken place. While further diversification in the composition of GDP and exports is certainly needed, an ongoing process is underway in this direction.

The fifth section discusses the standard economic threat that resource-rich countries must cope with, that is, the so-called Dutch disease. The Gulf oil countries have quite successfully dealt with it through a strategy of investment in infrastructure and almost unlimited importation of expatriate labor. The latter has caused its own problems, social and economic—notably a pattern of declining productivity, which is discussed in Sect. 6. This pattern must be reversed because, in the longer run, it is incompatible with a growing GDP per capita. This, I argue, is by far the most important challenge that these economies must now face.

In the seventh section, I discuss excessive dependence on hydrocarbons in the generation of aggregate value added, and argue that it is not an intractable problem. The Gulf economies have abundant opportunities to decarbonize domestic energy consumption as well as the production and transformation of hydrocarbons. Systematically embracing the capture of

$CO_2$ to avoid emissions in the atmosphere, and using captured $CO_2$ in enhanced oil recovery (EOR), possibly including also the use of imported $CO_2$ captured elsewhere in the world, makes the production of negative-emissions hydrocarbons technically possible (albeit expensive and currently uneconomic). The progressive transformation of crude oil and gas into non-fuel products or hydrogen, and the possible advent of a hydrogen economy, may eventually eliminate emissions from burning hydrocarbons directly.

It is therefore quite possible that hydrocarbons will retain economic value for decades to come, while their use evolves in ways that make it compatible with maintaining global warming within the limits mandated by the Paris Agreement and supported by climate science. This, however, would certainly entail an increase in the cost of production and correspondingly a reduction of the net rent accruing to the treasury of the major producing countries. In the eighth section, I argue that, combined with growing population, gradually declining access to rent implies that states whose financial wherewithal continues to be highly dependent on the oil rent must urgently develop fiscal tools to extract revenue from the domestic economy. This entails the progressive reversal of the rentier state paradigm, whereby the state economically supports society, to a more common situation of society paying for the state. This reversal is unlikely to happen without concomitant evolution of the political order away from patrimonial absolute monarchies to a situation in which, while monarchies may well remain in place, some form of accountability will need to be enforced. Hence, if one assumes that no political evolution is possible, then surely the Gulf oil states are fiscally unsustainable—but no political order can endure unchanged in the face of changing circumstances.

The need to raise taxes mandates a degree of separation between the State and its taxed subjects. Taxes cannot be raised exclusively or primarily on government employees and state-owned enterprises: a genuinely independent and competitive private sector must develop, which is capable of walking on its own legs and not depend solely on government procurement. As discussed in the ninth section, government policies toward the private sector have so far allowed for the accumulation of large fortunes but have fallen significantly short of truly empowering the private sector. Priority for state control of strategic assets has hindered privatization of all but marginal activities, and obsession for control and loyalty to the regime prevents the emergence of a truly entrepreneurial bourgeoisie.

Finally, social and political cohesion requires that greater attention be paid to income distribution. In the absence of any income or wealth taxation, oil-fueled economic growth has disproportionately favored a minority of citizens, and, after an early period characterized by significant social mobility, all opportunities now seem reserved for the elite. In the last section, I thus discuss the aspect of social sustainability, meaning that the development path must be such that some opportunities for upward mobility are offered also to the many that have been left behind.

## 2 No Economy Is Sustainable Without Change

Economic life is a process of continuous creative destruction, and no economy is sustainable if it does not continuously adapt. Changing consumer demand—and especially evolving technology and the constant introduction of new products—imposes continuous adaptation on all economies. No economy is sustainable unless it adapts continuously to changing circumstances.

Sustainability is therefore closely related to adaptability. Existing comparative advantage may be lost and new comparative advantage acquired: it all depends on investment decisions, based on starting circumstances. Protecting existing competitive advantages or acquiring new ones is more of an art than a science, and it is normal that some competitors will lose ground while others will gain. The losers may eventually disappear, if we are talking about individual companies; but countries do not disappear.

Countries whose economy loses competitiveness will experience a worsening of their trade balance. If not compensated by an inflow of foreign capital or other transfers, their currency will eventually depreciate relative to other currencies. In recent times, many countries have experienced dramatic devaluations until they found a new equilibrium in external accounts.

It should be noted that the Gulf countries enjoy trade surpluses in most years, and their imports include non-essential and non-productive items such as notably weapons systems. They also host large numbers of expatriate workers who generate large flows of remittances, which burden the respective balance of payments (and contribute to inflating imports). That said, the possibility of allowing a devaluation of the respective national currencies has repeatedly been considered (and rejected) in the Gulf Cooperation Council (GCC) countries. Devaluation would discourage imports and improve the competitiveness of some local products.

Devaluations restore competitiveness but damage the purchasing power of all wages paid in the local currency. Hence, countries that lose ground in competitiveness generally experience lower growth in income per capita, while countries whose competitiveness is improving experience higher growth in income per capita. Differential growth in income per capita is a common experience, and major changes have taken place since the end of World War II in the relative position of individual countries—generally in the sense that countries that had lower income per capita have been able to reduce the gap separating them from countries with the highest income per capita. Thus, it may be said that the latter have lost competitiveness, but no one would conclude that they are not sustainable.

The question of sustainability for the GCC economies therefore means being able to adapt to changing circumstances, preserving past gains in income per capita for their citizens, and possibly further closing the gap with the richest countries.

## 3 Identifying Needed Change

In this discussion, it is important to exactly identify the nature of the problem in order to suggest the right policy decisions. The economies of the GCC countries are certainly not sustainable if the assumption is that no change will take place—all economies must continuously adapt. Thus, the mantra that GCC economies are not sustainable has limited use in guiding expectations and policies unless the exact nature of the required change is identified.

The literature oscillates between different definitions of non-sustainability, and these do not necessarily converge in a reasonably clear policy agenda. Is the problem linked to the high share of GDP that is generated by the oil and gas sector? Or to the dominant share of oil and gas in total exports? If we conclude so, we should say that sustainability is linked to improved economic diversification. Or is it linked to the belief that oil and gas may lose economic value in a world that needs to rapidly decarbonize? In this case, sustainability may rather be linked to investment to preserve the economic value of hydrocarbons in a decarbonizing world.

Is non-sustainability linked to excessive dependence on expatriate labor combined with unemployment among the nations' youth? Is it related to the phenomenon of declining productivity of labor, which is closely related to dependence on expatriates? In this case, the issue should primarily be viewed as one of reforming labor policies.

Or is non-sustainability primarily linked to government finance—the fact that governments have not developed a system of taxation that may progressively wean them from excessive dependence of a possibly declining source of revenue, that is, hydrocarbon rents? In this case, the question should be primarily one of making sure that a taxation base is created and the required imposition system established. It is perfectly possible for the economy to remain competitive and for the state to be unable to extract sufficient revenue from it with appropriate taxes.

## 4 Diversification Has Taken Place

Contrary to the frequent assertion that the Gulf economies have not diversified as evidenced by their continuing high reliance on oil and gas as shares of both GDP and exports, statistics tell us that significant value-added diversification has taken place. This is hardly surprising for any observer that is aware of the impoverished and extremely simple status of the economies in question before the advent of the oil era.

The discussion of diversification in oil economies is complicated by the impact of variable oil prices. Whenever prices are high, value added from oil and gas upstream is also increased, more so than value added in the rest of the economy, and it will appear that the economy is losing diversity. Whenever prices are decreasing, the opposite will be the case, and it will show that the economy is now more diversified. Higher oil prices are a good thing for oil-producing countries, notwithstanding the fact that they will appear to be less diversified. In fact, it is not just appearance: the higher prices are a reality and reduced diversification an inevitable consequence. That said, it does not make sense to discuss diversification trends unless price variations are also taken into account. There are no other commodities as important as oil whose price is as volatile as that of oil, otherwise this would be a common problem for many countries.

We therefore need to resort to more complex measures of diversification, as extensively discussed in the chapter by Joerg Beutel in this volume. Although even the more sophisticated indices aimed at measuring the structural interrelations of an economy, such as those based on input-output tables, are not immune from reflecting changes in relative prices, their behavior tends to be more stable. Beutel shows that several diversification indices do not tell us that the economies of the GCC countries, in particular Saudi Arabia, are much less diversified than the comparator country Norway.

A converging conclusion is arrived at if we use the Economic Complexity Index (ECI) developed by Ricardo Hausmann and César Hidalgo at

Harvard and MIT (Hausmann et al. 2014). The index is based on the mapping of what the authors call the products space, in which each product category is mapped in proximity of other products requiring similar technology and know-how, meaning that the producer of one product could relatively easily shift to producing a proximate product. This creates an image in which some products, mostly primary products, appear in isolated positions at the margins, while others, mostly manufactured products, are located in several interlocking clusters at the center. The economic complexity of a country is:

> a measure of the knowledge in a society as expressed in the products it makes. The economic complexity of a country is calculated based on the diversity of products a country exports and their ubiquity, or the number of the countries able to produce them (and those countries' complexity). Countries that are able to sustain a diverse range of productive know-how, including sophisticated, unique know-how, are found to be able to produce a wide diversity of goods, including complex products that few other countries can make. (CID, Atlas, Glossary)

The Economic Complexity Index ranks countries based on how diversified and complex their export basket is. Countries that are home to a great diversity of productive know-how, particularly complex specialized know-how (and including know-how originating from abroad but attracted and implemented locally), will be able to produce a great diversity of sophisticated products.

Table 2.1 ranks key Gulf exporters' ECI scores for 1996 and 2016. The table includes several comparator countries, to highlight the extent to which superficial impressions may be misguided.

There are several interesting points to be noted in this table. To begin with, all oil or commodity exporters, including advanced countries, have low economic complexity because of their relative specialization. Thus, the USA ranks lower than one may expect of the leading techno-economy in the world and will probably show a further decline in the coming years, thanks to Donald Trump's push for "energy dominance" (White House 2019; DOI 2017). It is striking that Mexico is ranked as more complex than Canada, and both are more complex than Norway, which is normally viewed as the success case *par excellence* among oil exporters. Russia and India are rated barely more complex than Saudi Arabia and the United Arab Emirates. The latter two are doing better than New Zealand, and all

**Table 2.1** Ranking of Gulf countries and selected comparator countries by Economic Complexity Index

| | *2016* | | *1996* | |
|---|---|---|---|---|
| *Country* | *Rank* | *ECI value* | *Rank* | *ECI value* |
| United States | 10 | 1.55 | 7 | 1.868 |
| Mexico | 21 | 1.11 | 25 | 0.817 |
| Canada | 35 | 0.696 | 23 | 0.898 |
| Norway | 39 | 0.638 | 32 | 0.592 |
| Russian Federation | 48 | 0.235 | 38 | 0.420 |
| India | 49 | 0.191 | 56 | 0.007 |
| Saudi Arabia | 50 | 0.171 | 65 | −0.153 |
| United Arab Emirates | 51 | 0.162 | 80 | −0.476 |
| New Zealand | 54 | 0.124 | 42 | 0.313 |
| Oman | 71 | 0.292 | 82 | −0.510 |
| Kuwait | 73 | 0.314 | 89 | −0.666 |
| Qatar | 76 | −0.396 | 109 | −0.955 |
| Australia | 86 | −0.592 | 60 | −0.026 |
| Iran | 87 | −0.611 | 106 | −0.915 |

Source: Center for International Development, Harvard University, *Atlas of Economic Complexity*, http://atlas.cid.harvard.edu/.

Gulf Cooperation Council (GCC) countries are doing better than Australia, while Iran does just a bit worse. We normally do not worry about the sustainability of any of the comparator countries (except maybe Russia)—so the Gulf economies may be unsustainable, but lack of economic complexity does not appear to be the reason for it.

Between 1996 and 2016, all Gulf countries, including Iran, significantly improved their ECI position, while all comparator countries except Mexico lost ground. So diversification is indeed happening, and economic complexity is increasing.

As well as other measures of diversification, the ECI is vulnerable to changes in oil prices and quite unstable for major oil exporters—for example, Saudi Arabia ranked 36th in 2004, a year of low oil prices, and 104th in 2008, a year of peak oil prices. However, oil prices were relatively low in both years compared here (1996 and 2016), so the improvement is clearly not just due to changes in oil prices. Furthermore, oil prices affect all oil exporters, so that GCC improvement relative to other oil exporters must be real. It remains true, however, that low oil prices sustain diversification and high oil prices hinder it. At times of growing oil prices, diversification may well be taking place, but it will be obscured by the inflation of value added in the oil sector. When prices decline again, diversification becomes visible.

Some further comments are in order. First, the structural transformation of an economy takes time. Global oil demand may peak in 20 years (at a level above today's), but there is no credible prospect of oil losing economic significance for at least the next 50 years (more on this in Fattouh and Sen's chapter in this book). Diversification may become more challenging after the early successes (it is difficult to gain in diversity when you are diversified to begin with), but we should certainly expect further improvement in the ranking of the Gulf oil exporters. Some massive investment projects that have been undertaken in the past 10 years (such as the Sadara petrochemical j.v. between Saudi Aramco and Dow, the aluminum project going from mining bauxite to rolling mills developed by Maaden in association with Alcoa, or Borouge's new plants in Ruwais) have yet to fully impact export statistics, because of their long gestation and the progressive ramp-up of production.

Second, there are major differences between Gulf oil exporters. The gap in 2016 between Saudi Arabia and Iran is significant. We do not have an ECI for Iraq, but it would surely be much worse than Iran's. And the gap between the Gulf economies and oil exporters elsewhere in the world is even larger: Nigeria ranked 125 (out of 127 countries), Azerbaijan 120, and Angola 116. Not all countries are doing equally well, and relatively few can be said to have made real progress in diversification.

My conclusion is that GDP and export diversification are not good reasons to conclude that the Gulf oil export economies are unsustainable. Of course, the global economy is constantly evolving. Demand, supply, and terms of trade of individual products change constantly, and adaptation is a never-ending task; however, the Gulf economies are diversifying and are today much more adaptable and competitive than they were three or four decades ago. Diversification is happening and likely to continue, progressively reducing dependence on the oil sector—even if oil remains a valuable and important internationally traded commodity.

## 5 Dutch Disease and Expatriate Labor

By "Dutch disease" we mean the appreciation of the real exchange rate as expressed by the relative terms of trade of non-tradable relative to tradable goods and services. The change in the relative terms of trade is caused by the emergence of new natural-resource-based exports, which leads to an increase in revenue and domestic expenditure. Because the supply of tradables is elastic, the price is established by international competitive forces and is not expected to change. In contrast, supply of non-tradables is limited in the short term, and increased demand will lead to an increase in their average price. So non-tradables become more expensive than tradables (which is the definition of appreciation of the real exchange rate), and capital is incentivized to move from the production of tradables to that of non-tradables. If a country has a developed tradables sector at the beginning of the process, it may de-industrialize: this is what was feared for the Netherlands when large resources of gas were discovered in the Groningen field. Typically, non-tradables consist of real estate, personal and hospitality services and maybe perishable food products, while tradables consist of all industrial products.

In the short run (defined as the time that it takes for a new investment to take place and increase productive capacity), there is nothing that can be done to prevent the Dutch disease, except refraining from spending the additional revenue from natural resources. This is what Norway has done, but no poor country would reasonably be expected to do the same. The narrative of the Dutch disease is therefore almost invariably supported by empirical observation in the aftermath of a natural resources bonanza: housing and services become very expensive, local craft industry and frequently also agriculture are wiped out, and real estate investment becomes the only game in town. The Gulf countries in the 1970s fully conformed to this pattern.

But in the longer run, when investment can take place and change the economy's production possibility frontier, the real appreciation of the exchange rate may be countered. This can be achieved by expanding the production of non-tradables, thus countering the expected increase in relative prices (Sachs 2007), and importing labor to overcome the limitations in the supply of domestic, especially skilled, workers. The Gulf countries adopted both strategies early on: the state stepped in to quickly increase the supply of non-tradables, and the door was flung open to essentially unlimited importation of expatriate labor (yet with discrimination based on the country of origin).

Coupled with very weak protection of workers' rights, this has meant that the Gulf countries have imported the wage level of their poorer neighbors in Middle East and North Africa (MENA) or South Asia. The end result has been that wage-push inflation has been contained, but dependence on expatriates has reached levels not seen anywhere else in the world. Most nationals are employed by the government, but just a few are employed in the private sector in white-collar occupations. Essentially all blue-collar jobs are filled by expatriates, but a significant number of foreigners, mostly skilled, also work for the government.

Is this a sustainable state of affairs? While there will always be foreigners willing to take employment in the Gulf, notwithstanding the sometimes oppressive conditions, such high reliance on expatriates creates significant problems. Expatriates are mostly interested in remitting their incomes to their families and countries of origin; they are very conscious of their precarious state and are reluctant to "take roots" and invest in their host countries, even when this is allowed—and normally it is not. Thus, much of the aggregate wage bill exits the local economy almost immediately, meaning that the Keynesian multiplier is very low.

Secondly, expatriates have an income target and are likely to react to income taxation by simply requesting higher wages or returning to their countries of origin. The latter has been confirmed by the response to increases in fees enacted after the decline in oil prices from 2014. As there is considerable turnaround in the expatriate population, and the wages on offer already are set at the minimum level required to attract them, newcomers will focus on net income if taxes are imposed. This means that taxes on expatriates will end up being paid by employers. In the absence of policies to consolidate the expatriate population (policies which all Gulf governments abhor), foreign residents offer an intermediate, not an ultimate, taxable base.

Thirdly, the abundant supply of low-wage workers is distorting the competitiveness of industry. Investment is attracted in labor-intensive and low-skill activities simply because it is so cheap to hire workers—but these lines of production do not conform to the underlying comparative advantage of the Gulf countries, which is in capital-intensive industries requiring more sophisticated skills. Therefore, their longer-term commercial viability is uncertain to say the least.

## 6 The Productivity Issue

Considering the functioning of the labor market and the almost complete openness to importation of expatriate labor, it is not surprising to see that labor productivity, that is, value added per worker employed, has been declining throughout the Gulf economies, in contrast with the experience of almost all other countries in the world (Hertog 2018). The private sector has primarily aimed at minimizing the cost of labor, and has been ready to accept lower skills and qualifications for the sake of lower wages. This has mostly taken the form of shifting hiring practices as in between countries of origin, because the minimum reservation wage of expatriates is frequently related to the average wage in the country of origin.

In economies where the workforce is stable, it is normal to expect a progressive improvement in skills simply due to the permanence in the same job for long periods of time. But the high turnaround rate, which is characteristic of expatriate labor in the Gulf, works in the opposite direction, undermining learning by doing, and actually facilitating the recourse to progressively less skilled laborers, if they come cheaper.

When labor is cheap, there is little incentive to invest in machinery that may improve its productivity. Operating a machine normally requires at least a modicum of training and higher skills, thus improving productivity, which opens the door to higher wages. It also makes the turnaround of workers more difficult, because some retraining might become necessary. Therefore, the attitude has been rather in favor of keeping things simple and basic, avoiding productivity-enhancing investment.

In fact, the cost of labor is so low, and the hassle of recruiting from abroad so significant, that many employers indulge in labor hoarding. Because expatriate workers sometimes desert their employers, either because they somehow find a better job—although this may be against the law—or simply give up and return home, employers want to make sure that they have a certain number of "spare workers" at hand, just in case. This is a recipe for permanently declining productivity.

The meaning of declining productivity is that economic growth is achieved exclusively, thanks to adding to the number of expatriates. While this may allow for expanding the share of GDP accruing to profits in the private sector, it is hardly compatible with improving economic conditions for salaried nationals. Only nationals that possess capital and employ expatriates can benefit from increasing profits. But profits are difficult to tax, and the creation of a taxation base will remain an elusive task.

An economy in which labor productivity constantly declines is therefore not sustainable, because it deprives the state of an easily taxable base. Sustainability requires moving away from extreme reliance on expatriate labor. Labor-saving investment is necessary, coupled with acceptance of progressively higher wages. This in turn will reduce the gap between the wages of nationals and those of expatriates, thus creating conditions for overcoming the paradox of high unemployment among nationals, while at the same time so many jobs are taken by expatriates. It may also open the door to the creation of a taxable income base.

## 7 Potential for Further Diversification and Defense of Hydrocarbons' Value

While the Gulf countries have, comparatively speaking, made good progress in diversifying their economies, the potential for further diversification exists, specifically in the direction of developing industrial activities that may allow the continuing use of hydrocarbons, even while eliminating greenhouse gas emissions and countering climate change. The frequently held opinion according to which only zero use of hydrocarbons may be compatible with zero $CO_2$ emissions may turn out to be unfounded. There are many ways in which the Gulf oil and gas producers may contribute to decarbonization while continuing to use oil and gas.

### *7.1 Decarbonize the Production of Oil and Gas*

Growing attention is being devoted to the fact that the production of oil and gas is not equally emission-intensive across regions and fields (Masnadi 2018; IEA 2018a, 73–76). Focusing on the abatement of emissions connected to upstream and midstream activities is a significant strategy in the context of fighting global warming. In the case of gas, special attention has been attracted to reducing or eliminating flaring and venting. In the case of oil, energy requirements connected with fracking for shale oil or use of other enhanced oil recovery methods, and the heat required for the production of heavy oils and oil from bituminous sands have negative consequences for the environment even before the oil or gas is burned. To a large extent, the heat and power for oil and gas production, which is now derived from burning hydrocarbons, could be derived instead from renewable or clean sources. In Oman, the Miraah solar thermal power plant produces steam that is injected in the nearby Amal oil field, and at 1021 MW of capacity, it is one of the largest solar power plants in the world (IEA 2018a, 76; PDO 2017). Steam-based enhanced oil recovery is not very common in the prolific fields of the Gulf, but is likely to be increasingly in use as fields mature and decline. Steam and high temperatures are also needed in refining and petrochemicals. Electric power is needed in field operations as well as in transportation by pipeline and in refining and petrochemicals—and it could be derived from renewable or clean sources. While burning hydrocarbons inevitably produces $CO_2$, it may be possible to produce zero-$CO_2$ hydrocarbon fuels in the sense that no $CO_2$ needs to be emitted in the production process.

### 7.2 *Carbon Capture and Sequestration*

The next logical step is engaging in systematic carbon capture and sequestration (CCS) in the oil and gas fields. The oil industry has extensive experience of capturing $CO_2$ and injecting or re-injecting it in oil and gas fields (IEA 2016; McGlade 2019). This also facilitates the production of oil and gas (KAPSARC 2018), and is therefore either a method of avoiding venting $CO_2$, which is present in the field in association with methane, or of capturing $CO_2$ produced from other sources and sequestering it underground. It is conceivable that all $CO_2$ emitted from burning hydrocarbons in stationary plants may be captured and either used in other industrial uses or sequestered underground in oil and gas fields. If the Gulf oil and gas exporters are keen to preserve the economic value of their resources in a carbon-constrained world, carbon capture and sequestration should be their technology of choice.

In fact, CCS has attracted considerable scholarly interest (FT 2018b), and examples of implementation are slowly coming to the fore. In Abu Dhabi, the Al Reyadah project takes $CO_2$ from a steel plant owned by Emirates Steel and pumps it to the Rumaitha and Bab oilfields, where it is injected in the ground for enhanced oil recovery (Gulf News 2016). This is just a first, but eventually systematic capture from all plants (power plants first and foremost—at least until oil and gas-fired power plants will still be in use) should be envisaged.

But considering the centrality of CCS for allowing some continued reliance on fossil fuels, the level of engagement of the oil-producing countries (and major international oil corporations) in pursuing deployment of this technology is patently insufficient. Notwithstanding the keen interest in CCS demonstrated by the International Energy Agency and some European oil-producing countries (the UK and Norway) (IEA 2018b), and notwithstanding the multiple studies that have been carried out by research institutions notably in Saudi Arabia, there is not a single major project in the process of being implemented.

The consequence is that public and even expert opinion does not consider CCS as a viable alternative for global decarbonization. The special report of the IPCC issued in 2018 (IPCC 2018) strongly insists on the need for CCS and negative emissions down the road, and the major producing countries could very well position themselves as potential suppliers of storage services for $CO_2$, including that which may be captured in other countries that might not have appropriate geological sites for this purpose. True, for the time being we do not have a price for carbon sufficient to justify the economics of capturing, transporting, and storing $CO_2$, but hydrocarbon producers need to think strategically and anticipate the market if they wish to convincingly argue that there is a future for their products in a world that needs to rapidly decarbonize

### *7.3 Further Integrate Downstream Toward Non-fuel Uses*

A parallel defense strategy consists in emphasizing the use of hydrocarbons to produce products that are not meant for burning as fuels. The palette of oil products has always included lubricants and bitumen, but the major case in this category clearly is that of petrochemicals.

Demand for petrochemicals is growing rapidly in the world, much more rapidly than GDP, and there is little expectation that it may slow down in the coming decades (IEA 2018c). The petrochemical industry is sometimes mistaken for a low-technology industry, and little more than a producer of plastic bags polluting the oceans—but this is a caricature. While production of basic petrochemicals may rely on proven technology, and indeed yield simple products that may damage the environment if not properly recycled (plastic bags and the like), the industry is extremely diversified and increasingly sophisticated, producing an incredible array of materials that find use in almost all manufacturing.

Although polymers can be produced from organic material rather than from fossil feedstock, the prevalence of the latter is unquestionable. Size and plant synergies being essential to competitiveness and product diversification, it is difficult to envisage that the petrochemical industry might move away from reliance on fossil hydrocarbons. To the extent that use of fossil fuels might decline, oil will tend to be cheaper, and petrochemical materials ever more attractive in a wide range of utilizations. Saudi Aramco's and SABIC's effort to develop a crude to petrochemicals refinery (which would essentially produce no fuels) evidently is an early manifestation of this strategy (Arab News 2018).

### *7.4 The Hydrogen Economy*

The final, decisive strategy to preserve the economic value of fossil fuels in a decarbonizing world is to separate them into hydrogen and $CO_2$, capturing the latter and using the former as fuel. Hydrogen is an energy carrier, not a primary energy source, and can be produced starting from several molecules containing it. The perspective of a hydrogen economy has been discussed for decades (IEA 2019), and is normally considered part and parcel of decarbonization strategies, especially in conditions when excess electricity might be available (from excess non-dispatchable renewables or nuclear) for electrolysis of water. The concept of a hydrogen economy thus is relevant even independently of the desire to preserve the economic value of fossil fuels; yet it has considerable synergy with the latter, because, on the one hand, hydrogen is easily produced from light hydrocarbons with steam reforming, a well-established technology("Natural gas accounts for around three-quarters of the annual global dedicated hydrogen production" IEA 2019, 38), and, on the other hand, hydrogen can substitute for methane in pipeline networks and can be used as transportation fuel in fuel cells—a technology that has considerable advantages in battery electric vehicles for extended range driving.

Oil and gas producers could then position themselves as pioneers of the hydrogen economy, through either exporting hydrogen (presumably in liquid form, unless export pipelines are available), or transforming hydrocarbons into hydrogen close to the final consumer and taking responsibility for the capture, sequestration, or utilization of $CO_2$.

Considered as a whole, these four interlocking strategies for maintaining the economic relevance of hydrocarbons in a decarbonizing world envisage the further profound transformation of the economic structure of the Gulf oil countries. This transformation has not even started yet, and it is not clear that it will be pursued at all. It requires a degree of anticipation and forward looking that the existing governance structures may not be able to provide. Obviously today, the reality of relative prices—notably the absence of a sufficiently high price for carbon and consequent lack of $CO_2$ valorization—does not support massive investment in the direction envisaged. The question remains whether the major Gulf oil producers should not anticipate future global decarbonization policies in view of demonstrating that it may not be necessary to move away from hydrocarbons, rather to change the use that is made of them.

## 8 Need for Developing Taxation

Even in the perspective that I sketched in the previous section, it is clear that decarbonizing the production of hydrocarbons or turning them into hydrogen for clean burning as fuel will entail significant additional cost, which will go to reduce the size of the rent. In other words, the oil and gas business will come to more closely resemble all other lines of business, where a very significant profit margin is sometimes possible, but not such a huge rent as to allow to almost completely pay for the state from this source alone.

Developing additional sources of revenue requires time and progressive strengthening of the administrative capabilities of the state. Some taxes are easier to collect than others, but the easiest ones—excise taxes on specific categories of goods or taxes on imports—are not sufficient to diversify state revenue to the extent needed. In fact, the Gulf oil-producing countries have recently introduced excises on several goods such as tobacco or sugary beverages, but fuels, which are subjected to large excises in many countries, are made available at relatively low prices (or "subsidized"), and alcoholic beverages—another frequent target of excises—are either restricted or prohibited altogether.

The GCC countries have decided to implement VAT starting at a low level of 5%, but only Bahrain, Saudi Arabia, and the UAE did so. Saudi Arabia further increased VAT to 15% in July 2020. VAT is a tax on consumption which is regressive, because higher incomes consume proportionately less, and weighs more on nationals than on expatriates, because the latter are likely to remit or consume abroad a larger share of their income. Other things being equal, a VAT will tend to increase the level of domestic prices and erode the purchasing power of all fixed income earners.

In the end, the imposition of taxes on income and/or wealth for individuals and/or corporations will be inevitable. Taxing corporations may be politically less problematic and administratively simpler, but will meet resistance from domestic entrepreneurs. Subsidiaries of foreign corporations are likely to have ways to shift income to other jurisdictions, or compensate taxes paid locally with deductions in their home countries or other jurisdictions. Local corporations will not enjoy the same financial agility.

Taxes on personal income will meet with the greatest resistance, and may trigger demands for accountability of power holders and calls for more democratic institutions. In countries where semi-democratic institutions exist, such as Kuwait, the probability of gaining the support of the parliament to enact income taxes is realistically close to nil. Furthermore, as the vast majority of nationals is employed by the state, imposing a tax on their salaries would be equivalent to reducing the same. The net gain would be limited to the revenue that the tax may raise out of income from investment or from salaries of employees of the private sector.

Taxing private sector salaries would contrast with the strategic priority that must be attributed to encouraging nationals to take employment in the private sector. It would however also encourage expatriates to ask for higher wages and eventually possibly help in reducing the wage gap between expatriates and nationals.

Imposing an income tax on expatriates only has been repeatedly considered and rejected. Imposing all wage earners and then offering exemptions or other benefits to nationals employed in the private sector may be a more productive approach. In any case, a personal income tax scheme would need to be strongly progressive to counter the growing inequality of incomes and wealth which is inevitable in the absence of redistribution mechanisms.

A strong argument may be put forward in favor of a wealth tax, both because it would at least conceptually be close to the traditional form of Islamic taxation, that is, zakat, and because a lot of wealth has been accumulated in the form of real estate, which could be easier to assess than financial or other forms of wealth. However, zakat already is levied, at least on paper, and the extent to which it might be increased or administered more effectively is uncertain.

## 9 Genuine Reliance on Private Sector

The need for nurturing a strong base for taxation is the main argument to conclude that economic sustainability eventually requires the consolidation of a competitive and independent private sector. The Gulf oil-producing countries have over decades implemented various converging policies to favor the birth of private entrepreneurs and allow them to amass very sizable fortunes, but the tendency always has been to consider the private sector as an appendix and a client (economically as well as politically) of the state. With the exception of a relatively short period of time in the very early years of the current century, when some of the GCC states, including most notably Saudi Arabia, seemed ready to accept a more active role of the local bourgeoisie (Luciani 2005), the state has mostly moved to limit the scope of private initiative in multiple fields that are considered strategic. There has been an almost total disjuncture between an official discourse that always insisted on the role of private enterprises, on privatization, on supporting small and medium enterprises; and the reality of a state profoundly suspicious and unappreciative of the qualities of the national bourgeoisie.

The latest episode in the difficult relationship between the state and holders of political power on the one hand and the entrepreneurs on the other has been the imprisonment of many leading Saudi business people in the Ritz Carlton Hotel in Riyadh and their significant dispossession. This process, which has been carried out with zero transparency and complete disregard of any accepted judicial procedure, has been justified as a cleanup of extensive corrupt practice; however, no specific case of corruption has been made public. The government authorities have repeatedly announced, as proof of success, the fact that assets supposedly worth 100 bn dollars have been sequestered: whether this is in fact the case or not is impossible to say, but the discouraging effect on private investment and initiative in the Kingdom is absolutely certain (FT 2018a).

A radical change of tack is required if the private sector is indeed to become the protagonist of the next phase of economic growth and provide the state with a taxable base capable of offering a source of income in alternative to the oil rent. The will to accept an arm's-length relationship between the state and private enterprises through adherence to the rule of law and promotion of a level playing field is currently not detectable. Under these circumstances, the state is postponing adaptation and again relying on income from oil, plus possibly other international financial placements. This may well work for a time, depending on oil demand and prices, and on how successful the investment decisions of sovereign wealth funds will turn out to be. It may work better for the very high rent per capita countries (Elbadawi and Makdisi in this volume) than for the others. For the latter, the danger is clearly that the shift in attitude may come very late in the day, only when the state will have lost its financial autonomy and will need to solicit the support of the private sector from a position of weakness.

## 10 Tackle Inequality

The combination of issues related to excessive dependence on expatriate labor, underdevelopment of fiscal tools, and lack of acceptance of a genuinely independent private sector results in a pattern of increasing inequality in the distribution of income and wealth which may translate into growing political discontent.

We have limited systematic information on the distribution of income and wealth in the GCC countries, including because of the absence of direct taxes, which deprives the state of crucial knowledge tools in this respect. But basic facts are very visible. We know that a majority of nationals work for the government, and their wages are not such that you can build a fortune with the savings. They are secure and guaranteed for life by generous retirement schemes that trigger early in life, but do not allow for much luxury. At the opposite extreme, we know that quite a few private entrepreneurs are independently assessed by Forbes to be billionaires or multibillionaires. Some members of the royal family are included in this list, but others would certainly belong to it if Forbes had the information to assess their wealth, or they cared to appear in it. At the same time, social mobility has stalled: stories of rags to riches were common in the 1970s, but are very unusual today. Only the scions of the wealthy families have a chance to occupy prominent positions in business or public service.

There is nothing exceptional in this trajectory. It has been well documented that income inequality is growing almost everywhere in the world (Atkinson 2015; Piketty 2013), and inheritance is again more important than earned income to define the economic status of the vast majority of individuals, even in more mobile and meritocratic societies. We also know that progressive income taxes are an essential tool to prevent ever-greater concentration of income and wealth, and the fact that taxes have become less progressive than they used to be is one of several causes of growing inequality.

Growing inequality is breeding discontent in many countries, and the dilemmas that this poses are a focus of growing political attention. The sustainability of inequality is a question mark for the future of many countries, and of democratic as well as authoritarian regimes. The Gulf oil-producing countries are no different—except that not much attention is paid to the issue.

Yet sustainability and inequality are very closely related. Sustainability requires reduced reliance on expatriate labor, increased national employment in the private sector, increased productivity and wages, more value-added creation in market conditions rather than in the government sector; in other words, more independent and economically active *citizens*. The political system would inevitably be impacted: will the transition be traumatic?

## References


Arab News (2018) Saudi Aramco, SABIC sign deal with Britain's 'Wood Group' to develop world's largest crude oil to chemicals project, Riyadh 11/03/2018.

Atkinson, Anthony n.d.(2015) Inequality: What Can Be Done, Harvard University Press, Cambridge, MA.

CID (n.d.) (Center for International Development, Harvard University), Atlas, Glossary, available at http://atlas.cid.harvard.edu/learn/glossary, last accessed 2019.06.26.

DOI (2017) 'Strengthening the Department of the Interior's Energy Portfolio', Order No. 3351, US Department of the Interior.

FT (2018a) Wealthy Saudis sit on cash as purge casts shadow over investment 24/07/2018.

FT (2018b) Scientists attempt to breathe new life into CCS 20/09/2018.

Gulf News (2016) Carbon capture company Al Reyadah launched in Abu Dhabi 05/11/2016.

Hausmann, n.d.Ricardo, César Hidalgo et al. (2014) The Atlas of Economic Complexity: Mapping Paths to Prosperity, MIT.

Hertog, n.d.Steffen (2018) "Challenges to the Saudi Distributional State in the Age of Austerity" in Madawi Al-Rasheed Salman's Legacy: the Dilemmas of a New Era in Saudi Arabia, Oxford University Press Oxford.

IEA (2016) 20 Years of Carbon Capture and Storage: Accelerating Future Deployment, Paris.

IEA (2018a) Outlook for Producer Economies—WEO Special Report, Paris.

IEA (2018b) IEA and UK kick-start a new global era for CCUS press release 28/11/2018.

IEA (2018c) World Energy Outlook 2018, Paris.

IEA (2019) The Future of Hydrogen: Seizing today's opportunities, Paris.

IPCC (2018) Global Warming of 1.5 °C Special Report October 2018, Geneva.

KAPSARC (2018) Decarbonizing Oil: The Role of $CO_2$-Enhanced Oil Recovery ($CO_2$-EOR), Riyadh.

Luciani (2005) "From Private Sector to National Bourgeoisie: Saudi Arabian Business" in Paul Aarts and Gerd Nonneman eds. Saudi Arabia in the Balance, NYU Press, New York.

Masnadi (2018) Mohammad et al. "Global Carbon Intensity of Crude Oil Production" in Science vol. 361 issue 6405.
McGlade, Christopher (2019) Commentary: Can $CO_2$-EOR really provide carbon-negative oil? IEA, Paris.
PDO (2017) Miraah Solar Plant Delivers First Steam Oilfield press release 11/2/2017.
Piketty, n.d.Thomas (2013) Le capital au XXI siècle, Seuil, Paris.
Sachs, Jeffrey (2007) "How to Handle the Macroeconomics of Oil Wealth?" in Macartan Humphreys, Jeffrey Sachs and Joseph Stiglitz eds. Escaping the Resource Curse, Columbia University Press, New York.
White House (2019) President Donald J. Trump Is Unleashing American Energy Dominance, fact sheet, May 19, 2019.

CHAPTER 3

# Fiscal Sustainability, the Labor Market, and Growth in Saudi Arabia

*Ishac Diwan*

## 1 Introduction

The drop in oil prices since mid-2014 has opened up a twin-deficit in the state budget and the balance of payment of Saudi Arabia. The specificities of the current shock are its extraordinary size (50 percent fall in oil prices), the fact that it comes on the heels of a long period of high prices (2000–2014), and the high probability that it will be permanent rather than transitory. As a result, the non-oil economy will have to be taxed to balance the books. But the implied tax rates are very high because of the narrowness of the non-oil economy and its dependence on the oil economy. Thus, unless the tax base grows over time, deep cuts in domestic consumption will have to be envisaged. But while the current shock brings more immediacy to the question of how to adjust to lower oil revenue, with a rising population, and oil revenue expected to fall over time as interest in climate change increases, the rentier mode of development followed by Saudi Arabia was clearly unsustainable to start with.

I. Diwan (✉)
Ecole Normale Supérieure, Paris Sciences et Lettres, Paris, France



G. Luciani, T. Moerenhout (eds.), *When Can Oil Economies Be Deemed Sustainable?*, The Political Economy of the Middle East,
https://doi.org/10.1007/978-981-15-5728-6_3

On this diagnostic, most analysts agree. But what should replace the current economic model? Here, pundits are deeply divided: some believe that the inadequacies of the current economic model will necessarily lead, sooner or later, to a doomsday scenario of instability and pauperization; others advocate a move from the current economy still dominated by hydrocarbons to a modern and diversified knowledge-based economy, OECD-style. What seems to be missing in the current discourse on Saudi Arabia is a more reasonable vision for the future, one that not only resolves the problem of non-sustainability of the current path but that, at the same time, would also be desirable and feasible.

This confusion in the economic discourse about the kingdom has its source in the unusual structure of its labor market. The steep dualism between expatriates and national workers is unique among large countries. The main problem with the current economic system is that under it, nationals are simply not employable in large numbers in the private sector. In this chapter, I argue that the main challenge to building a better future is to find ways to employ them productively. Doing so would greatly boost growth: indeed it may be the only source of growth that is currently readily available. It would also save on the foreign exchange now being remitted by expatriates abroad, reducing the need to produce tradables besides oil. But because the current division of labor between nationals and expats is deeply inscribed in the current social contract, the Kingdom will have difficulty in balancing the interests of businesses and society at large.

This chapter explores these issues in turn—the size of the current shock and the magnitude of the fiscal challenge, the need to find a realistic and effective new growth model, the challenge of increasing the national labor force participation, the implied policy challenges at the macro and corporate levels and, finally, the type of political economy constraints that will be encountered. Its value added is not so much with highlighting each of these issues, which have been somewhat explored in the past, as much as it is with its claim that addressing the current challenge requires a strategy that "normalizes" the Saudi economy. The discussion also highlights the main areas of economic and social behavior that require a much better understanding for effective policies to be devised in the years ahead.

## 2 Sizing-Up the Macro Shock

The first specificity of Saudi Arabia (KSA thereafter) today is the extraordinary size of the macro shock that has hit its economy. According to the IMF (2017, 2018), oil exports dropped by half, from around $300 billion/year during 2012–2014, to $140–160 billion/year during 2015–2018. Oil revenues that accrued to the state budget represented close to 90 percent of government revenues in 2014, and they have dropped from SAR 913 billion in 2014, to an estimated SAR 454 billion in 2017–2018.

By most accounts, the current depressed level of oil prices, which dropped from about $100/barrel during the decade that ended in 2010, to about $50 afterwards, is unlikely to fully recover in the medium term, given demand and supply projections. This means that there is a good chance that the old level of prices will never be recovered. Although this assertion can be debated, we will consider this hypothesis as our starting point. The implication is that the fiscal deficit resulting from this massive negative shock cannot be financed forever, and that there is therefore a need, sooner or later, to adjust the budget to the new level of oil prices.

The burden of the adjustment is made harder by the fact that by 2014, a year with record oil revenues, the budget was already running a (small) deficit. This is unlike the situation of the previous oil shock of the 1990s, when the kingdom was running a large surplus, and when the needed expenditure cutback was therefore more limited. It is also unlike the situations of most other GCC countries today (except Bahrain and Oman), which were running large surpluses in 2014, and where the negative shock only reduced this surplus, or turned it into a small deficit.

The size of the shock can be appreciated in several ways. In 2015–2016, the fiscal deficit was about SAR 400 billion/year (or 16 percent GDP), in spite of the initiation of some budget cutting measures. By 2017, the deficit was cut by nearly half, to about SAR 240 billion (9.3 percent GDP), largely through deep cuts in public investment, which were running at a record high level to start with. In addition, a long-standing taboo was broken with the introduction of several new taxes (excise, VAT), and the initiation of increases in the domestic prices of energy and water, sectors that have been heavily subsidized to date. By 2017, public investment has fallen to a third of its 2014 level, to 7 percent of GDP, and non-oil revenues rose from 4.6 to 7.5 percent of GDP (see Table 3.1). So, by the end of 2017, what has been achieved is already a large fiscal adjustment by

**Table 3.1** Central Government Budget 2014–2022 (SAR, billions)

| | | | | *Prelim.* | *Projected* | | | | |
|---|---|---|---|---|---|---|---|---|---|
| | *2014* | *2015* | *2016* | *2017* | *2018* | *2019* | *2020* | *2021* | *2022* |
| Revenue | 1044 | 612 | 519 | 649 | 757 | 851 | 982 | 1019 | 1053 |
| Oil | 913 | 446 | 309 | 455 | 454 | 463 | 479 | 500 | 519 |
| Non oil | 131 | 166 | 210 | 194 | 303 | 388 | 503 | 519 | 534 |
| o/w taxes | 38 | 41 | 54 | 63 | 116 | 155 | 187 | 194 | 199 |
| o/w energy/water | | | 20 | 35 | 81 | 119 | 198 | 203 | 210 |
| Expenditures | 1141 | 999 | 936 | 890 | 936 | 982 | 1023 | 1056 | 1077 |
| Wages | 335 | 394 | 439 | 412 | 422 | 433 | 444 | 455 | 466 |
| Goods and services | 286 | 229 | 229 | 234 | 236 | 239 | 244 | 250 | 252 |
| Interest | 4 | 3 | 5 | 11 | 17 | 26 | 32 | 35 | 36 |
| Subsidies | 45 | 42 | 54 | 60 | 75 | 81 | 86 | 89 | 91 |
| Investments | 471 | 331 | 209 | 173 | 186 | 203 | 217 | 227 | 232 |
| Deficit | −97 | −387 | −417 | −241 | −179 | −131 | −41 | −37 | −24 |
| Notes | | | | | | | | | |
| GDP | 2836 | 2454 | 2424 | 2596 | 2713 | 2795 | 2896 | 3018 | 3095 |
| o/w non oil GDP | 1615 | 1768 | 1797 | 1859 | 1966 | 2047 | 2136 | 2239 | 2308 |

Source: IMF article IV (2017)

global standards. But the adjustment so far has been relatively easy, as there is not more room left for reductions in the investment budget.

In spite of these considerable efforts, the fiscal deficit remained sizable at the end of 2017, as there was still about SAR 241 billion of excess of expenditure over revenues (9.3 percent GDP). Since the deficit has to be financed from the non-oil part of the economy, one gets a better sense of its magnitude by expressing it as a share of non-oil GDP. By this measure, it stood at 13 percent of non-oil GDP at the end of 2017, a large figure (down from 22 percent of non-oil GDP in 2015 and 2016). At the same time, the need to finance the deficit during the past 4 years has led public debt to rise from 1.6 to 19 percent GDP already, and external reserves to be reduced from $724 to $473 billion between 2014 and 2017.

The challenge ahead will be more difficult, as the type of expenditure left on the budget are more difficult to compress. Wages and compensation of public servants now approach 50 percent of expenditure, and interest payments have started to rise. Moreover, as new taxes and new tariffs start hurting poorer households, the need for some compensation will arise. The government is finalizing a cash transfer program (called the citizen account), which should start disbursing during 2018.

The government is thus left with only two choices in the short term: to increase revenues and to finance (the remaining) deficit.[1] The government has taken an aggressive stance at deficit reduction, signaling that it intends to increase taxes and cost-recovery rapidly. Whether this will happen, however, remains to be seen. To get a sense of the magnitude of the needed adjustment, government plans foresee to eliminate the deficit by 2020. To achieve this ambitious goal, tax revenues are projected to rise to SAR 200 billion by 2022 (VAT initially at 5 percent and rising afterwards, excise on tobacco and drinks, and higher expat fees), from a base of SAR 38 billion in 2014, and water and energy cost recovery are projected to rise by SAR 210 billion. In addition, non-tax revenues are expected to also rise due to an ambitious program of privatizations. This financing effort represents, therefore, about SAR 400 billion of additional revenues that will have to be paid by national households, corporations, and expatriate workers.

The important question is whether these expectations are realistic. The risk is that the planned increase in revenues meets serious social resistance and is rolled-back, and that the gap will continue to be filled up with debt. By some estimates, the kingdom can borrow abroad and sell assets to theoretically finance at least 10 years of deficits at the current level, before going bankrupt. This allows it to kick the can down the road rather than find more sustainable solutions, as has often happened to regimes built on oil (Karl 1997).

## 3 The Stabilization Challenge

To gage of the realism of the kingdom's fiscal objectives, it is important (but difficult) to get a sense of the incidence of the planned fiscal effort on the various parts of the population.

Expats do create a buffer for the economy but are unlikely to be part of the tax base. The predominance of expats in the construction sector (Fig. 3.1) meant that, unlike other countries, the recent and ongoing closing-down of major infrastructure projects has not created large unemployment among nationals, as instead, the expat labor force has shrunk (Mahroum 2016). One presumes however that expats are being paid their reservation wage, and while they can be fired at will, it is not possible to

[1] Ending the war in Yemen would also lead to savings—but these costs are not detailed in the government budget as presented in IMF and other public documents.

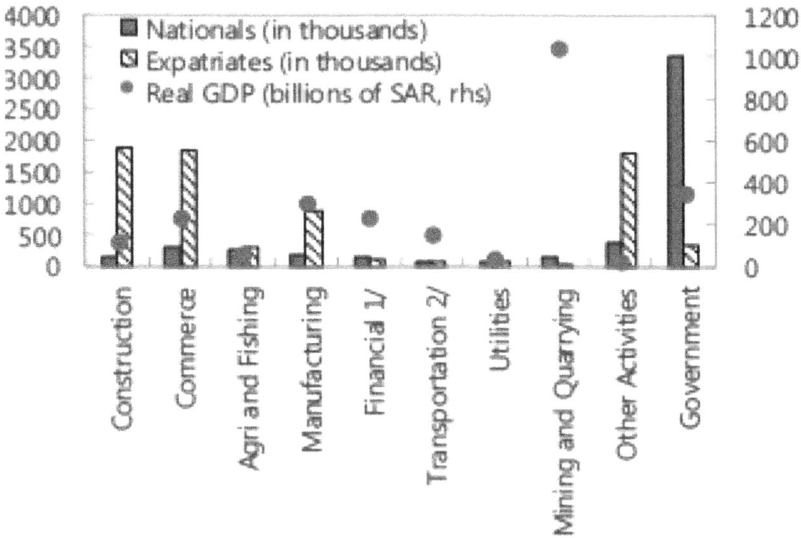


**Fig. 3.1** Employment and Real GDP by sector (2016). (Source: IMF Article IV (2017))

compress their net incomes by much and keep them employed. Thus, while several fees are levied on them already (work permit, residence permit, training charges), and while these fees are expected to rise more in the future (by a total 2.2 percent of GDP by 2022, according to IMF projections), one expects that such additional costs will have to be borne by their employers rather than through a compression of their net income.[2] Some more research on this issue, taking advantage of recent increases in these fees, will be useful to understand this central set of issues better.

The burden of taxation and cost recovery will, therefore, have to fall largely on Saudi households and firms. Concerning firms, there is little knowledge of how large their profits are—there is actually no such data, since corporate revenues are not taxed. Thus, it is not clear to what extent they would be willing to absorb the new taxes and higher input prices through reductions in their profits, as opposed to choosing to either close down their operations (especially for those in the traded sector), or to pass on the increased burden to consumers (especially for those in

[2] The same logic applies to the VAT and rise in utility tariffs that apply to them.

non-tradables). There are indications that, in the aggregate, profitability is on the high side: the non-oil corporate sector has been growing fast in the past decade, and its investment rates can be estimated roughly from the national account to be about 25 percent of non-oil GDP, a pretty high figure which suggests overall high levels of profitability. Large firms traded on the stock exchange, and listed in the Orbis database, tend to report large profits.[3] Moreover, the private sector as a whole has relatively high access to credit, as total credit to the private sector is reportedly at about 90 percent of non-oil GDP, again, a large figure by international standards.

But it seems plausible to think that the corporate sector is highly dualistic, with large corporations benefitting from a privileged access to the credit market, and facing less competitive pressures than smaller firms (Hanieh 2016). This would imply that large firms are able to absorb more losses than SMEs. It should be noted however that state ownership in the largest firms is quite sizable—directly at about 45 percent, and overall even more, given state stakes in banks and insurance companies that own another 15 percent of these firms (Coksun et al. 2018—see Fig. 3.2). This

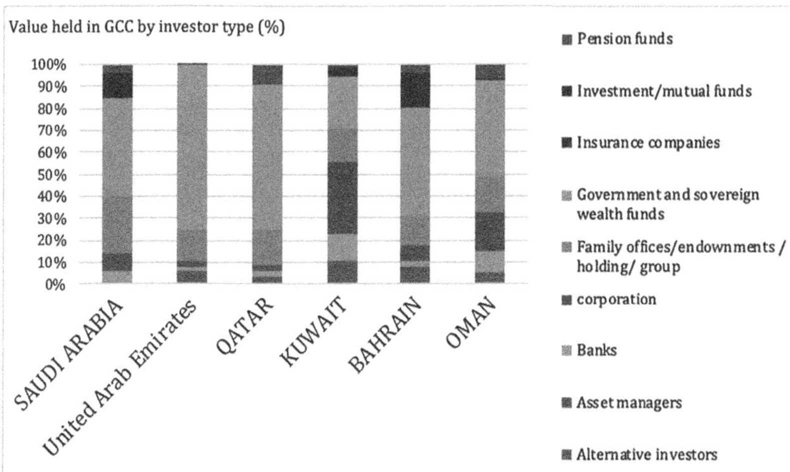


**Fig. 3.2** Ownership of traded companies in the GCC. (Source: Coskun et al. 2018)

[3] Excluding the petrochemical companies and energy utilities, their profit rates (relative to revenues) were of the order of 17 percent in 2014, falling to 15 percent in 2016 (IMF 2017). This can be considered high by international standards.

limits the net financial gains of passing the burden of adjustment on such firms. At the same time, there are indications that smaller firms operate in a highly competitive environment, and that they are already hurting from the deteriorating market conditions. More research is needed to understand in more detail the performance of the corporate sector and its ability to sustain more taxation, as this will help devise better policies in the years ahead.

There is a possibility of taxing wealth, such as bank accounts, and other forms of assets. There has been to date no indication that this may take place formally, although the episode of the Ritz-Carleton hotel suggests that this may take place informally, at least at the top of the corporate pyramid. There have been reports of large amounts of capital flight in recent years, which suggests that capital-holders are trying to shield their assets from potential taxation and state predation.

To get a sense of magnitude of the burden on households and corporations implied by the IMF projections, I will try below to heroically estimate a lower and upper bound aggregate tax rate. As described above, the new taxes announced are essentially a VAT and excise taxes, which will fall directly on households; higher expat fees, which indirectly affect profits and consumer prices; and increased utilities prices, which fall directly on consumers and firms, but that may end up falling more on consumers via their indirect influence on output prices.

One extreme assumption one can make is that all of the burden will fall on Saudi households. If that were the case, the burden would be unbearable. To get a feel of magnitudes, 70 percent of working Saudis work for government, and their total compensation in 2017 was SAR 412 billion (Table 3.1). Even if one accounts generously for income earned by Saudis in the private sector (which is unlikely to be more than a third of this amount), and from non-labor sources (such as return on savings and other assets), it is apparent that the burden of the extra taxes and utility costs will be a large proportion of their income—possibly more than half!

A lower bound would consider a larger tax base that includes corporate profits and self-employment earnings. Non-oil GDP stood at SAR 1.8 trillion in 2017. One can estimate based on ministry of labor data that total expats income is about SAR 600 billion. Taking this out leaves about SAR 1.2 trillion of non-oil national income, which, together with the public sector wage bill, adds to a total tax base of about SAR 1.6 trillion. With this larger tax base, a lower bound estimate of the average tax burden would stand at about 25 percent, a lower, but still quite a large burden.

We can conclude that the current financing goals are not realistic. For the burden of adjustment to become bearable in the medium term, non-oil GDP needs to rise in ways that boosts the tax base. More realistically then, a strategy to eliminate fiscal deficits that can work would be more gradual, and it would rely centrally on pro-growth reforms that increase the size of the tax base over time. As long as this does not happen, public debt is likely to continue to rise. There are moreover financial limits to the extent of gradualism, since both slow fiscal adjustment and growth-inducing reforms require financing. There is thus a central trade-off: if adjustment is too fast, it would be resisted socially; but if it is too slow, it would consume much of the fiscal space, leaving little financing resources available to invest sufficiently in new sources of growth.

## 4 What Growth Strategy?

The government has announced ambitious plans to boost private sector growth, including improvements in the business climate, in trade logistics, and in the application of the rule of law. In the IMF scenario—which is reportedly more conservative than government expectations—these reforms are expected to increase non-oil GDP growth by 3 points by 2020, for a total of 3.5 percent growth per year. This seems quite optimistic, given the large weight of government spending in fostering growth in the past; and it says little about the potential for medium term growth, because the growth model itself remains fuzzy and poorly defined. The question of interest is, therefore, what growth model can deliver a larger tax base? From a longer-term perspective, oil rents are not sufficient anymore to finance anything close to the current consumption levels of the population, and this can only get worse over time in the absence of a new source of growth. With its current population of 23 million, and oil revenues of only $6000 per capita, Saudi Arabia has clearly outgrown its current model. A new growth model would also allow consumption to grow again after a necessary adjustment, and a credible model could even justify a temporary smoothing of consumption until growth picks up again, fueled by a new economic reactor.

In the short term, growth will be made harder by the headwinds created by fiscal stabilization itself. Cuts in government expenditure create a drag on non-oil GDP growth, whose size depends on the size of the "fiscal multiplier". The multiplier depends on how much of government spending ends up as imports and remittances—the larger these, the smaller the "multiplier". It has been estimated, based on historical data that the short-term multiplier is quite low at about 0.5 (Espinoza et al. 2013). This means that large leakages reduce the growth impact of government expenditures. Still, one would expect a cumulative fall in the growth of non-oil GDP on account of a stabilization effort of about 10 percent of GDP to be about 5 percent points. Already, non-oil GDP growth, which was growing fast before the 2014 shock, has halted to about zero. The stabilization drag is expected to continue to operate for several years as firms adjust their level of operation to lower levels of government spending.

But what will drive economic growth in the medium and long terms? There are several reasons to believe that the main source of growth available in the medium term is the increase in the labor force participation of Saudi nationals.

From a tax collection perspective, GDP growth per se does not tell us much about the economy's ability to absorb increased revenue collection, which depends not just on growth rates, but on the type of growth. A path of labor intensive growth leveraged on expat labor can allow for larger taxation of corporate profits, but if it does not manage to increase employment among Saudis, it would require more social spending to preserve social peace, and thus, would not be conducive to correcting the macro-imbalances. On the other hand, growth based on the expansion of Saudi labor would lead to a broader tax base over time.

More broadly, only 40 percent of working-age nationals participate in the labor force (but only 35 percent work, as the rest are unemployed). This compares to labor participation rates of about 60 percent in the OECD. Low Saudi participation rates are largely due to very low participation by women (19 percent), but men's participation (55 percent) is not high by international standards either. Huge gains could be made if they were instead encouraged to do so, because Saudi nationals are both grossly underemployed and increasingly well educated, thus increasing the opportunity cost of low participation. To give a sense of the potential gains if national labor was employed more effectively, a simple projection model suggests that, with participation rates growing from 40 to 60 percent of the working-age population and unemployment dropping to its natural rate, non-oil national income would more than double if the additional workers join the non-oil sector at current productivity levels. Improvements in labor productivity would add to this growth rate further. Altogether, it can be estimated that this addition to national wealth is comparable in magnitude to the kingdom's current oil wealth.

The extent of economic gain that can be obtained by fixing other inefficiencies, or by removing other constraints to growth, pales in comparison. The total cost of energy subsidies is about 7 percent GDP—there is certainly much over-consumption given low prices, but the extent by which the economic pie can actually be increased on this front is limited. Importing technology to increase productivity, the classical catch-up development policy, is desirable, but the reason for the relatively low labor productivity in the past has not been the lack of openness of the Saudi market to outside influences, but rather, the incentive of firms to use labor-intensive technologies, which is itself connected to the widespread availability of cheap (imported) labor.

This poses the question of the employability of Saudi citizens. Jobs cannot come from an expansion of the civil service. The government is no longer hiring all Saudis who are willing to work. And in the current economic environment, nationals are simply not employable in large numbers in the private sector. Already, unemployment is officially at 11.6 percent overall, 32.8 percent for women, and 29.4 for youth, and rising. Moreover, there are about 200,000 new entrants to the labor market every year.

One can envision a worst case scenario, where dwindling oil revenues continue to be shared among the nationals, cheap labor continues to be freely imported, and Saudi reservation wages only fall slowly over time as oil rents per capita decline. In such a scenario, the kingdom will turn into an increasingly impoverished welfare state, with rising unemployment and lower labor force participation. Moreover, income inequality would rise fast, as business-owners continue to enrich themselves by exploiting cheap expat labor, while the rest of the population gets poorer over time. Governance, which has relied largely on the co-optation of citizens, could become more repressive, as has happened in the countries of the region with smaller oil endowments and larger populations, such as Iraq or Algeria (Cammett et al. 2019).

Besides the no-reform slow-impoverishment scenario, an equally undesirable strategy is one that uses the existing fiscal space to finance a costly pie-in-the-sky national project that does not deliver sustainable growth. Some elements of the Vision 2030 can be read in this manner, and especially those that project Saudi Arabia to become a sort of Dubai on steroids, with Saudi youth managing a large population of migrant workers in a super-competitive knowledge economy. The challenge for such a strategy is to make middle-skill Saudi workers complementary to expat workers. Such a vision seems too ambitious at the macro level for a country the size of KSA. One can think at best of particular sectors, where such a production strategy can make sense, but it cannot cover the whole economy. But even then, insulating sectors with cheap foreign labor from others that employ higher skill national workers runs again in the same problem as with the current quantitative restrictions on labor use—the existence of cheap unskilled expats will reduce the incentives of firms to increase capital investment and improve efficiency and wages in the sectors selected for upgrades.

While the more pessimistic scenario is the more likely of the two, both fail to present a reasonable vision for the country's next 20 years, when oil revenues are likely to remain sizable, but not sufficient to sustain the current model of development. It must be evident to many Saudis that it is high time for the productive structure to adapt to the new realities. The country's human and real assets have changed profoundly in the past 50 years. While importing labor to build the country made sense in the past, there are now large cohorts of educated Saudis graduating and aspiring to productive employment. The situation is thus profoundly different, and it requires profoundly different economic incentives. The massive import of foreign labor was a response to an exceptional situation, unseen in these proportions in any other country at any other time. This period has now to come to an end.

To employ its youth gainfully, Saudi Arabia now needs to become a "normal" oil economy—like, for example, Norway—that exports mostly, if not only, oil, but that derives national income from the work of its own population, primarily in the service sector. In this "normal" model, Saudi workers would replace expatriates, largely in service jobs. The economy would remain dominated by a large oil sector (which includes backward- and forward-linked sub-sectors), employing specialized national workers (e.g., about 50,000 oil engineers). A large share of the Saudi labor force (say half) would remain employed in government; and many public-sector firms would continue to play an important economic role, employing specialized Saudi workers in the oil, health, academic, telecom, and finance industries.

The new jobs would largely be in high-productivity occupations in the service sector. Except in a few areas of comparative advantage, not many firms would produce globally competitive tradables; those that do so now would be unlikely to survive given that unskilled wages would rise, capital would come at a higher cost, subsidies would be cut, and taxes would be introduced. At best, a few tradable sectors could thrive, such as religious tourism and sectors with close linkages with petroleum—at least initially.

At the end of this transition, millions of expatriates would have returned to their home countries, having provided a vital contribution to the task of building up a modern country at record speed. The Saudi economy would become smaller, but it would employ a large share of its own population productively. At the end of the day, the 6 million expats working in private sector firms may be replaced by less than 1 million Saudi workers, employed at higher productivity and at higher wages. The private sector may shrink considerably—even as the tax base grows. The economy may end up with a lower GDP, but it would produce a larger national income. Oil would remain central, but it would have a much larger multiplier effect in terms of the domestic employment of nationals.

In sum, the main economic challenge of the transition to a normal economy is to create highly productive jobs for nationals. It is easy enough to create high-paying jobs—in the public sector, including the security forces, or by replacing migrants in labor-intensive private-sector occupations. But the first option (creating new public-sector jobs) would expand fiscal deficits unsustainably. The second option (replacing expatriate workers with nationals) can deliver high wages in the non-tradable service sector if the total number of migrant workers is reduced sharply. But unless productivity rises too, this would be reflected in higher non-tradable prices, eroding standards of living.

In spite of the relative complexity of the task ahead, one should keep in mind that the overall challenge remains modest—to find productive jobs for about 1 million Saudis over the next five years, in an economy whose GDP is close to $1 trillion. What is needed is a three-prong strategy that tweaks the time-tested Saudization strategy further, by moving faster and more ambitiously on increasing the cost of expat labor, and that supplements labor policies with an industrial policy that helps domestic firms adapt faster to a new set of input prices, and by a conducive macro environment that makes room for the necessary investment needed to raise labor productivity. The rest of the chapter explores each of these themes in greater detail.

## 5 The Central Challenge: Saudis Employability

The key challenge then is to provide incentives for Saudi women and men to join the labor market in larger numbers, and for Saudi firms to hire them.

As a background for this discussion, lets us summarize the well-known duality of the Saudi labor market. About 70 percent of nationals work for the public sector, while expatriates fill 80 percent of private sector jobs. About 4 million Saudis work for the public sector, and 1 million in the private sector.

According to data from the Central Authority for Statistics, the wage ratio between Saudi men that work in the public sector over that of those that work in the private sector is about 2 (Fig. 3.3). This means that Saudi current reservation wage is at around half the public sector wage. Unemployment benefits, close family ties and support, and the hope to get a *wasta* (connection) to join some part of the public sector will keep reservation wages from falling rapidly, even as the prospect of getting a public sector job fades away.

At the same time, a small proportion of Saudis already work in the private sector, where the men among them earn on average twice more than expats (and higher multiples for those with lower levels of education). Firms are thus willing to hire Saudis at wages above those of expats. This may reflect in part constraints on firms imposed by the Saudization program and, in parts, the comparative advantages of nationals in some sectors. But clearly, the demand by firms for national labor must be severely constrained by the size of the wage differentials.

Unemployment in such circumstances reflects both demand and supply constraints—Saudis with reservation wages too high given private sector

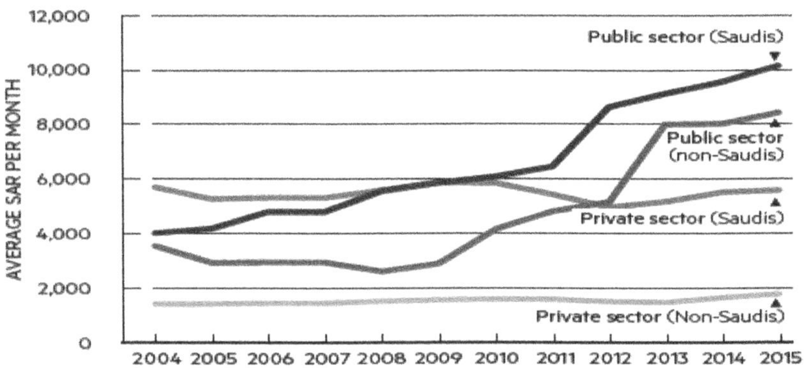


**Fig. 3.3** Average month wages, 2004–2015. (Source: Ministry of Labor Report (2016))

wages for nationals, and firms finding it more profitable to hire cheaper expats. To improve the national labor force participation, expat wages need to go up in order to improve the attractiveness of national workers to firms, and/or national reservation wages need to go down to expand labor supply. The more movement in these prices, the faster Saudi employment would expand. The balance in the use of the two approaches will determine the terms of the burden sharing of reforming the labor market, that is, the welfare loss of workers versus that of firms. While cultural norms of low participation may initially create a resistance to the expansion of work, it is highly unlikely that these norms will not adjust to economic opportunities as they get created, as is the case around the world.

The labor market would therefore have to become much tighter, with constraints placed on the total number of migrants in the country—and not just firm-level quotas that are poorly applied, as is the case now. So far, the Saudization effort has largely focused on a system of quota for Saudi nationals in various sectors. But it has become clear over time that a system of growing quotas cannot work effectively unless the cost of employing expat rises at the same time, so that the (rising) quotas remain close to the quantity of national labor effectively demanded by the market. As long as it is cheaper to employ expats, rent-seeking would ensure that the quota system will leak.[4] The current policy stance also includes efforts to shrink the Saudi-expat wage gap. The government-stated goal is to reduce it by 20 percent by 2022. This goal seems however too modest. The main reason for the modest approach is that Saudization policies are naturally resisted by the corporate sector, since they lower profitability in the short term.

The alternative of squeezing the population into accepting lower reservation wages generates social discontent and resistance leading to delays and to policy reversals—so far, the government has backed down on several reforms that hurt the population (including wage cuts). Given the scale of the challenge, there is therefore a need to broaden the policy toolkit. There are three other ways to improve the labor force participation of Saudis in the non-oil sector besides taxing the expats and pushing Saudis to accept lower reservation wages: (1) to subsidize the private

[4] This will be exacerbated by the highly crony nature of the private sector, whose heights remain dominated by royals and families close to the court, and who had, until recently, a decisive influence over the application of policies and regulatory norms to their business concerns (Mazaheri 2016).

sector wages of Saudis with a negative income tax; (2) to subsidize their in-firm training so as to improve their productivity; (3) to influence social norms and work condition in ways that encourage Saudi women to work more in the private sector. More likely, a successful strategy would use all these instruments.

The discussion above assumes that the challenge of creating jobs for Saudis is a problem of riches and that all that KSA needs to do is to substitute Saudi for foreign workers in existing positions. But simple substitution will not do. As stressed by various analysts, the 6 million jobs currently occupied by expats in the non-oil sector are either overly low skill or high skill intensive (Hertog 2014; Kabli 2014). This dualistic production structure has been molded by economic incentives—whenever possible, firms have tended to use production methods that take advantage of the low cost of unskilled labor, which saves on the necessity to invest in machinery and other productive assets.[5] At the same time, the generous energy subsidies, and the large supply of loanable funds have also advantaged capital intensity among large firms, resulting overall in a two-peak aggregate production function, focused around the employment of low and high skills workers, to the detriment of technologies using middle levels of skills more intensely.

The problem with this two-peak dual production structure is that it does not correspond to the relative scarcity of national factors, as the Saudi labor force is now predominantly of a medium skill level (see Fig. 3.4). Thus, the type of needed structural change is one that upgrades labor-intensive sectors, by adding capital, skills, and technology. On the positive side, such middle-skill jobs would be at higher productivity, allowing private firms to pay the higher wages needed to attract national labor without generating too much inflationary pressures. Such job upgrading should over time eliminate much of the menial positions occupied presently by expats.

The good news is that the education level among young Saudis has risen, which makes it possible for them to occupy jobs at relatively high productivity levels. The bad news, however, is that existing incentives have pushed firms in the private sector to create jobs that require either very low skills (especially in services) or very high skills (especially in the energy sector). To create jobs that are attractive for Saudis, who have mostly

[5] This is also reflected in a decline in aggregate labor productivity over the past decade, even as non-oil GDP rose fast.

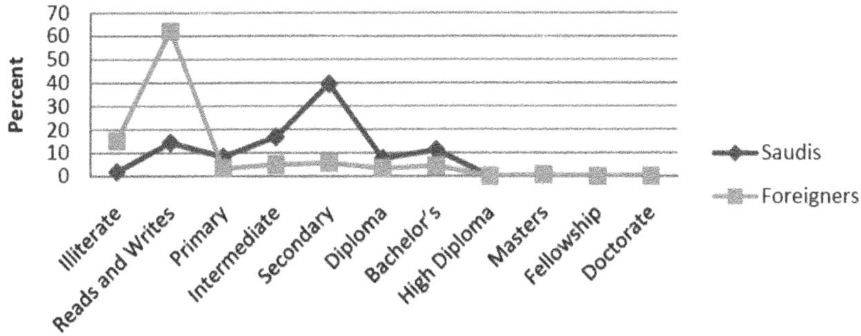


**Fig. 3.4** Nationals and Expats employment in the non-oil sectors in relation to educational levels, 2010. (Source: Kabli 2014, using Ministry of Labor statistics)

mid-level skills, there is therefore a need for structural change in the production methods used by the private sector in response to changes in factor endowments and prices.

## 6 The Policy Agenda Ahead

There are two types of additional economic challenges to transitioning to a "normal" economy, besides the tightening up of the labor market (and a multitude of political-economic ones discussed in Sect. 6). These are related to the need: (1) to improve the dynamism of the private sector so as to support structural change; and (2) to make the right macroeconomic choices as related to speed in order to leave enough financial headroom for the financing of a technological upgrading of the private sector.

The first issue about the need for a more dynamic private sector is related to the need for structural change discussed above. The Vision 2030 document puts a lot of emphasis on the development of new small and medium-size enterprises (SMEs) to make the market economy more dynamic. According to national statistics, there are however more than 1.1 million firms with less than 50 employees, which is already a large figure.[6] Indeed, a recent survey of SMEs by the General authority on Statistics reveals that an important constraint facing SMEs in KSA was too much competition and the implied lack of profitability (General Authority

[6]This can be compared to 1.5 million such firms in Turkey, a country with a population four times that of KSA.

of Statistics 2017). What is needed then is not a net rise in the number of SMEs, but rather, the replacement of the old labor-intensive firms by new skill intensive ones.

Successful SMEs should afford to pay sufficient wages to attract Saudi workers by investing in more capital- and skill-intensive production methods, and by training their workers to improve their productivity. To encourage their entry will require however a much more ambitious effort to improve the business climate, which remains opaque and constraining, and to open up access to finance, which is now severely restricted for SMEs. These efforts will only help if, at the same time, the less efficient firms face higher labor costs and are pushed to close down and exit. Efforts should thus also be needed to facilitate the exit of firms that cease to be profitable, including a reform of the bankruptcy code.[7]

To help SMEs quickly adjust to the evolving input price structure (more expensive labor and other input costs), more targeted industrial policies would also be needed. The key challenge is to support the transformation of whole sectors—reforming the construction industry so that it can move from its current labor-intensive techniques to more capital- and skill-intensive methods, or helping to create new SMEs that cater to a world where housework is three to four times more expensive, by creating substitute services in the transport, cleaning, child-care, and food sectors. In this transformation, individual firms face several externalities: (1) their (costly) discoveries will be imitated by others; (2) they may need to rely on inputs (say machinery) that are not yet produced (or serviced); and (3) they risk losing the workers that they train at great cost to their competitors. As with industrial policies more generally, the question is not as much whether such policies are needed, but rather how to implement them in ways that work. Success here would be judged by their ability to elicit innovation by the private sector, while avoiding rent-seeking (on this, see Rodrik 2009). It has been noted that Saudization is more achievable in sectors where the wage gap is smaller (Hertog 2014). In sectors with large wage gaps, a decision needs to be made about whether to sacrifice or rescue the sector. Construction and commerce are two sectors that would almost certainly need to be rescued given their important size (see Fig. 3.4). These sectors attract at present nearly 4 million uneducated expats and use labor-intensive methods to produce about 20 percent of

[7] Moreover, policies that help firms survive the coming hard times—such as support to compensate for higher energy costs—should carefully discriminate between firms that should, and those that should not survive.

non-oil GDP (see Fig. 3.1). They need to be incentivized to increase their labor productivity by investing in machinery and technology.

In the absence of supporting policies, there is a risk of migrant jobs disappearing, but being replaced with only a limited number of jobs for nationals, as has been observed in the Saudization policies followed so far (Hertog 2014; Peck 2017). The recent work of Jennifer Peck (2017) on the effects of the Nitakat program between 2011 and 2012 illustrate well the risks of a half-baked strategy. Using a detailed dataset at the firm level, she finds that the new quotas imposed by the program managed over a period of 16 months to increase the number of Saudi jobs by 93,000. But, at the same time, the destructive effect of the program was considerable: 11,000 firms (mostly small ones) preferred to exit rather than have to respect the new quotas (or pay the fine); and 934,000 expat jobs were eliminated. These stark results owe much to the precipitous nature of the program—if firms were given more time, and had better access to finance, more of them might have adjusted in ways that would have rescued more jobs.

The second major set of policy issues is related to the broad macro framework and, specifically, to that of the speed of adjustment. As discussed earlier, the required new investments would generate large new financing needs. In order to pay sufficiently to attract Saudi workers and use their skills most productively, the new economy will need to invest in more capital- and skill-intensive production methods. These new financing needs are likely to be large. At the aggregate level, to create about one million jobs every 5 years, they can be of the order of $0.5 trillion over the next ten years. These funds will have to come from the national banking and financial sectors, FDI, or from public funds. At the same time, overly capital-intensive firms and sectors, which rest on massive energy and capital subsidies, may need to be scaled down or eliminated, as financial capital becomes more scarce and in higher demand in medium-skill intensive sectors.

At the macro level, the need for large amount of investment finance creates a trade-off with the speed of adjustment. Large amounts of public financing of deficits would crowd out funds needed to finance private-sector investment. Given that the private investment required for a successful structural reform strategy is large, there is also a global finance trade-off. In the back-of-the-envelope calculations cited above, KSA's fiscal space can be evaluated at about $1 Trillion. This makes it possible to wait 10 years to adjust, if nothing goes to investment. Alternatively, half of the space would be needed to create one million good new jobs. Thus, choices will have to be made. Slowing adjustment too much would constrain how much can be invested to upgrade jobs and productivity.

## 7 The Political Economy Challenge

Important elements of the reforms needed for KSA to become a "normal" economy are already in place. Vision 2030 focuses on many aspects of this agenda. Taxes are rising and subsidies are coming down slowly. Saudization policies, which were started a decade ago, are becoming more binding, and expats are becoming more expensive and are starting to leave in droves. And policy signals have been sent to encourage more innovative SMEs to enter domestic markets. But overall, the program projected by Vision 2030 is not sufficiently focused on the creation of jobs for nationals and is overly concerned with an unattainable diversification agenda. As such, it remains blurred and lacks credibility. To send an unmistakable signal that productive jobs are the priority, Saudization policies, and private sector development efforts, would need to become more ambitious. Mega-projects (such as Neon city) that could easily turn into white elephants should also be replaced by pragmatic industrial policies that help whole sub-sectors to upgrade rapidly.

In developing these policies, and beyond the important economic complexities discussed above, the deeper constraints to progress need to be found in the particular political economy of the Kingdom. It is only a bit of a caricature to state that the current growth model rests on a two separate deals: one deal with businesses for a free hand at importing labor from abroad, and one with citizens for guaranteed public sector jobs and lifelong support. With lower oil rents to share, the domestic settlement comes under strain. Cutting support for businessmen, the population, or both risks undermining the current national pact. The economic elites would oppose policies that lead to more restrictions on their ability to hire expats at low cost. They would claim that most Saudis' education and attitude are not favorable to their employment. National workers will resist a reduction in their reservation wage, claiming that it is the responsibility of the state to protect their consumption level.

To build support for reforms, the perception that the costs of reform are fairly shared by all groups will be essential. The main challenge for a successful transition to a "normal" economy is to avoid distributional fights. Often, reforms that have significant distributional implications end up pitting different groups against each other, with each trying to shift the burden of adjustment by engaging in a "war of attrition" in an attempt to wait the other group out (Alesina and Drazen 1991). This happens more when actors have low trust that costs and gains will be fairly shared among themselves.

Moreover, in order to reduce opposition to change, it does make sense for the state to try to smooth out the initial consumption drop by only lowering the fiscal deficit gradually, at the speed at which the private sector picks up steam in creating jobs for nationals.

On the political side, there are three main risks. First, a populist reform program that advantages labor over firms will have a hard time generating a supply response. The risk here is that expat jobs will be eliminated, but they would not be replaced by new Saudi jobs. But second, a program that puts firms ahead of labor will not manage to increase the size of the tax base, because as long as the current incentive system of firms does not change fundamentally, the creation of more jobs will only attract more migrants rather than employ more Saudis. There is thus a fine line to walk in order to ensure that the (largely front-loaded) burden and (mainly long-term) benefits of the reforms will be fairly balanced among the population. A failure to ensure such a balance brings in the third type of risk, which is that of ending up doing nothing, kicking the can down the road, and borrowing to avoid reform as long as possible. Such a path would only lead to slow decay, and the need for harsher adjustment down the road.

Finally, beyond economic plans that manage to reduce opposition to stabilization and reform by sharing the burden of reforms more fairly across groups and time, one cannot avoid wondering how politics will adjust in the long term to the demands of a more autonomous population and private sector. How the current political regime will choose to adjust the ways the country is governed once circumstances change remains highly uncertain. In the long term, rent distribution will become less dominant and it will become unable to neutralize political voice as much as it did in the past. KSA's truest exceptionalism—the richest non democratic large country in the world—will surely come under stress as incomes start shifting from a rentier logic, to one based on private initiative.

## References

Alesina, Alberto, and Allan Drazen. "Why Are Stabilizations Delayed?" The American Economic Review 81.5 (1991): 1170–1188.

Cammett, Melani, Ishac Diwan, and Andrew Leber: Is Oil Good or Bad for Private Sector Development? Economic Research Forum, working paper, March 2019. Cairo, Egypt.

Coskun, Ali, Serhat Cevikel, and Vedat Akgiray. The role of the state in MENA capital markets. In Diwan, Malik, and Atiyas: Crony Capitalism in the Middle East. OUP, forthcoming, 2018.

General Authority of Statistics, Kingdom of Saudi Arabia: SME survey, 2017.

Espinoza, Raphael, Ghada Fayad, and Ananthakrishnan Prasad. The Macroeconomics of the Arab States of the Gulf. International Monetary Fund, 2013.

Hanieh, Adam. Capitalism and class in the Gulf Arab states. Springer, 2016.

Hertog, S. (2014). Arab Gulf States: An Assessment of Nationalization Policies. Gulf Labor Markets and Migration—GLMM. European University Institute & Gulf Research Center.

International Monetary Fund. Article IV for Saudi Arabia, 2017, and 2018. Washington, DC.

Kabli, Saud. Unemployment in Saudi Arabia. AUC master's thesis, 2014.

Karl, Terry Lynn, The Paradox of Plenty, University of California Press, 1997.

Kingdom of Saudi Arabia. Vision 2030. Retrieved at: http://vision2030.gov.sa/en.

Mahroum, Sami. The Gulf States Expats Dividend. Project Syndicate, July 2016.

Mazaheri, Nimah. Oil Booms and Business Busts: Why Resource Wealth Hurts Entrepreneurs in the Developing World. Oxford University Press, 2016.

Ministry of labor and social development. Saudi Arabia Labor Market Report 2016. 3rd edition, Kingdom of Saudi Arabia, July 2016.

Peck, Jennifer R. "Can hiring quotas work? The effect of the Nitaqat program on the Saudi private sector." American Economic Journal: Economic Policy 9.2 (2017): 316–347.

Rodrik, Dani. "Industrial policy: don't ask why, ask how." Middle East Development Journal 1.01 (2009): 1–29.

CHAPTER 4

# Outlook for Producer Economies

*Ali Al-Saffar*

## 1 Introduction: Risks to Producer Economies—Current Problems and Emerging Pressures

The oil price cycle of the last decade has exposed once more some of the enduring structural economic weaknesses in "producer economies", those countries for whom oil and gas revenue constitutes a significant proportion of exports and fiscal revenue.[1] As the most prominent producing region in the world, the impact of the change in oil prices has been particularly acute in the Middle East, where average net incomes from oil and gas fell by 40% in 2015–2018 compared to their highs in 2010–2014 (International Energy Agency 2018). The public debate on the need to

The contents of this chapter are based on the Outlook for Producer Economies, published by the International Energy Agency in October 2018. Ali Al-Saffar was the lead author.

[1] For the purpose of this chapter, it is assumed that producer economies are those countries in which oil and gas exports make up at least one-third of total goods exports, and where revenues from oil and gas contribute to at least one-third of total fiscal revenue.

A. Al-Saffar (✉)
International Energy Agency, Paris, France
e-mail: Ali.AL-SAFFAR@iea.org



G. Luciani, T. Moerenhout (eds.), *When Can Oil Economies Be Deemed Sustainable?*, The Political Economy of the Middle East,
https://doi.org/10.1007/978-981-15-5728-6_4

diversify oil- and gas-dependent economies has once again gathered steam, with several countries announcing ambitious reform programmes aimed in large part at reducing their dependence on hydrocarbons revenue.

In the past, the imperative to reform has waxed and waned, following closely the oil price cycle, but the combination of pressures from shale oil in the United States, which accounted for 7 in every 10 incremental barrels since 2008, and the long-term uncertainty over demand for oil and gas are intensifying the imperatives for change (ibid.). Fundamental changes to the development model in resource-rich countries look unavoidable. This chapter will enumerate how pressures on producers across the Middle East could evolve, using the results of the International Energy Agency's New Policies and Sustainable Development Scenarios. While acknowledging that the reform agenda will necessarily be much broader than just energy, it will propose ways in which the energy sector can be leveraged to help promote a more balanced economy.

## 2 How Do the Economic Pressures in Producer Economies Evolve Going Forward?

There are multiple uncertainties affecting the outlook for hydrocarbons and the revenue flows that sustain producer economies, not only because of fluctuating prices, but also because of questions concerning long-term demand. How the economic pressures on producer economies evolve in the future depends largely on what the energy sector looks like going forward. To get a sense of the range of possibilities, we will examine two distinct trajectories.

The *New Policies Scenario* provides a measured assessment of where today's policy frameworks, and policies that have been announced but not yet implemented, together with the continued evolution of known technologies, could take the energy sector. In this scenario, equilibrium prices for oil and natural gas (in the context of rising global demand for both commodities) rise gradually from today's levels. The pace of oil demand growth slows markedly, but there is no peak in global consumption, and demand is expected to reach 106 million barrels per day (mb/d) by 2040. Demand for natural gas increases strongly and accounts for a quarter of global primary energy demand by 2040.

The combination of continually growing oil demand and the attendant robust prices may at first appear to offer a relatively benign outlook for major producers—by the late 2020s, annual net income from oil and gas is higher for all Middle East producers than it was in 2010–2017 (Fig. 4.1). But several important factors need to be considered that would suggest that even here, the incentive to diversify oil-dependent economies remains strong.

The first is that the short- and medium-term market dynamics differ considerably from the long-term ones. The majority of the growth in oil production until 2025 continues to come from countries outside of the Middle East, particularly from the United States. It is only after 2025, once the US tight oil plateaus and then starts to fall back, that the world again becomes more reliant on the major conventional oil resource-holders to balance the market. This is why, in most cases, the pick-up in oil net income takes place only in the latter years. These latter years are the ones where there is the most uncertainty about the outlook for demand.

Secondly, while the broad pattern may show absolute levels of income growth, this masks the volatility imposed by price cycles, which, in the absence of new policy tools in many producers, could translate into uncertain fiscal spending and unpredictable economic growth.

Lastly, it is crucial to incorporate the impact of population growth. For sizeable producers with relatively small populations (e.g., the United Arab Emirates, Kuwait and Qatar), this is less of a concern. However, in the region's two largest producers Iraq (+28 million increase in population anticipated by 2040) and Saudi Arabia (+10 million), population growth

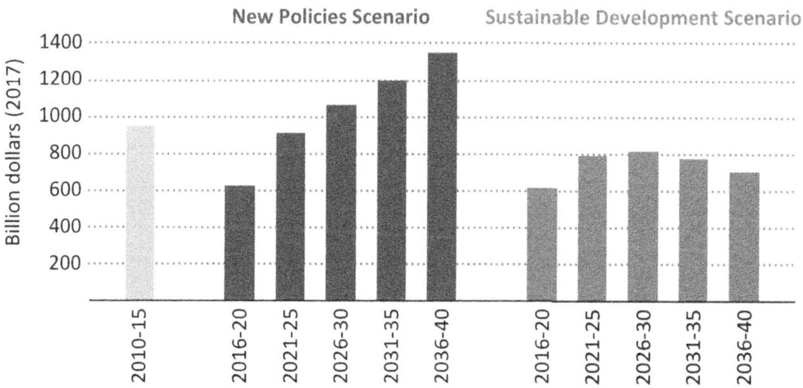


**Fig. 4.1** Average annual net income from oil and gas in Middle East producers. (Source: International Energy Agency)

has significant implications for net income from oil and gas when calculated on a per-capita basis. This indicator will decline in Saudi Arabia over the period until 2025 and increase only modestly overall until 2040, despite a large increase in total net income (Fig. 4.2).

Population growth also has implications for patterns of employment. If we assume that the public sector employs the same proportion of the labour force in 2030 as it does today, Iraq's public sector wage bill would increase by almost 150%, even without any real increase in average salaries, reaching over $70 billion in 2030 (equivalent to 40% of its net income from oil and gas in 2030). The public sector wage bill in Saudi Arabia would increase even more, reaching almost $200 billion (or 60% of its net income from oil and gas in 2030). Pressure for reform in some key producer economies remains high, even in a world where demand for oil continues to increase and oil prices remain robust.

The possibility that the energy sector moves in another direction is examined in the *Sustainable Development Scenario*. This scenario starts from some key energy-related components of the Sustainable Development Goals[2] and then works back to the present to see how

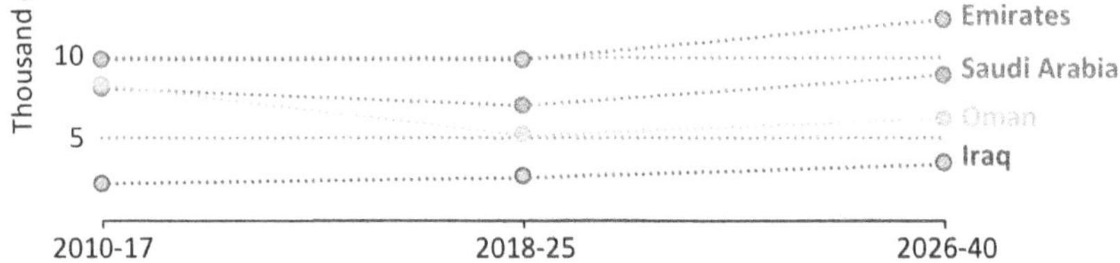


**Fig. 4.2** Total net income from oil and gas in the New Policies Scenario. (Source: International Energy Agency)

[2] These goals include achieving universal access to modern energy by 2030, reducing dramatically the premature deaths due to energy-related air pollution and delivering on the Paris Agreement—this scenario is therefore fully aligned with the Agreement's goal of limiting the increase in the global average temperature to "well below 2 °C".

they might be achieved. It sets out the major changes that would be required to deliver these goals simultaneously. As such, it incorporates not only price effects for oil and gas, but also a major change in the volumes demanded. In this scenario, oil demand peaks in the near term and will then decline to around 70 mb/d by 2040, while natural gas consumption rises by only one-quarter the amount projected in the New Policies Scenario.

An energy sector that is set on a trajectory that complies with the Paris Climate Agreement looks significantly different to the one that prevails today, and indeed from the New Policies Scenario discussed earlier. For producer economies, the effects of a smaller market for oil are compounded by lower prices for it, causing cumulative net incomes across the Middle East to fall until 2040 by over one-quarter relative to their levels in the New Policies Scenario. In such a scenario, constrained government revenues make it harder to manage immediate budgetary requirements while simultaneously complicating the task of making changes to diversify the economy (Fig. 4.3).

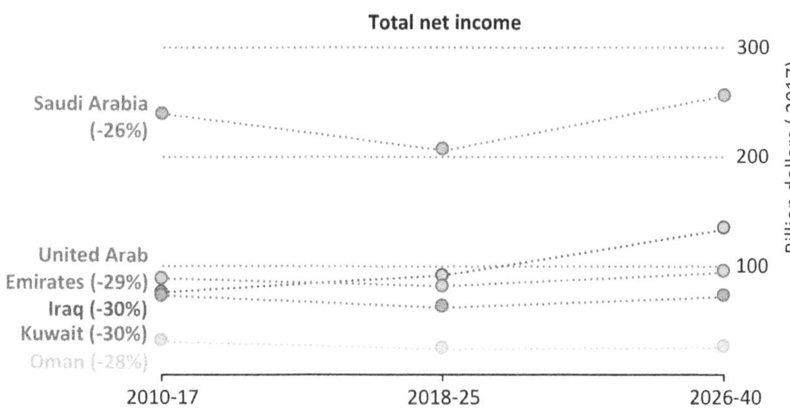


**Fig. 4.3** Total net income from oil and gas in the Sustainable Development Scenario. Note: The change in average annual net income until 2040, relative to the New Policies Scenario, is shown in brackets. (Source: International Energy Agency)

## 3 The Macro-Economic Impact of the Two Scenarios

Beyond the decrease in net income, what could lower oil prices mean for the economic outlook in producer economies?

To help answer this question, we coupled the results of the International Energy Agency's World Energy Model with the Organisation of Economic Co-operation and Development's (OECD) computable general equilibrium model, ENV-LINKAGES, to generate results for the Middle East. This modelling does not assume that major structural economic changes take place among the world's major hydrocarbon producers, but instead assumes a degree of continuity with the past, in which diversification efforts have made steady, rather than transformative, progress.

The results of this economic modelling show that as export revenues decline in the period until 2025 in a lower oil price environment, the trade balance falls into deficit for the Middle East as a whole and stays in deficit through to 2040. For countries that have a floating exchange rate, the resulting currency depreciation makes imports ever more expensive. For countries that have pegged regimes, such as the Gulf producers and Iraq, maintaining the link becomes an increasingly difficult and expensive proposition. Furthermore, reduced revenues impinge on the ability of countries to pursue expansionary fiscal policies to stimulate growth. This hits all sectors of the economy, with the business services and public sector output each around 20% smaller in 2040 than in the New Policies Scenario, and construction output reduced by more than 15%. This has implications for jobs in the Middle East: there will be 8 million fewer jobs in 2040 than in the New Policies Scenario.

The net impact of these effects is that wider economic output is severely curtailed compared with the New Policies Scenario. In the Middle East alone, the cumulative economic losses between 2018 and 2040 amount to $11 trillion, five times the current GDP of the region, and the economy is one-fifth smaller by 2040 than in the New Policies Scenario. This has a clear impact even at the household level where the average person in the Middle East has $1500 less disposable income per year than they would in the New Policies Scenario to 2040.

## 4 The Future Role of a Competitive Energy Sector

The reform agenda for producers looking to diversify their economies and make them more resilient to commodity price movements is, fundamentally, much broader than energy. However, the overall reform efforts could benefit from a well-functioning energy sector. If channelled efficiently, it could provide a platform, both in terms of revenue and comparative advantage, for achieving broader social and economic objectives. We identify six key areas where producer economies can leverage their energy sectors to provide long-term advantage and increase their resilience to a range of future market and policy possibilities.

### *4.1 Capturing More Value from Hydrocarbons*

Across the Middle East, producers have already made significant efforts to move downstream in an effort to capture additional value from hydrocarbons resources. The Middle East currently accounts for around 10% of global refinery runs, and this share is set to increase as several large-scale projects, such as the Jubail, Yanbu and the Jizan complex in Saudi Arabia, are already under construction across the region. Countries are not limiting this increase in capacity to their home territories, and several are pursuing growing markets, notably in Asia. Saudi Aramco recently reached agreements to invest in refineries in China, India, Indonesia, Malaysia and the United States, while the Kuwait Petroleum Corporation is looking at investments in India's Bina Refinery.

There are several motivations behind this downstream drive, including a wish to extract more value from the oil the region produces and to secure outlets for crude exports. The expansion increases revenues for each barrel produced, and thereby also risks increasing dependency on oil revenue. However, downstream earnings typically move in a different direction from upstream earnings—they tend to be higher when crude oil prices are low, and vice versa—so they also provide a hedge against lower oil prices.

Producers across the Middle East are also pursuing large investments in petrochemicals complexes. Beyond the attraction of potentially higher and more resilient margins, the likelihood of a robust outlook for petrochemicals products in all scenarios means that these offer a degree of hedging against the possibility of a contraction of oil demand as a result of a rapid uptake of electric vehicles or higher levels of efficiency improvements in a transitioning energy world. Indeed, even though there is growing attention on reducing single-use plastics and increasing plastic recycling, especially in advanced economies, the impact of these trends is more than offset by a surging demand in developing economies and the increasing use of plastics in place of other materials such as wood and metal. As such, oil demand for petrochemicals is expected to increase by around 5 mb/d, accounting for the largest share in incremental oil demand growth until 2040.

Middle East chemical production is expected to double between today and 2040—allowing the region's share in global chemical production to increase by four percentage points, reaching 17% by 2040 on the back of feedstock cost advantage and the high level of efficiency of newly built facilities. A corollary of this increase in downstream activity is that out of a 6.5 mb/d incremental increase in oil supply until 2040 from the region, only 800 kb/d is exported as crude (with additional 2.1 mb/d passing through refineries, and 3.6 mb/d being used in petrochemicals production).

### 4.2 *Using Natural Gas Strategically in Support of Diversification Goals*

In several prominent oil producers in the Middle East, natural gas has been considered as being a sometimes-convenient by-product of oil extraction. Rising consumption, particularly in power generation, and the limited availability of associated gas (with the exception of Qatar) now require that gas be sought and produced as a commodity in its own right, and often imported. This requires a reassessment of pricing policies to incentivise upstream activity, as well as a review of the priority sectors for gas consumption.

There is also a pressing case for a fundamental rethink of the strategic importance of natural gas, including where it is likely to bring the best value within the energy system, especially in countries where there are strains on the gas balance. In many cases, gas can bring value by displacing oil in domestic energy use, especially where oil is combusted to generate electricity. But there is a strong economic case across the Middle East for faster deployment of solar PV to displace gas as well as oil in power generation, which would augment gas availability for use in value-added industries.

The potential significance of natural gas is broader than the direct incomes that it can bring; it can underpin an industrial strategy in a way that oil cannot, and in this sense can be an important conduit to economic diversification. At present, most industrial applications of gas are capital- but not labour-intensive. For example, the Qatar Petrochemical Company generates more than $1 billion in annual revenues, but employs just 1000 people. These large industrial complexes, however, can act as anchor consumers to help underwrite large infrastructure investments that bring benefits to smaller industrial consumers, supporting the emergence of lighter manufacturing and small and medium enterprises, which thereby contribute to economic diversification and job creation. Indeed, in the projections of the International Energy Agency's (IEA)s World Energy Outlook, industry is a key driver of natural gas consumption growth in the Middle East, increasing by 65 bcm until 2040, reaching 170 bcm.

## 4.3 Tapping the Large, Underutilised Potential of Renewables

The anticipated growth in demand for electricity in a number of producer economies raises questions about the economic viability of the current mode of electricity supply. Across the Middle East, for example, a 5.7% per year increase in electricity demand has translated to a doubling in the region's oil consumption for power generation over the last 20 years, reaching around 1.8 mb/d in 2017. This diverts oil away from exports towards inefficient domestic consumption and incurs a significant opportunity cost—especially significant in periods when global spare production capacity look thin.

At present, peaking capacity in many parts of the Middle East is provided by oil-fired plants, often burning crude oil directly or using heavy fuel oil. In Saudi Arabia, for example, the daily load curve in summer reaches almost twice its peak in winter months because of air conditioning use. This means that 20–25 gigawatts out of a total of 88 gigawatts of capacity are used only for around half the year. These are mostly oil-fired plants, and they increase daily liquids burn by as much as 500 kb/d in peak summer periods relative to winter months. In the future, without a significant improvement in efficiency over time, and considering the large anticipated increase in the use of air conditioners across the Middle East (demand for space cooling alone could skyrocket from 135 terawatt-hours (TWh) today to over 300 TWh in 2040), the peak will grow significantly, giving a measure of the imperative for a more efficient electricity system going forward.

Solar resources are abundant and are ideally suited to meeting this peak (daily demand for cooling peaks in the early afternoon, matching the normal peak in solar PV output). At present, this potential is almost entirely untapped, with the 1.2 GW of solar capacity making up less than 0.5% of total generation capacity in the Middle East (compared to over 90 GW of oil-fired capacity). But fast-falling costs for solar PV mean that, even if oil were priced for generation at $40/barrel, unsubsidised solar would be displacing it quickly on a cost competitive basis (Fig. 4.4).

Although the economic case for renewable power is compelling, reaching deployment levels that reflect this will depend on removing barriers to their uptake. However, at current levels of deployment, concerns about the impact of variable renewables on grid stability in the Middle East are limited, although care will be needed to ensure that network planning matches plans for new utility-scale renewable projects. Most GCC countries in particular have generation fleets that are flexible enough already to enable a much higher penetration of renewables, and the rise in electricity demand for desalination could provide a further synergy for renewables, providing the option of being used as a demand response facility, helping to ensure an outlet during periods of excess electricity production.

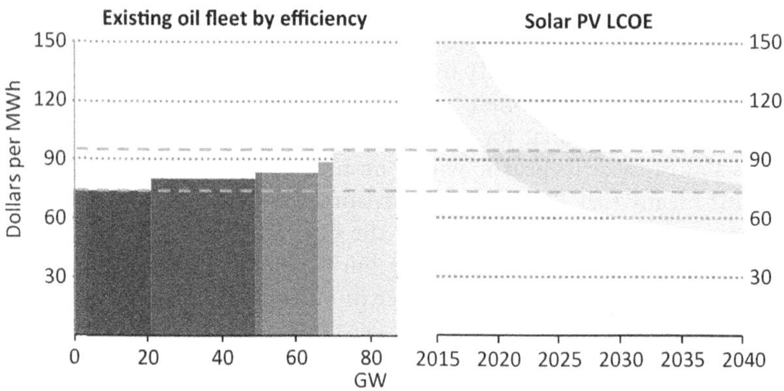


**Fig. 4.4** Levelised cost for solar PV compared with existing oil-fired generation at $40 per barrel oil price. (Source: International Energy Agency)

### 4.4 *Phasing Out Subsidised Use of Energy to Improve Its Efficiency*

According to estimates by the International Energy Agency, fossil-fuel consumption subsidies totalled around $105 billion across the Middle East in 2017 (Fig. 4.5). Prices for oil products and natural gas in most producer economies are well below the value that could be obtained for these commodities on the international market, even after transportation costs are taken into account. Eleven major producer economies in the world—Bahrain, Iran, Iraq, Kuwait, Nigeria, Oman, Qatar, Russia, Saudi Arabia, UAE and Venezuela—account for around half of the International Energy Agency estimate for worldwide fossil fuel consumption subsidies, which, in most cases, represent foregone revenue rather than actual budgetary payments.

Artificially cheap energy encourages wasteful consumption—primary energy demand in the Middle East has grown at 4.4% per year since 2000, a rate that is more than double the world average. Among other things, this has meant that two in every five new barrels of oil production have been consumed domestically in this time. Economies across the region are

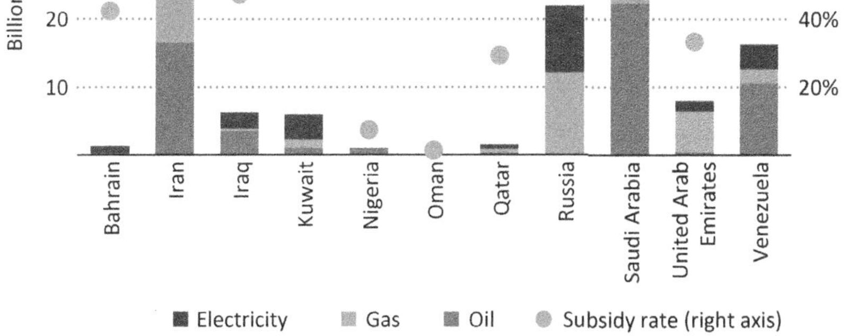


**Fig. 4.5** Estimated value of fossil fuel subsidies in selected producer economies, 2017. (Source: International Energy Agency)

now among the most energy-intensive in the world—the United Arab Emirates, the least intensive in the region, requires 10% more energy to generate a dollar of economic output than the world average.

Beyond the fiscal burden and their impact on consumption, subsidies also distort broader investment incentives across the energy sector. Low natural gas prices, for example, have reduced the incentive for private companies to invest in new exploration and production projects in parts of the Middle East. As well as accommodating for the fact that low-cost energy is deeply embedded in the social contract in many producer economies, successful reform must also reconcile the need to reform prices with the imperative of sustaining or even enhancing industrial competitiveness. Across the Middle East, even without subsidies, most oil and gas producers would still have a comparative advantage in energy, since a low production cost base can provide a stable low domestic price.

The implications of pricing reform for energy consumers can be mitigated substantially if reform is paired with enhanced energy-efficiency measures. Raising fuel and electricity prices reduces the payback period for products with higher efficiency, and helps raise public awareness of the links between efficiency and the cost of the energy they consume, but a push is typically required on the supply side to ensure that more efficient products are available on the market.

## 4.5 *Ensuring Adequate Investment for a Dynamic Upstream*

The ability to maintain oil and gas revenues at reasonable levels provides an important element of stability for the economy as a whole, especially when market conditions are tough. In this regard, though it may sound counter-intuitive in the narrative surrounding economic diversification, it remains crucial for producers to attract investment and maintain or improve the productivity of their upstream sectors. In a scenario where oil demand does not peak before 2040, all key producer economies will need to increase investment over the next decade just to maintain production levels close to those in 2017 (Fig. 4.6).

The need to step up investment comes at a time when there are many competing priorities for domestic spending, potentially limiting the capital available to national companies. The improving prospects of a range of new production areas mean there is also likely to be strong competition for international investment capital. The situation is, of course,

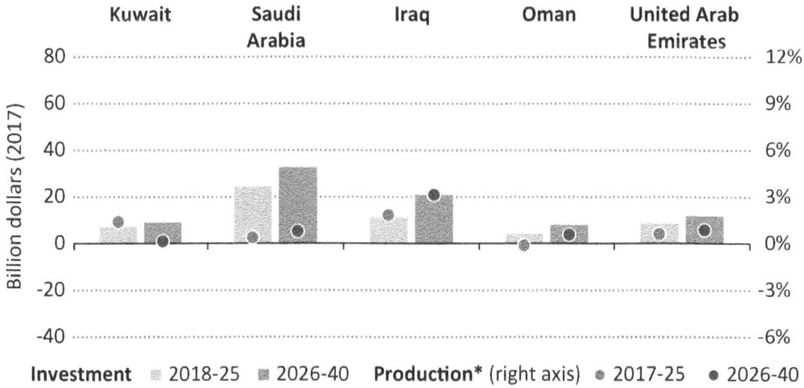


**Fig. 4.6** Average annual upstream oil and natural gas investment and production in selected Middle East producers. (Source: International Energy Agency)

different across the producer economies, but it is clear that oil revenues will be crucial to national solvency well into the future, and will be a vital ingredient for any government wishing to reorientate the economy going forward.

Occupying the bottom end of the oil supply cost curve, Middle East producers could remain integral producers even in a Paris-compliant energy landscape where oil demand peaks imminently and falls to around 70 mb/d by 2040. Some producers, led by Saudi Arabia and the United Arab Emirates, have already shown that through intensified efforts to ensure elimination of gas flaring and methane leakages, they are also extremely competitive on a greenhouse gas emissions intensity basis, one factor that could differentiate suppliers of oil in the future.

## 4.6 *Supporting the Development of Cleaner and More Efficient Energy Technologies*

Many producers have world-leading expertise in energy technologies; in addition to their potential in renewables, they are also well positioned to develop new approaches that reduce or minimise the life cycle emissions of oil and gas.

Saudi Arabia and the UAE have made significant efforts to capture produced natural gas for either domestic use or export and so they flare and vent only very small volumes. This is also the case in Qatar and Kuwait. Despite the fact that the crude oil they produce is relatively sour, oil from these countries has among the lowest emission intensity of any source globally. Saudi Arabia and the UAE therefore point the way for other producer economies to step up their endeavours to eliminate flaring and venting.

There are a number of other options to reduce the GHG emissions intensity of oil and gas extraction. One possibility is to use solar energy rather than oil and gas to provide the energy needed to extract oil from the subsurface. This is particularly attractive in many Middle East countries, as they have high-capacity factors for solar PV and solar thermal plants. Although Saudi Arabia already has some of the least emission-intensive oil in the world, it is continuing to explore options to reduce this further.

Another possibility is to enhance oil recovery (EOR) using $CO_2$. $CO_2$-EOR is one of the few options available today to monetise carbon capture, utilisation and storage (CCUS), since it can lead to higher levels of oil extraction while permanently keeping injected $CO_2$ in the ground. There are important issues to be resolved—most notably ensuring that $CO_2$-EOR leads to an overall reduction in the level of $CO_2$ in the atmosphere—but this represents an important opportunity for the producer economies to make a substantive contribution to the energy transitions. Not only could this help to reduce $CO_2$ emissions directly, but as more capture, transport and $CO_2$ storage facilities are deployed, it could help reduce the cost for future facilities, improving the economics of CCUS more generally. Saudi Arabia has expressed interest in these technologies and has a demonstration project that takes $CO_2$ from a gas-processing plant for injection into the Uthmaniyah area of the super-giant Ghawar field.

Some producer economies are also considering possible alternative uses for oil and gas in the future which would minimise overall GHG emissions. One option is to increase the levels of hydrocarbons in non-combustion uses such as petrochemicals (discussed in the previous section). Another is to convert the hydrocarbons to a zero-carbon-vector such as hydrogen. This can be produced from methane using steam. If the $CO_2$ from this process is stored underground, this zero-carbon hydrogen could help decarbonise a number of end-user sectors across the global energy system while still providing a mechanism to monetise the producer economies' hydrocarbon resources.

The economic argument for such measures becomes particularly compelling when synergies are found to couple industries in a mutually beneficial way. This is already happening, to some extent, for example, in the United Arab Emirates, where over 40 million standard cubic feet per day of carbon dioxide are being captured at the Al Reyadah steel plant before being piped to be used in enhanced oil recovery. This has the added benefit of freeing up the much-needed natural gas that would otherwise be used for the same purpose. Oman is pioneering the use of large concentrating solar projects for enhanced oil recovery, which could displace natural gas in providing the thermal energy needed in its oil production, thereby reducing the emission intensity of upstream production while at the same time freeing up increasingly scarce and valuable gas for domestic consumption or export.

## 5 Implications and Conclusions

The objective to diversify hydrocarbon-dependent economies and reduce their reliance on oil and gas is not new, and the reform discussions unfolding in many countries today have many parallels with previous efforts. The longstanding nature of these efforts underscores that diversifying away from commodity dependency is genuinely challenging and raises a complex set of policy questions that do not have a straightforward solution.

Even in a relatively benign market and policy environment that assumes continued demand growth and relatively robust prices, the current economic model that recycles oil and gas incomes into jobs is unsustainable for some of the world's major producers. The challenges facing producer economies in the Sustainable Development Scenario are more profound—the strong policy push for reductions in emissions, accompanied by rapid technological change, would imply persistent pressure on oil markets and consequently on the producers as well. Here, there is an inescapable imperative to prepare for a world in which hydrocarbons are no longer the main source of revenue, even if there may be alternative ways to monetise hydrocarbon resources that do not contribute to global emissions.

The starting conditions among even the Middle East's producers are very different, and risks are not evenly distributed. Collectively, however, these countries constitute an integral component of global supply, both in terms of how they can influence the future trajectory of the energy system and how they will be affected by it. In all scenarios, the Middle East's low-cost suppliers retain a significant share of global oil and gas supply.

Periods of higher prices caused by a supply crunch brought on by a period of under-investment may offer the possibility of temporary financial relief for producer economies, but they also present a clear downside. The imperative for reform has, in the past, subsided during periods of higher prices, and here too, they risk easing the pressures for change at just the time when higher prices accelerate the policy momentum behind alternatives to oil and gas, especially in some of the emerging global energy demand giants in Asia that are particularly sensitive to price swings. Higher prices would also have implications on the supply side, encouraging new, higher cost production in other parts of the world, and setting the stage for prices to fall again. Such volatility would be commensurately reflected in revenues, and in the absence of pro-cyclical fiscal policies in producer economies, can lead to diminished long-term economic growth.

A successful transformation of producer economies would have profound implications for the energy outlook. By reducing the "social costs" of production, that is, by diminishing the dependence on hydrocarbon revenues to finance areas such as education, health care and public sector employment, it would also lower the oil price at which these countries can manage without persistent fiscal deficits (Dale and Fattouh 2018). In such a case, oil markets could find an equilibrium at a lower price, with major low-cost producers taking a larger share of the market based on their position at the lower end of the global supply cost curve.

The implications of lower oil prices for energy transitions are more ambiguous. On the one hand, low prices could facilitate some policy shifts, such as introducing an effective or actual price on $CO_2$ emissions. On the other hand, experience since 2014 has underlined that lower prices can provide a boost to global hydrocarbon consumption. Payback periods for many efficiency measures increase, and the growth of renewable technologies, outside the power sector, becomes much more challenging. The case for electric vehicle technologies would be particularly disadvantaged. To achieve the same outcomes, policies would need to be strengthened to counteract the effect of lower oil prices on transport and industrial demand.

This raises the question of the strategy that producer economies might follow in a lower demand scenario, and whether the prospect of accelerated energy transitions might invite producers to accelerate extraction of hydrocarbons (the so-called green paradox). The answer again is contingent upon the pace and extent of their broader economic transformation. If investments were to be mobilised to maximise production in such a case, this would bring down prices. Yet oil prices that cause producer economies to run into significant fiscal difficulties are unlikely to be maintained for long.

Ultimately, in a world where demand for energy services is only going to increase, resource-rich countries will always seek value from their endowments, hydrocarbon or otherwise. The task ahead is to make this quest compatible with the gathering pace of change in global energy.

## References

Dale, S. and B. Fattouh (2018), Peak Oil Demand and Long-run Oil Prices, Oxford Institute for Energy Studies, Oxford, United Kingdom.

IEA (International Energy Agency) (2018), *Outlook for Producer Economies*, OECD/IEA, Paris.

CHAPTER 5

# Economic Diversification in Arab Oil-Exporting Countries in the Context of Peak Oil and the Energy Transition

*Bassam Fattouh and Anupama Sen*

## 1 Introduction

Economic diversification has been a key developmental goal for the Arab oil-exporting countries since the 1970s, as evidenced in their various national development plans. Some have undoubtedly made progress over the last few decades in diversifying their economic base, but despite these efforts, several indicators of economic complexity, diversity and export quality continue to be lower in Arab oil-exporting economies than even in many emerging market economies, including commodity exporters in

The authors are grateful to the Kuwait Foundation for the Advancement of Sciences for funding support.

B. Fattouh • A. Sen (✉)
Oxford Institute for Energy Studies, Oxford, UK
e-mail: bassam.fattouh@oxfordenergy.org; anupama.sen@oxfordenergy.org

 
G. Luciani, T. Moerenhout (eds.), *When Can Oil Economies Be Deemed Sustainable?*, The Political Economy of the Middle East,
https://doi.org/10.1007/978-981-15-5728-6_5

other regions (IMF 2016, 9–11).[1] For the Arab oil exporters, the biggest challenge has been to diversify the sources of government income (for instance, through raising additional revenues by taxing individuals and businesses) and to generate non-oil export revenues (through building export-oriented industries) (Luciani 2019; Shehabi 2019). Following the oil price fall in 2014, the weakening of macroeconomic indicators, deterioration in fiscal and current account balances and sharp decline in private sector activity indicate that government spending fuelled by oil revenues still remains the main engine for economic growth.

However, there is a renewed sense of urgency around diversification in these countries, driven by a paradigm shift around the future prospects of global oil demand, and whether the oil industry will continue to generate sufficient rents into the future to sustain oil exporters' economies and their extensive welfare systems. There is a growing consensus that the pace of oil demand growth is likely to slow over time, eventually plateauing or declining, as efficiency improvements, technological advances, climate change and environmental policies and changing social preferences lead to substitution away from oil in its traditional sectors (such as transportation), which have historically driven oil demand growth (Dale and Fattouh 2018, 3). For example, the International Energy Agency predicts that even if oil prices remain within the range of \$60–\$70/barrel, net oil income across the Middle East oil exporting economies will not recover to the levels seen in 2010–2015. Without far-reaching reforms, this would translate into large current account deficits, downward pressure on currencies and lower government spending. The downside economic risk equates to a \$1500 drop in average annual disposable income per person (IEA 2018b, 11).

The concept of 'peak oil demand' is gaining consensus, with many scholars, company executives and policymakers predicting an imminent peak, as soon as before 2030. Increased uncertainty about the prospects of global oil demand is already influencing the behaviour of all oil market players, including the international majors, National Oil Companies (NOCs) and oil-exporting countries—which are intensifying their efforts to diversify their economies and sources of income. Regardless as to when

[1] Recent measures include the Economic Complexity Index which measures the number of products made by an economy and controls for the likelihood that the same product is also made by others, IMF Export Diversification Index which is a combined measure of the 'extensive' and 'intensive' dimensions of diversification using trade data, IMF Export Quality Index which measures the average quality within any product category and Manufacturing Value-Added Gini which is constructed based on the relative value-added of different manufacturing industries within an economy (IMF 2016, 9–11).

oil demand will eventually peak, the current debate matters as it signifies a shift of perception from scarcity to abundance. Consequently, the concepts of scarcity premiums, the effectiveness of rationing oil supplies in an inter-temporal framework and the idea that oil kept underground today will command a higher price in the future need to be critically assessed, especially in the light of the region's massive reserves (Dale and Fattouh 2018). Fattouh et al. (2018, 19) highlight that this issue is far more serious for oil-exporting countries than it is for international oil companies—while for the latter, reserve to production ratios typically range from 8 to 10 years, for the former, they run into decades (e.g. 63 years for Saudi Arabia; 90 years for Kuwait).

Given these economies' long-standing efforts to diversify, one might ask: what is different this time round? The renewed sense of urgency in fact constitutes a break with history—when the concern was mainly over the macroeconomic consequences of heavy dependence on a single export commodity with a highly volatile price—to the possibility that as demand slows down, global oil markets become increasingly competitive and oil industry margins decline, Arab oil exporters can no longer rely indefinitely on oil export revenues for their future economic prosperity. In the bigger picture, it is argued that the world is on the brink of an 'energy transition'[2] in which hydrocarbons will eventually be substituted away in the global energy mix in favour of low or zero carbon energy sources, alongside a paradigm shift in energy technology, institutions and infrastructure (Fattouh et al. 2018, 5; Sovacool and Geels 2016).

Against this changing context, this chapter addresses some key questions: how soon can we expect 'peak oil demand' to occur? How are diversification efforts in key oil exporters linked to the ongoing global energy transition? Will the hydrocarbon sector play any role during the energy transition? And will the emergence of renewables as competitive energy source impact economic diversification strategies in these countries?

We make three main arguments in this chapter. First, the speed of the energy transition is highly uncertain and heavily driven by government policies, implying that it will vary across regions and be unpredictable on a global scale. Second, the diversification strategy adopted by oil-exporting countries will be conditioned by the speed of the energy transition, during which the oil sector will continue to play a key role in these economies, including in their diversification efforts. Oil producers will need to be far

[2] Previous energy transitions have involved wood being substituted by coal, and coal being substituted by oil on a mass basis.

more strategic in developing their energy sector, including the renewables sector, and strengthening forward and backward linkages to help diversify their economies. Finally, there is a co-dependence between the success of diversification efforts in oil exporters and the global energy transition. While the transition is already shaping the political and economic situation of the Arab oil exporters, the success (or failure) and the speed at which Arab oil exporters transition to a more diversified and more resilient economies will shape the global energy transition.

The next section reviews the debate on peak oil demand and the energy transition, and Sect. 3 summarises their implications for oil-exporting Arab countries. Section 4 considers the role of the oil sector in economic diversification and energy transition, and Sect. 5 discusses the implications of this changing context for traditional cooperation between oil producers. Section 6 concludes.

## 2 The Energy Transition and Perceptions of Peak Oil Demand

The current debate on peak demand tends to be dominated almost entirely by the time or point at which global oil demand is expected to peak. Consequently, most analyses of peak demand contain a wide range of projections, some suggesting that it could peak around the mid-2020s and others expecting it to grow beyond 2040. Figure 5.1 depicts these variations based on projections published by organisations such as BP, International Energy Agency (IEA) and US Energy Information Administration.

The IEA 2018 New Policies scenario, for instance, predicts that global oil demand will grow by around 1 million barrels per day (mb/d) on average each year until 2025; thereafter, average annual demand growth slows to around 0.25 mb/d, but global demand does not peak before 2040. All of this growth occurs in developing economies, whereas demand in advanced economies drops by over 0.4 mb/d on average each year until 2040. In contrast, in its 2018 Sustainable Development Scenario, demand falls by 25 mb/d between 2017 and 2040 (IEA 2018c, 133–134). The IEA's 2018 Oil Market Report, on the other hand, sees demand growing strongly until 2023 (expanding by 1.4 mb/d in 2018 alone), but its pace slowing down to 1 mb/d thereafter (IEA 2018a). BP's 2018 Energy

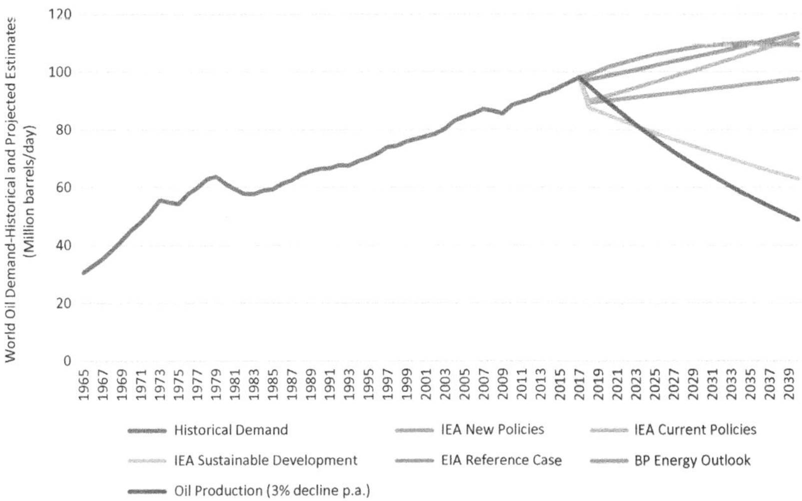


**Fig. 5.1** Projections of world oil demand (million barrels/day). (Source: IEA 2018c; EIA 2017; BP 2018)

Outlook contains three differing scenarios[3] of which the Evolving Transition (ET) scenario envisages oil demand continuing to grow until the 2030s, and plateauing thereafter (BP 2018).[4] The Energy Information Administration (EIA) in its 2017 International Energy Outlook 'reference case' scenario sees oil demand growing from 95 m/d in 2015 to 113 mb/d by 2040, but in a 'low oil price' scenario reflecting abundant supplies, it projects higher demand (5.5 mb/d more than the reference case) by 2040 (EIA 2017). In 2018, the EIA published three different 'side cases' on demand, focusing in detail on the way in which different macroeconomic scenarios might play out in three major world regions (China, India and Africa) and postulating different impacts on world primary energy demand as a consequence. While a full discussion of the latter is beyond the scope of this chapter, it highlights the range of uncertainty around various peak demand projections and the multitude of variables affecting the same.[5]

[3] Evolving Transition (ET), Faster Transition (FT) and Even Faster Transition (EFT).
[4] Demand in 2040 is projected at 109 mb/d.
[5] Refer to EIA (2018) for a full discussion.

An obvious question that arises from the multitude of peak demand forecasts is as follows: what kind of a scenario should Arab oil-exporting countries be preparing for? While exercises in forecasting peak demand contribute towards providing a renewed motivation for diversification, they should not be the sole factor upon which these strategies are undertaken. This is because the speed of the energy transition matters. At the same time, the speed of the transition will also be influenced by the manner in which key actors in the energy sector, including the Arab oil-exporting economies, choose to respond to it. The risks associated with the energy transition may be felt by investors much faster than the timescale within which the transition is completed. For example, this could be because of government policy signals supporting new low-carbon energy technologies vis-à-vis hydrocarbons, which contribute to improving the future competitiveness of these technologies. Investors in the hydrocarbon sector may therefore adjust their current risk preferences in relation to new investment and operational decisions (Fattouh et al. 2018, 2).

The wide range of forecasts on peak demand allows us to draw some key observations to set the context of this discussion. First, the range of uncertainty is high. Peak demand forecasts are highly dependent upon their underlying assumptions. For instance, while the IEA's New Policies scenario is based on government policies that have already been announced or implemented, its Sustainable Development Scenario assumes a tightening of climate and environmental policies that are sufficiently aggressive for carbon emissions to decline at a rate that is broadly consistent with achieving the goals of the Paris Agreement reached at COP21 in 2015 (Dale and Fattouh 2018, 4). However, many of the more ambitious commitments made by the countries that are likely to pose the largest share of rising emissions in the future (e.g. China and India) are voluntary rather than binding, and therefore partially dependent upon the resolve of policymakers to fulfil them.[6]

It is also difficult to entirely discount the possibility of countries reneging on some of their commitments, as some governments have recently acquiesced to domestic political pressures and rising rhetoric around trade

[6] We state that thats is partially dependent, as stringent environmental policies in developed countries have encouraged innovation, driving down the costs of clean energy technologies to make them cost-competitive on a global basis.

protectionism.[7] Another illustration of the complexity of assumptions around peak demand projections relates to the consensus that the majority of oil demand growth until 2040 will come from non-OECD Asia, on the back of rising incomes and car ownership. The EIA, for instance, has predicted that China's use of liquid fuels for transportation could increase by 36 per cent from 2015 to 2040, while India's use over that period could increase by 142 per cent (EIA 2017, 36). At the same time, most forecasting organisations also allow for a scenario in their forecasts which takes into account the impact upon transport oil demand of vehicle electrification, automation and shared mobility. These, in turn, entail very different sorts of impacts in different combinations. For instance, in its 2018 Energy Outlook, BP did not see electrification on its own impacting oil use in transport significantly. But automation—which reduces the cost of shared mobility by 40 to 50 per cent—was seen as having a bigger impact.

BP estimated that electric vehicles (EVs) would account for 15 per cent of about 2 billion cars on the road in 2040, but for 30 per cent of all passenger car transportation measured by distance travelled (due to the uptake in shared mobility). A prominent academic paper by Fulton et al. (2017, 15–29) estimated a 50 per cent reduction in energy use in transport in an automation-electrification-shared mobility scenario over an automation-electrification scenario, by 2050. On the other hand, the BP Energy Outlook predicted that oil use in transport would remain almost unchanged in 2040, from 18.6 mb/d in 2016, with gains from efficiency and electrification offset by a doubling of overall demand in car travel (FT 2018). These variations in forecasts from different organisations show that the confidence intervals around any peak demand forecast are likely to be large, reflecting the high uncertainty about the speed of the transition and the factors shaping the transition.

The second key observation is the possibility of not one, but multiple peaks. An important consideration while assessing peak demand scenarios is the 'rebound effect'—that is, the premise that a peak in oil demand could cause oil prices to fall, triggering higher demand from consumers and a potentially more than one peak. For instance, the EIA in its 2017 Outlook predicted that in a 'low oil price' scenario, global oil demand in 2040 would be 4.5 mb/d higher than in the reference case, as low prices stimulate higher consumption (EIA 2017, 34). Dale and Fattouh (2018,

[7] For instance, the USA pulled out of the Paris Agreement in June 2017, although the consensus is that this will not derail the Agreement between the remaining signatories.

5) argue that this effect can lead to multiple peaks. Their examination of data on US oil and gasoline demand shows that US oil demand peaked in 2005, declining at an average rate of 1 per cent per annum over the next 8 years—but since oil prices fell in 2014, US demand has begun to grow again, and a continuation of low prices[8] could result in US demand exceeding the 2005 peak.[9] The 'rebound effect' is also evident in past energy transitions, in which new energy sources have unlocked new energy demand. For instance, Fattouh et al. (2018, 8) argue that the advent of coal in the nineteenth century opened up new forms of transportation, and its first use in the steam railway engine around 1829 was followed by an acceleration in global energy demand that was 1 per cent per annum higher than that of global population growth for nearly 50 years thereafter.

The third key observation that can be drawn from the range of peak demand forecasts is that oil will continue to be an important part of the energy mix for the foreseeable future. The incumbent advantages of oil as an energy source, including its high energy density and an existing infrastructure ecosystem geared around it, imply that even if oil demand peaks, it is unlikely to 'fall off a cliff'. Instead, the broader characteristics of the current energy transition from hydrocarbons to low-carbon energy sources and the speed of transition are of more relevance to economic diversification, rather than predictions of when oil demand will peak. But while there is unlikely to be a sharp discontinuity in oil use, it is also unlikely that the current transition will exactly mirror the speed of past transitions, as it has some fundamentally different characteristics.

Historical data and evidence indicate that past energy transitions have been slow. Fattouh et al. (2018, 10) state, for instance, that fast transitions have rarely happened, and, when they have, they have been anomalies that are related to countries or specific contexts with little scope for replicability.[10] The scale and complexity of energy transitions tend to create path dependency and 'lock-in' of infrastructure that has been developed over long periods of time. Energy transitions also involve 'fighting' against an entire infrastructure ecosystem based on the incumbent energy source which represents massive sunk investments—this is visible, for instance, in

[8] This was not evident at the time of writing.

[9] US gasoline consumption reached its highest ever level in 2016 after falling for much of the previous 10 years.

[10] Examples of fast transitions include Sweden's shift to efficient lighting in nine years; Indonesia's substitution of kerosene with LPG in 3 years; Brazil's substitution of petroleum with ethanol in 90 per cent of all new vehicles in six years (Fattouh et al. 2018, 12).

efforts to substitute internal combustion engine vehicles (ICEVs) with EVs. Fattouh et al. (2018, 10) show that the key empirical evidence for a slow transition is past inter-fuel competition, which led to the substitution of coal for pre-industrial biomass and muscle power, and oil for coal. The market share of coal increased from 5 per cent to 60 per cent between 1830 and 1914, peaking in the year that the First World War broke out. Oil increased from 1 per cent to 40 per cent between 1900 and 1973, peaking in the year of the first OPEC oil shock (2017). Gas increased from 4 per cent in 1945 to 24 per cent in 2018, while nuclear rose from zero per cent in 1954 to 2 per cent in 2000. There are also very few examples of major energy sources disappearing from the global energy mix. BP's 2018 Energy Outlook, for instance, sees most oil demand growth after the 2030s coming from non-combustible uses, although this could also face a slowdown if measures to curb environmental pollution from materials (e.g. plastics) are tightened/enforced.

However, the possibility of a fast transition cannot be entirely discounted either, as the current transition bears characteristics that represent a clear break with the nature of past transitions. Historical transitions were more about developing technologies in an age of scarcity based on markets and innovation, whereas low-carbon transitions are more about adjusting the environmental selection in an age of abundance, via policies, regulations and incentives (Sovacool and Geels 2016). Therefore, while past transitions were opportunity-driven, the current transition is solution-driven. At the same time, since the current transition is heavily driven by government commitments on mitigating their emissions and consequently by national policies, its speed could differ across regions as well as sectors, making it difficult to draw firm conclusions on a global scale.

The discussion therefore underscores the futility of adopting an approach whose starting point is 'oil will no longer be demanded by a certain date', or, in other words, preparing for some 'distant future'. Instead, it is necessary to adopt a more dynamic analysis that moves beyond the potential role that the oil sector and oil rents would play in the transition phase.

# 3 The Implications for Arab Oil-Exporting Countries

How should Arab oil-exporting countries respond to this changing context? The aforementioned discussion suggests that, on balance, oil-exporting countries should adapt to the energy transition which is already underway, but given that the speed of the transition is uncertain, in doing so they could take into account the consolidation of three key trends.

## *3.1 Oil Demand Is Unlikely to Increase Strongly Over the Next Two Decades*

Government oil substitution policies point in the direction that oil demand is not likely to increase strongly over the next two decades, although the timing of when oil demand growth will start slowing down and turn negative is still highly uncertain. These policies are prevalent in OECD and non-OECD countries. Some countries[11] in OECD Europe have for instance announced bans on ICEVs by 2040 as part of their carbon reduction targets which are among the world's most ambitious, while in non-OECD Asia, China and India have both announced ambitions to scale up EVs in the vehicle fleet[12]—with China attempting to integrate the EV sector into its overall industrial strategy. The two countries also have localised (state or province level) restrictions on ICEVs, through lottery sales and bans of older models. Further, both countries are moving towards the adoption of stricter fuel-efficiency standards by the early 2020s.[13] The drive to reduce urban air pollution could also accelerate substitution. The 'ET' scenario in BP's 2018 Energy Outlook sees China's energy demand growing by just 1.5 per cent per annum (less than a quarter of its historical 20-year growth rate), with renewables overtaking oil to become the largest energy source by 2040. India, on the other hand, is seen as overtaking China as the main engine of oil demand growth, although it remains smaller in terms of absolute volumes consumed. This trend has implications for future market share strategies of the Arab oil exporters, which should be taken account of in their diversification strategies.

[11] The UK and France, for example.

[12] For instance, India is targeting 30 per cent of the vehicle fleet by 2030.

[13] India plans to leapfrog from Euro IV to Euro VI standards by 2020.

### 3.2 Large Investments Will Still Be Needed in the Oil Sector to Fill the Gap in Supply

Even in the event of peak demand and in the absence of investment in the oil sector, the decline in supply will be faster than the decline in oil demand. Dale and Fattouh (2018, 5), in their analysis of peak scenarios for instance, assume a relatively conservative 3 per cent decline rate in production and no new investments until 2040[14] to illustrate that the gap between demand and supply could range from anywhere between 35 mb/d and 75 mb/d.[15] The IEA (2018c) estimates a near-term supply deficit of 26 mb/d by 2025 even under the Sustainable Development Scenario. Upstream investment has not recovered since the 2016–2017 oil price fall; it was flat in 2017, with a modest rise in 2018 mainly from USA's use of light tight oil (Fattouh et al. 2018, 7). A survey by Fattouh et al. (2018) indicates that risk preferences of institutional investors (mainly in the USA and Europe) are already changing in response to the energy transition, reflected in much higher 'hurdle rates' required by investors now, relative to the last few years, to consider new long-term investments in oil exploration and development projects. Instead, there is a strong preference to concentrate oil and gas companies' conventional activities in the 'harvesting' phase and away from the 'exploration and appraisal' phase.

As the world's lowest-cost oil producers, Arab oil-exporting countries will most likely be required to fill this gap, but any expansion in productive capacity will require massive investments running into billions of dollars. These investments in productive capacity will need to be funded by sufficient revenues, largely from oil exports. At the same time, Arab oil producers face competing needs on their revenues given the social welfare measures funded by these revenues, which underpin their societies. Thus, although the physical cost of production and developing their reserves is low, one should add some measure of social costs in addition to the physical cost, as producers need higher prices for their economies to function and to expand productive capacity.[16]

[14] We also show this in Fig. 5.1.

[15] The lower bound represents the difference between projected production (assuming the 3 per cent decline rate) and projected demand in 2040 under the IEA Sustainable Development Scenario, while the upper bound represents the difference with projected demand in 2040 under the EIA's scenario.

[16] In the age of abundance and greater competition, prices would ideally tend towards marginal cost, but MENA producers require a premium to finance their socioeconomic models.

Countries also require relatively stable political environments to make these investments. In the absence of such stability, it is possible that some countries with cheap reserves will be unable to develop these reserves. In other words, it is not necessarily true that low-cost producers will develop their reserves at the expense of high-cost producers with more stable environments. In the long run, as economies diversify, the 'social costs' of production embodied by the fiscal breakeven price should naturally decline as economies are supported by a broader economic base (Dale and Fattouh 2018, 8). The major caveat to this point is that if climate change emission reduction policies lead to a fall in oil demand at the same pace as, or faster than, production declines from existing fields, then there would be no supply-demand gap (Fattouh et al. 2018, 7). The realisation of a constrained oil demand trajectory however depends on how determined policymakers in the developing world are (especially China and India) about moving away from fossil fuels—which we have discussed earlier in this chapter.

### 3.3 *Renewables Are at an Inflection Point*

While there are many uncertainties induced by the energy transition, there is almost a consensus among forecasts provided by various organisations that the share of renewables in the energy mix will rise (Fig. 5.2). As described in Fattouh et al. (2018, 4), renewable energy's recent cost deflation has been nothing short of revolutionary for the global energy industry. Five years ago, US wind costs were $11 c/kWh (US cents per kilowatt hour) and solar costs were $17 c/kWh, on a fully loaded basis, including the capital costs of construction. Neither was commercial without subsidies. In 2019, the International Renewable Energy Agency (IRENA) estimated that solar PV costs in the Gulf Cooperation Council (GCC) had declined to less than $3 c/kWh, leaving behind natural gas, LNG, coal, oil and nuclear. In Saudi Arabia and Oman, wind has emerged as another cost-effective option. The four bids submitted for the 400 MW Dumat Al Jandal wind project were reported to be between 2.13 US cents/kWh and 3.39 US cents/kWh (IRENA 2019, 86).

As a result, on a plant-level basis and excluding the cost of dealing with intermittency, wind and solar have emerged as very competitive sources of energy globally (Fattouh et al. 2018, 4). The 2018 BP Energy Outlook

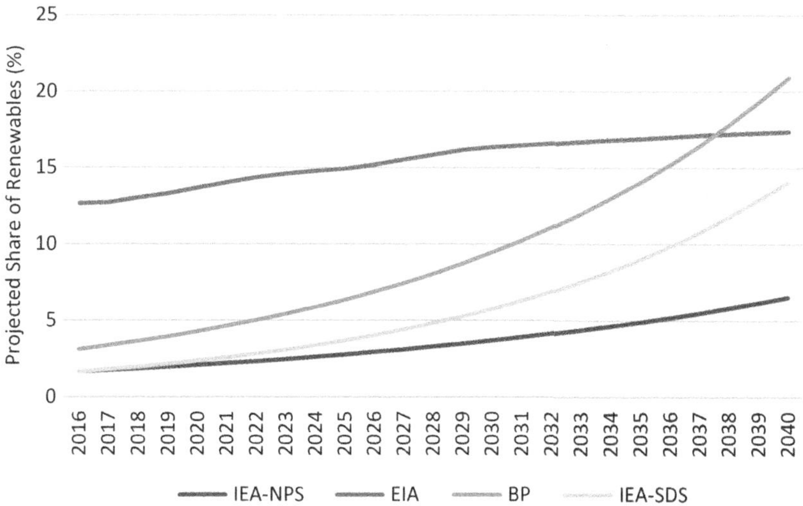


**Fig. 5.2** Projected rising share of renewables in total primary energy demand (%). (Source: Calculated by Authors from BP 2018; IEA 2017; EIA 2017)

stated that the share of power generation gained by renewables over the Outlook period will be faster than any other energy source over a similar period—it saw renewables increasing fivefold and capturing 40 per cent of new demand growth (BP 2018). The EIA's International Energy Outlook 2017 similarly saw renewables as the fastest growing source of overall energy consumption and power generation in 2015–2040—renewables will more than double their share in energy consumption by 2040, with their share in power generation rising by an average of 2.8 per cent a year and reaching over 30 per cent by 2040, as technological improvements and government incentives in many countries support their increased use.

In the next section, we discuss how the aforementioned three trends feed into an economic diversification agenda for the Middle East and North Africa (MENA) which makes far more strategic use of the oil sector as the energy transition progresses, rather than one that simply envisages a sudden 'switch' away from oil and the redundancy of the oil sector.

## 4 The Strategic Role of the Oil Sector in the Energy Transition

As low-cost producers with some of the largest reserve bases, Arab oil producers are expected to fill the gap by heavily investing in their oil sector. Therefore, even in a world where oil demand growth is expected to slow down, the oil sector will continue to play the dominant role in these economies for the foreseeable future. However, as leaders develop new visions to transform their countries, the energy sector will be under increasing pressure to show that it can contribute to the diversification efforts, not only by generating the rents that could be used to create new industries and sectors, but also by extending the value chain and creating new industries within the energy sector through fostering backward and forward linkages. In the following section we discuss three important elements of this strategic role, which complement each other.

### *4.1 The Oil Sector Will Continue to Dominate the Economy, But it Needs to Play a More Active Role in the Diversification Process*

Although the oil sector remains very profitable and enjoys higher margins than any new industries that government are aiming to establish, from a developmental perspective, it suffers from two major shortcomings. First, it does not generate a stable source of income as oil prices fluctuate considerably, and in some countries the rents are not big enough to provide sufficient income for growing populations and an extensive welfare system. In a world where demand is expected to peak, these challenges become more acute. Second, the oil industry is capital-intensive in nature and does not generate enough jobs for the hundreds of thousands entering the labour market each year.

Extending the value chain beyond simply producing crude oil and exporting it to international markets could in principle address some of these challenges. The Middle East currently accounts for around 10 per cent of global refinery runs, and countries are increasingly looking at petrochemicals projects as well as refinery expansions as a way of pursuing higher and more resilient margins. By extending the value chain, Arab oil producers can create new industries with different types of jobs and whose products' prices are not highly correlated with oil prices. In the past, the focus has been on exporting basic petrochemicals (for instance, converting ethane to ethylene), which did not generate much of the expected benefits for two reasons. First, the prices of basic petrochemical products are highly correlated with oil prices. Second, refining and petrochemicals are also highly capital-intensive industries and do not generate many jobs. Therefore, the recent emphasis in some of the Gulf Cooperation Council (GCC) countries has been on extending the value chain to more complex petrochemical products and even finished products manufactured in industrial parks that attract the private sector and foreign direct investment. The move towards non-combustible uses of oil also offers a degree of hedging against the possibility of a drop in oil demand (IEA 2018b).

For instance, Saudi Arabian Oil Company and Dow Chemical established a joint venture in 2011, with an investment of $20 billion, which incorporates 26 integrated large-scale manufacturing plants with over 3 million metric tonnes of capacity per annum. It has introduced many new products to Saudi Arabia (e.g. the first isocyanates and polyurethane plants), enabling many industries for intermediate products that either did not exist in the Kingdom or only existed through imports of raw materials, potentially opening up a range of new downstream opportunities. The Saudi Arabia Basic Industries Corporation (SABIC) has similarly established a number of industrial parks specialising in intermediate raw material inputs and offering easy access to banking facilities, research resources, skilled workforce, logistic services and other inputs—it has a particular focus on the automotive industry which offers opportunities down the value chain.

According to the IEA (2018b), Middle East chemicals production is expected to double between now and 2040—allowing the region's share in global chemicals production to increase by four percentage points, reaching 17 per cent by 2040 on the back of feedstock cost advantage and the high level of efficiency of newly built facilities. A corollary of this increase in downstream activity is that, out of a 6.5 mb/d incremental increase in oil supply until 2040 from the region, only 800 kb/d is exported as crude (with additional 2.1 mb/d passing through refineries and 3.6 mb/d being used in petrochemicals production). Adding more stages to the oil value chain in this way generates not only more jobs, but also different types of jobs including jobs in the service sector such as trading, marketing and sales, procuring and logistics, as well as supporting services such as accounting, finance and human resource management.

In this respect, the local content requirements (LCRs) that give priority to nationals in terms of employment, domestic companies in terms of contracts and locally produced goods and services will only increase in importance. The objectives of such policies are to create a level playing field of local industries, create jobs for locals and enhance transfer of technology and technical expertise and skills. However, experience shows that such policies, if not properly implemented, can have unintended consequences such as increasing the cost of projects, misaligning the interests between the government and investors and hence dis-incentivising private investment and even encouraging corruption.

Arab oil producers, as owners of large low-cost resources, have rationed their supplies in order to preserve their resources for future generations—a strategy which made sense during the age of scarcity (Dale and Fattouh 2018). However, in carbon-constrained world, the main challenge for these producers in addition to maintaining the high levels of revenues upon which their economies are dependent is to monetise their massive reserve base in order to avoid to a greatest extent the possibility that significant amounts of recoverable oil may never be extracted.[17] This requires that Arab oil exporters play a key role in providing low-carbon solutions through developing and investing in low-carbon technologies that could extend the life of oil and gas.

Thus, rather than treating the oil industry as sunset industry in a world of heightened uncertainty about the prospects of oil demand, these countries will need to be much more strategic in terms of how the oil sector can further contribute to economic diversification. While the oil sector will continue to play a key role and to generate most of the income during the transition, it is expected to play a bigger role in the diversification and the economic transformation agenda. Business as usual for the oil sector and those who are leading it is no longer an option. Particularly, its success in extending the value chain and effectiveness in implementing LCRs will determine the weight the sector plays in the decision-making process and in influencing the transition. But, as noted by Fattouh and Shehabi (2019), the extent to which the energy sector can play such a role depends on governments removing a wide array of structural barriers such as reducing the role of the public sector in employment generation and capital employment, diversifying the sources of income through serious tax reform and dismantling the oligopolistic market structure. After all, 'the real problem lies in the economic and political structures and in policies surrounding the energy

[17] In other words, that countries are not effectively left with 'stranded assets'.

sector; these factors not only constitute barriers to meaningful diversification, but also limit this sector's contribution to broader and deeper diversification' (Fattouh and Shehabi 2019).

### *4.2 The Use of Demand-Side Measures to Optimise the Resource Base*

In addition to building new production capacity, these countries need to undertake other measures to optimise the use of the resource base. These include policies such as diversifying the energy mix and implementing energy efficiency measures and pricing reforms that liberate hydrocarbons that are currently used to meet soaring domestic demand at low (subsidised) prices for export, which would add economic value (as long as the international price is above the fiscal breakeven price). Across the Middle East, for example, a 5.7 per cent annual increase in electricity demand has translated to a doubling in the region's oil consumption for power generation over the last 20 years, reaching around 1.8 mb/d in 2017 (IEA 2018b). This necessitates the implementation of energy-efficiency measures including subsidy reforms or price incentives. Indeed, subsidy reforms implemented in the GCC following the oil price fall of 2014–2016 have largely continued despite the price recovery in 2018. Further, such measures are complementary to an overall economic diversification strategy, which entails structural changes and fiscal reforms. For instance, the fact that all GCC countries (with the exception of Kuwait which has faced hurdles) have implemented energy pricing reform, constitutes an important step towards meaningful economic diversification in the long term.

Matar et al. (2017), for instance, explore the impact upon energy demand of various alternative policies to induce a transition to a more efficient energy system in Saudi Arabia, including immediately deregulating industrial fuel prices, gradually deregulating fuel prices and introducing investment credits or feed-in tariffs for efficient fuels. They find that the alternative policies result in nuclear and renewable technologies becoming cost-effective and producing 70 per cent of the Kingdom's electricity in 2032 in contrast with hydrocarbons at present. They can also reduce the consumption of oil and natural gas by up to 2 million barrels of oil equivalent per day in 2032, with cumulative savings of 6.3 to 9.6 billion barrels of oil equivalent. They estimate a net economic gain up to half a trillion US dollars from increased oil exports, even with investments in nuclear and renewables. However, given the rigidity of existing political structures, institutions and the implicit social contract through which limited political participation is compensated for by distribution of hydrocarbon rents, gradual and small-scale reforms combined with mitigation measures can be implemented, but one should also not expect 'speedy transformations' of oil-exporting economies (Fattouh et al. 2018, 21).

### 4.3 Renewables as a Complement to Economic Diversification Strategies

Middle Eastern and in particular GCC oil exporters should not miss on the renewable 'revolution'. They have great potential for renewable energies, owing to high levels of irradiation throughout these countries, and wind potential in some. Many countries in the region also have fewer limitations on the use of land for construction of wind and solar farms. Furthermore, their locations are often close to the regions' main energy markets. Collectively, these conditions create a unique opportunity for these countries to exploit their renewable resources to their full potential to serve rising domestic demand, whilst also harmonising with the changing global energy landscape in which renewables are fast becoming mainstream (Poudineh et al. 2016, 6). At present, this potential is almost untapped, with the 1.2 GW of solar capacity making up less than 0.5 per cent of total generation capacity in the Middle East (compared to over 90 GW of oil-fired capacity) (IEA 2018b).

However, given the uncertainty in the speed of transition, Arab oil exporters need to adopt a strategy that is likely to be successful under a wide set of future market conditions (Fattouh et al. 2018, 4). Renewables may replace hydrocarbon resources in the domestic energy mix, but not immediately in the government budget, because investments in renewables still do not generate the high returns that the oil and gas industry does. Furthermore, while the renewable energy industry is part of the diversification strategy, it alone cannot deliver the real needs of these economies, such as job creation and welfare improvements.

Therefore, these countries need to gradually 'extend' their energy model rather than completely 'shift' from hydrocarbons to renewables and integrate renewables into their hydrocarbon assets (i.e. oil-exporting countries cannot simply 'transform' into renewable exporting countries) (Fattouh et al. 2018, 4–5). Indeed, these countries have unique characteristics that make the rationale of investment in renewables for them quite compelling. These countries have rising energy demand and are at a stage of development where economic growth is tied up with energy consumption. The rise in energy demand is expected to strain the export capability of these countries. Countries such as Kuwait and the UAE are already net importers of natural gas. Investment in renewables could help boost the short-term revenues of oil-exporting countries as it frees up their hydrocarbon resources for export (as long as international prices are above the breakeven price).[18] In short, for Middle Eastern oil exporters, investment in renewables addresses, to some extent, the government's short-run revenue maximisation objective by freeing exports of hydrocarbons, but does not guarantee their long-term sustainability. In the long run, diversification of their economies remains the main adaptation strategy that these countries need to pursue (Fattouh et al. 2018, 21).

## 5 The Role of Oil Policy and Producer Cooperation

While diversification should remain the ultimate objective of Arab oil-exporting countries, the process is complex and fraught with challenges and potential setbacks and requires broad and deep structural reforms. Many oil producers may take a long time to develop alternative industries and activities that are as profitable as extracting low-cost oil. During the transition, the oil sector will continue to generate most of the income and therefore in addition to diversification, the oil exporting countries should

[18] The economics of renewables in exporting countries depends on the opportunity cost of domestic oil and gas consumption, which is reflected in international price of hydrocarbon resources. According to the Energy Information Administration (EIA) (2016), generating 1 MWh of electricity requires 1.73 barrels of oil or 10.11 mcf of natural gas. The record low auction prices for solar photovoltaics (PV) in Dubai, Mexico, Peru, Chile, Abu Dhabi and Saudi Arabia have shown that, under the right circumstances, an LCOE (levelised cost of electricity) of $0.03/kWh for solar is achievable. IRENA also estimates the global average cost of solar PV to be around $0.06/kWh. If the lower band is considered (which is closer to costs of solar at the region), the break-even prices of oil and gas would be $17.34/bbl and $2.96/mcf, respectively, which are well below international prices. If we consider the global average costs of solar instead (which is pretty much conservative for the region), the break-even prices will increase to $34.68/bbl and $5.93/mcf, which is still below the international price for oil but slightly higher than the average price for natural gas. Even if we add the costs of dealing with solar intermittency (at around $5/MWh) to these numbers, the economics of renewables is still winning over traditional resources in these countries. The economics of renewables will be boosted if we account for the gain that these countries will make from liberated oil and gas for exports. This highlights the importance of integrating renewable into the existing fossil fuel-based generation mix of oil-exporting countries (Fattouh et al. 2018, 20).

aim to maximise the income from their hydrocarbon assets. This implies that oil policy and oil monetisation strategies will remain key in shaping these countries' economic strategies for the foreseeable future.

Faced with the possibility of significantly lower oil demand, some suggest that these countries have no option, and indeed it is even rational that they monetise their reserves as quickly as possible and squeeze out high-cost producers and gain market share—just as with any other competitive market. However, this argument ignores the significant challenges that a shift to a competitive market poses for major oil-producing countries. If most low-cost producers adopt a similar strategy and increase supplies in the face of expected slowing demand growth, this will result in massive fall in oil prices and oil revenues, putting the political, social and economic stability of these countries at risk and derailing the entire economic diversification agenda. In other words, the heavy reliance on oil revenues places a constraint on how fast Arab oil exporters can shift to a more competitive world where prices converge to the marginal cost of physical production. There is also the question as to whether low-cost producers can sharply increase their production capacity, especially in an environment of low oil prices. This is a major undertaking, which requires huge investments, and cannot be implemented in countries with unstable political and economic environment.

This discussion implies that even as we shift to more competitive markets, oil policy and managing producer-producer relations will continue to matter. Rather than simply pursue a policy of non-cooperation and competition with low- and high-cost producers, it is most likely that producers would continue to cooperate and restrain their output in an attempt to maximise revenues. This is despite the fact that the challenges for producers pursuing a cooperative approach are immense, especially in a more competitive oil market.

To start with, the existing framework for cooperation is not well developed to deal with producers with different revenues needs and different degree of financial resilience. It also requires that producers constantly manage the market, based on newly developed criteria. Further, any cooperative action must go beyond output to include long-term investment plans; rapid investment and bringing on new capacity beyond what is needed in the market create problems similar to the high-output/low-price strategy. With many countries within OPEC and non-OPEC having ambitious plans to increase productive capacity, coordination on investment will be extremely difficult, if not impossible. Also, stabilising expectations around a higher oil price will not only encourage US shale producers to increase output, but would also encourage investment in the long-term capital-intensive cycle (Curtis and Montalbano 2017). Above all, long-term cooperation requires unprecedented exercise of leadership to maintain the coalition among producers during good and bad times.

All these issues suggest that maintaining cooperation in a more competitive world is very challenging, and while producers have the incentive to cooperate, the cooperation between producers has to take a different shape to what has existed in the past. For instance, producers should not only be concerned with low oil prices, but also be proactive when prices are too high, as high oil prices induce strong supply and demand responses and speed up the energy transition.

The cooperative solution, which results in a higher oil price, is also not without its costs, and those costs need to be managed by ensuring that prices don't rise too high. It is also true that this cooperative strategy will be less effective over time in a carbon-constrained world. However, this does not imply that cooperation is not possible or sustainable for a prolonged period: as long as these economies are not diversified, the alternative of non-cooperation is also not sustainable. In a world where the prospects of oil demand and the speed of energy transition are highly uncertain, the immediate benefits from pursuing cooperation are more visible and certain than the benefits of pursuing the alternative strategy of fast monetisation of reserves.

Finally, it is worth stressing the co-dependence between diversification in oil exporters and the global energy transition. The transition is already shaping the political and economic outcomes in the Arab oil exporters, but the transition in the major Arab oil exporters to a more diversified and more resilient economies will also shape the global energy transition. In other words, this is a two-way street. If the transition in Arab countries does not go smoothly and countries fail in their diversification efforts, this could result in lower investment in the oil sector, output disruptions and more volatile oil prices. Also, in the absence of diversification, oil exporters will continue to push for higher oil prices. These have the effect of speeding up the global energy transition (Fattouh et al. 2018, 21). In contrast, if these countries succeed in their diversification objectives, they will not only increase the resilience of their economies, but this would allow them to pursue a more flexible and proactive oil policy and adopt long-term strategies that could influence the speed of global energy transition and secure long-term oil demand.

## 6 Conclusions

This chapter has added context to the debate on economic diversification by analysing it against the key arguments around peak oil demand and the energy transition. While exercises in forecasting peak demand contribute towards providing a new motivation for diversification as compared with the past, they should not be the sole factor upon which these strategies are undertaken. The starting point of any analysis of the Arab oil exporters should not be based on an approach solely predicated upon the timeline within which 'oil will no longer be in demand'. Instead, it should take into account the consolidation of three key trends: first, that the range of uncertainty surrounding peak demand is high, and the speed of energy transition is highly uncertain and will not be uniform across the world. Second, that a single peak is not a certainty, particularly when taking into account the possibility of rebound effect. And third, that although an energy transition may be in progress, it is difficult to assume that the transition will result in a sudden and sharp discontinuity in oil demand, implying that oil will continue to be an important part of the energy mix (including in non-combustible uses) for the foreseeable future and that the oil sector will continue to play a key role in oil exporters' economies.

The broader characteristics of the current energy transition from hydrocarbons to low-carbon energy sources are of more relevance to economic diversification rather than predictions of when oil demand will peak. In this regard, although there is unlikely to be a sharp discontinuity in oil use, it is also uncertain whether the current transition will exactly mirror the slow speed of historical transitions which were more about developing technologies in an age of scarcity based on markets and innovation, rather than being problem- and policy-driven.

The diversification strategy adopted by oil-exporting countries will be conditioned by the speed of the transition, which could be decades, during which the oil sector will continue to play a key role in these economies, both as a means to diversification (for e.g. extending the value chain beyond simply producing crude oil and exporting it to international markets) and as a generator of income. Arab oil producers will at the same time need to adopt complementary cost-effective strategies to optimise the use of their resource domestically to increase hydrocarbon exports. These strategies include taking advantage of the 'inflection point' in renewables, which has made renewable technologies competitive to liberate hydrocarbons used in the domestic economy at subsidised prices, for export, as well as adopt demand-side measures to encourage more efficient energy use. These are more cost-effective strategies than simply increasing productive capacity, which is costly and requires massive investments.

At the same time, Arab oil exporters will need to be far more strategic in their use of the oil sector to diversify their economies. In a more competitive world, oil policy will also continue to matter; cooperation between producers will be imperative, yet challenging, as a cooperative strategy could be less effective in a carbon-constrained world. The global energy transition will not only shape the political and economic outcomes in the Arab oil exporters, but the transition in the major oil exporters and the key choices they make will also shape the global energy transition.

## References

BP (2018) Annual Energy Outlook. https://www.bp.com/content/dam/bp/en/corporate/pdf/energy-economics/energy-outlook/bp-energy-outlook-2018.pdf [Accessed January 30, 2019].

Curtis, T. and Montalbano, B. (2017) 'Completion Design Changes and the Impact on US Shale Well Productivity', Energy Insight 21, Oxford Institute for Energy Studies.

Dale, S. and Fattouh, B. (2018) 'Peak Oil Demand and Long Run Oil Prices', OIES Energy Insight 25, Oxford Institute for Energy Studies. https://www.oxfordenergy.org/publications/peak-oil-demand-long-run-oil-prices/. [Accessed January 30, 2019]

EIA (2017) International Energy Outlook, US Energy Information Administration.

EIA (2018) International Energy Outlook, US Energy Information Administration.

Fattouh, B., Poudineh, R. and West, R. (2018) 'The Rise of Renewables and Energy Transition: what adaptation strategy for oil companies and oil-exporting countries?', OIES Paper MEP19, Oxford Institute for Energy Studies. https://www.oxfordenergy.org/publications/rise-renewables-energy-transition-adaptation-strategy-oil-companies-oil-exporting-countries/ [Accessed January 30, 2019].

Fattouh, B. and Shehabi, M. (2019) The Gulf Economies' Long Road towards Better Diversification, WE Energy.

FT (2018) 'BP Says Oil Demand to Peak by Late 2030s', *Financial Times*, February 20.

Fulton, L, Mason, J., and Meroux, D. (2017) 'Three Revolutions in Urban Transportation', UC Davis Sustainable Transportation Energy Pathways of the Institute of Transportation Studies. https://www.itdp.org/wp-content/

uploads/2017/04/UCD-ITDP-3R-Report-FINAL.pdf [Accessed January 30, 2019].

Sovacool, B.K. and Geels, F.W. (2016). 'Further reflections on the temporality of energy transitions: A response to critics', *Energy Research & Social Science*, 22, 232–7.

IEA (2017) World Energy Outlook, International Energy Agency, Paris.

IEA (2018a) 'Oil 2018: Analysis and Forecasts to 2023', International Energy Agency, Paris.

IEA (2018b) 'Outlook for Producer Economies: What do changing energy dynamics mean for major oil and gas exporters?' International Energy Agency, Paris.

IEA (2018c) World Energy Outlook, International Energy Agency, Paris.

IMF (2016) 'Economic Diversification in Arab Oil Exporting Countries', Annual Meeting of Arab Ministers of Finance, April 2016, International Monetary Fund. https://www.imf.org/external/np/pp/eng/2016/042916.pdf [Accessed January 30, 2019].

IRENA (2019) 'Renewable Energy Market Analysis—GCC 2019', *International Renewable Energy Agency*, Abu Dhabi. https://irena.org/publications/2019/Jan/Renewable-Energy-Market-Analysis-GCC-2019 [Accessed January 30, 2019].

Luciani, G. (2019) 'Unsustainable, but why?', *Oxford Energy Forum*, March, Oxford Institute for Energy Studies.

Matar, W., Murphy, F., Pierru, A., Rioux, B. and Wogan, D. (2017) 'Efficient industrial energy use: the first step in transitioning Saudi Arabia's energy mix', *Energy Policy*, 105, pp. 80-92.

OPEC (2017) World Oil Outlook, Organisation of Oil Exporting Countries.

Poudineh, R., Sen, A. and Fattouh, B. (2016) 'Advancing Renewable Energy in Resource-Rich Economies of the MENA', OIES Paper MEP15, Oxford Institute for Energy Studies. https://www.oxfordenergy.org/publications/rise-renewables-energy-transition-adaptation-strategy-oil-companies-oil-exporting-countries/ [Accessed January 30, 2019].

Shehabi, M. (2019) 'GCC economies do not need diversification; they need better diversification', *Oxford Energy Forum*, March, Oxford Institute for Energy Studies.

CHAPTER 6

# Economic Diversification and Sustainable Development of GCC Countries

*Joerg Beutel*

## 1 Introduction

For decades now, exports and imports of most countries have been growing more rapidly than domestic production. This is a strong indication that, besides foreign trade in final products, trade in intermediates is also becoming increasingly important. Globalisation in production is changing the way in which nations interact, and any analysis of diversification should therefore also encompass the worldwide exchange of intermediates in production. For this reason, an input-output approach, which accounts for the role of intermediates, is more appropriate for any analysis of diversification than a traditional approach based purely on macroeconomic data.

This chapter analyses economic diversification in the GCC using data from input-output tables to compare the performance of these economies with that of a 'reference case', Norway, which is considered as having successfully diversified its economy despite having a large oil resource

J. Beutel (✉)
Konstanz University of Applied Sciences, Konstanz, Germany
e-mail: Joerg.Beutel@htwg-konstanz.de



G. Luciani, T. Moerenhout (eds.), *When Can Oil Economies Be Deemed Sustainable?*, The Political Economy of the Middle East,
https://doi.org/10.1007/978-981-15-5728-6_6

base. It also assesses their relative progress on sustainable development using a new measure, adjusted net savings, which measures the 'true' rate of savings in an economy after accounting for investments in man-made and human capital, depletion of natural resources and damage from environmental pollution. This view of sustainable development requires that the nation passes on an *aggregate stock* of physical capital, human capital and natural capital to the next generation that is not smaller than the one that currently exists. This requires that the loss of depleting resources be offset by increasing the stock of physical and human capital.

The article concludes that GCC countries have (contrary to expectation) collectively performed relatively well on diversification, but their performance on sustainable development varies between countries.

The required information for sustainable development and diversification can be deducted from the national accounts in great detail. The main sources for analysis are monetary and physical supply and use tables, sector accounts, extended input-output tables and social accounting matrices.

## 2 Trends in Production and Foreign Trade

Since the foundation of the World Trade Organization (WTO) in 1995, trade boosted thanks to reduced barriers. Since then, exports and imports of most countries have been growing more rapidly than domestic production. Globalisation in production is changing the way in which nations interact, and any analysis of diversification should therefore also encompass the worldwide exchange of intermediates in production. This is the main reason why we promote an input-output approach for the analysis of diversification and sustainable development (Beutel 2012).

### *2.1 Exports and Imports*

If exports and imports are growing faster than GDP, the shares of exports and imports in GDP are also increasing. Furthermore, if net exports grow faster than GDP, the purchasing power of the nation is increasing and imports also tend to grow faster than GDP. In Table 6.1, GDP and foreign

**Table 6.1** GDP and foreign trade at current prices of the world's 10 largest economies

| | *USA* | *China* | *Japan* | *Germany* | *United Kingdom* | *France* | *India* | *Italy* | *Brazil* | *Canada* | *G10* |
|---|---|---|---|---|---|---|---|---|---|---|---|
| | Gross domestic product (millions of US dollar) | | | | | | | | | | |
| 1995 | 7,664,060 | 736,870 | 5,449,118 | 2,591,447 | 1,335,286 | 1,609,794 | 358,024 | 1,170,824 | 778,053 | 604,014 | 22,297,488 |
| 2000 | 10,284,779 | 1,214,915 | 4,887,520 | 1,949,952 | 1,647,874 | 1,368,437 | 453,578 | 1,141,759 | 652,360 | 742,288 | 24,343 464 |
| 2005 | 13,093,726 | 2,308,800 | 4,755,410 | 2,861,339 | 2,520,709 | 2,203,624 | 812,059 | 1,852,616 | 891,634 | 1,169,393 | 32,469,311 |
| 2010 | 14,964,372 | 6,066,351 | 5,700,098 | 3,417,095 | 2,441,173 | 2,646,837 | 1,650,635 | 2,125,058 | 2,208,838 | 1,613,463 | 42,833,921 |
| 2015 | 18,120,714 | 11,226,185 | 4,379,869 | 3,375,611 | 2,885,570 | 2,433,562 | 2,132,755 | 1,832,347 | 1,803,650 | 1,552,808 | 49,743,071 |
| 2016 | 18,624,475 | 11,218,281 | 4,936,212 | 3,477,796 | 2,647,899 | 2,465,454 | 2,259,642 | 1,858,913 | 1,795,926 | 1,529,760 | 50,814,358 |
| | Exports of goods and services (millions of US dollar) | | | | | | | | | | |
| 1995 | 812,813 | 140,675 | 488,884 | 570,285 | 335,955 | 360,689 | 40,196 | 289,773 | 57,122 | 218,079 | 3,314,471 |
| 2000 | 1,096,835 | 279,471 | 519,274 | 601,161 | 409,394 | 385,536 | 61,702 | 292,950 | 65,854 | 328,370 | 4,040,547 |
| 2005 | 1,308,901 | 773,092 | 666,349 | 1,079,882 | 622,240 | 581,070 | 160,991 | 456,701 | 135,919 | 430,280 | 6,215,425 |
| 2010 | 1,852,335 | 1,602,483 | 857,110 | 1,443,735 | 688,755 | 689,322 | 372,898 | 535,264 | 237,189 | 469,066 | 8,748,157 |
| 2015 | 2,264,916 | 2,431,269 | 773,034 | 1,582,312 | 790,107 | 722,100 | 425,341 | 548,573 | 232,476 | 490,372 | 10,260,500 |
| 2016 | 2,214,566 | 2,197,922 | 797,532 | 1,603,933 | 739,195 | 721,408 | 433,320 | 554,276 | 224,280 | 474,344 | 9,960,775 |
| | Imports of goods and services (millions of US dollar) | | | | | | | | | | |
| 1995 | 902,571 | 128,718 | 420,001 | 558,153 | 332,299 | 335,901 | 44,596 | 246,757 | 73,196 | 199,279 | 3,241,470 |
| 2000 | 1,472,630 | 250,686 | 449,416 | 595,914 | 439,771 | 370,654 | 66,047 | 283,347 | 82,121 | 286,674 | 4,297,259 |
| 2005 | 2,030,086 | 648,505 | 594,571 | 935,107 | 686,991 | 590,200 | 184,026 | 458,748 | 105,596 | 384,903 | 6,618,733 |
| 2010 | 2,364,992 | 1,380,082 | 773,860 | 1,266,126 | 752,366 | 739,135 | 447,311 | 577,110 | 260,184 | 499,993 | 9.061,158 |
| 2015 | 2,788,958 | 2,045,768 | 791,424 | 1,312,470 | 839,561 | 758,772 | 474,641 | 494,893 | 253,714 | 527,476 | 10,287,680 |
| 2016 | 2,735,805 | 1,948,009 | 749,700 | 1,326,711 | 797,270 | 769,431 | 466,266 | 491,307 | 217,764 | 510,595 | 10,012,858 |

*(continued)*

**Table 6.1** (continued)

| | *USA* | *China* | *Japan* | *Germany* | *United Kingdom* | *France* | *India* | *Italy* | *Brazil* | *Canada* | *G10* |
|---|---|---|---|---|---|---|---|---|---|---|---|
| | Net exports of goods and services (millions of US dollar) | | | | | | | | | | |
| 1995 | −89,758 | 11,957 | 68,883 | 12,132 | 3656 | 24,788 | −4400 | 43,016 | −16,075 | 18,801 | 73,000 |
| 2000 | −375,795 | 28,785 | 69,858 | 5247 | −30,377 | 14,882 | −4345 | 9603 | −16,267 | 41,696 | −256,712 |
| 2005 | −721,185 | 124,587 | 71,778 | 144,776 | −64,751 | −9130 | −23,035 | −2047 | 30,323 | 45,377 | −403,309 |
| 2010 | −512,657 | 222,401 | 83,250 | 177,609 | −63,611 | −49,813 | −74,413 | −41,846 | −22,994 | −30,927 | −313,002 |
| 2015 | −524,042 | 385,501 | −18,391 | 269,843 | −49,454 | −36,672 | −49,301 | 53,680 | −21,238 | −37,104 | −27,180 |
| 2016 | −521,239 | 249,914 | 47,832 | 277,223 | −58,076 | −48,024 | −32,946 | 62,968 | 6515 | −36,252 | −52,083 |
| | Share of exports in GDP (%) | | | | | | | | | | |
| 1995 | 10.6 | 19.1 | 9.0 | 22.0 | 25.2 | 22.4 | 11.2 | 24.7 | 7.3 | 36.1 | 14.9 |
| 2000 | 10.7 | 23.0 | 10.6 | 30.8 | 24.8 | 28.2 | 13.6 | 25.7 | 10.1 | 44.2 | 16.6 |
| 2005 | 10.0 | 33.5 | 14.0 | 37.7 | 24.7 | 26.4 | 19.8 | 24.7 | 15.2 | 36.8 | 19.1 |
| 2010 | 12.4 | 26.4 | 15.0 | 42.3 | 28.2 | 26.0 | 22.6 | 25.2 | 10.7 | 29.1 | 20.4 |
| 2015 | 12.5 | 21.7 | 17.6 | 46.9 | 27.4 | 29.7 | 19.9 | 29.9 | 12.9 | 31.6 | 20.6 |
| 2016 | 11.9 | 19.6 | 16.2 | 46.1 | 27.9 | 29.3 | 19.2 | 29.8 | 12.5 | 31.0 | 19.6 |
| | Increase of share of exports in GDP (%) | | | | | | | | | | |
| 95–16 | 1.3 | 0.5 | 7.2 | 24.1 | 2.8 | 6.9 | 7.9 | 5.1 | 5.1 | −5.1 | 4.7 |
| | Share of imports in GDP (%) | | | | | | | | | | |
| 1995 | 11.8 | 17.5 | 7.7 | 21.5 | 24.9 | 20.9 | 12.5 | 21.1 | 9.4 | 33.0 | 14.5 |
| 2000 | 14.3 | 20.6 | 9.2 | 30.6 | 26.7 | 27.1 | 14.6 | 24.8 | 12.6 | 38.6 | 17.7 |
| 2005 | 15.5 | 28.1 | 12.5 | 32.7 | 27.3 | 26.8 | 22.7 | 24.8 | 11.8 | 32.9 | 20.4 |
| 2010 | 15.8 | 22.7 | 13.6 | 37.1 | 30.8 | 27.9 | 27.1 | 27.2 | 11.8 | 31.0 | 21.2 |
| 2015 | 15.4 | 18.2 | 18.1 | 38.9 | 29.1 | 31.2 | 22.3 | 27.0 | 14.1 | 34.0 | 20.7 |
| 2016 | 14.7 | 17.4 | 15.2 | 38.1 | 30.1 | 31.2 | 20.6 | 26.4 | 12.1 | 33.4 | 19.7 |
| | Increase of share of imports in GDP (%) | | | | | | | | | | |
| 95–16 | 2.9 | −0.1 | 7.5 | 16.6 | 5.2 | 10.3 | 8.2 | 5.4 | 2.7 | 0.4 | 5.2 |

Source: UNdata—National Accounts Estimates by Main Aggregates 1995–2016

trade are shown for the 10 largest economies of the world (G10) for the last two decades. The most striking examples of an increase in the share of exports in GDP between 1995 and 2016 are Germany (+24.1 per cent), India (+7.9 per cent), Japan (+7.2 per cent), France (+6.9 per cent), Italy (+ 5.1 per cent) and Brazil (+5.1 per cent).

Substantial increases in the share of imports in GDP during the period 1995–2016 are observed for Germany (+16.6 per cent), France (+10.3 percent), India (+8.2 per cent), Japan (+7.5 per cent) and Italy (+ 5.4 per cent). For the 10 largest economies combined (G10) the increase of the export share in GDP during the period 1995–2016 was +4.7 per cent and of the import share in GDP +5.2 per cent.

In Table 6.2, the same information on GDP, exports and imports was collected for the GCC countries. If net exports of a nation grow more rapidly than GDP, the purchasing power of the nation is increasing. In consequence, imports tend also to grow more than GDP. In the GCC countries, the share of exports in GDP increased by 12.9 per cent during the period 1995–2016, while the corresponding share of imports in GDP increased by 14.0 per cent. The most rapid increase of the export share was observed for the United Arab Emirates (UAE) (+57.9 per cent) and Oman (+14.0 per cent). Similar results were observed for the import share in GDP for UAE (+41.3 per cent) and Oman (+17.0 per cent).

### 2.2 *Intermediate Consumption, Value Added and Output*

The numbers in the table suggest that during the last 20 years, the globalisation of economic activities has caused increasing worldwide interdependencies in production, leading to the intermediate consumption of goods and services becoming a key element in the intensification of economic diversification. If the consumption of intermediate products is growing above its GDP growth rate, an economy is moving towards more complex participation in inter-industrial production.

For the future, the challenge for many countries is to become a successful member in the international chain of value added. As shown on Table 6.3, the share of intermediates in total output for G10 countries increased by 5.9 per cent in the period 1995–2011. As a consequence, the corresponding share of gross value added in output declined in the same period by 5.9 per cent. In other words, in this period, the production

**Table 6.2** GDP and foreign trade at current prices of GCC countries

| | *Bahrain* | *Kuwait* | *Oman* | *Qatar* | *Saudi Arabia* | *United Arab Emirates* | *GCC* |
|---|---|---|---|---|---|---|---|
| | Gross domestic product (millions of US dollar) | | | | | | |
| 1995 | 6787 | 26,554 | 13,650 | 8041 | 143,152 | 66,603 | 264,786 |
| 2000 | 9063 | 37,718 | 19,450 | 17,548 | 189,515 | 105,701 | 378,994 |
| 2005 | 15,969 | 80,798 | 31,082 | 43,998 | 328,461 | 182,978 | 683,286 |
| 2010 | 25,713 | 115,416 | 58,641 | 123,627 | 528,207 | 289,787 | 1,141,392 |
| 2015 | 31,126 | 114,059 | 69,832 | 164,641 | 651,757 | 358,135 | 1,389,551 |
| 2016 | 32,179 | 110,346 | 63,171 | 152,452 | 639,617 | 348,744 | 1,346,509 |
| | Exports of goods and services (millions of US dollar) | | | | | | |
| 1995 | 4798 | 14,234 | 4757 | 3613 | 53,450 | 30,591 | 111,442 |
| 2000 | 7176 | 21,301 | 10,474 | 11,964 | 82,259 | 53,889 | 187,063 |
| 2005 | 13,397 | 51,692 | 19,633 | 29,017 | 187,389 | 122,071 | 423,199 |
| 2010 | 17,880 | 76,952 | 38,500 | 78,069 | 261,831 | 225,275 | 698,507 |
| 2015 | 26,326 | 62,024 | 39,166 | 92,291 | 218,010 | 359,401 | 797,218 |
| 2016 | 23,782 | 55,530 | 30,845 | 72,397 | 195,169 | 362,069 | 739,793 |
| | Imports of goods and services (millions of US dollar) | | | | | | |
| 1995 | 4122 | 11,409 | 4009 | 3526 | 39,659 | 24,663 | 87,388 |
| 2000 | 5132 | 11,371 | 5054 | 3966 | 46,926 | 36,003 | 108,452 |
| 2005 | 10,287 | 22,839 | 11,173 | 13,208 | 81,798 | 83,671 | 222,975 |
| 2010 | 13,097 | 35,034 | 24,166 | 29,717 | 174,203 | 184,221 | 460,439 |
| 2015 | 22,302 | 51,627 | 36,667 | 59,271 | 253,555 | 265,950 | 689,371 |
| 2016 | 21,088 | 51,922 | 29,301 | 63,475 | 194,169 | 273,274 | 633,230 |
| | Net exports of goods and services (millions of US dollar) | | | | | | |
| 1995 | 675 | 2825 | 749 | 86 | 13,791 | 5928 | 24,054 |
| 2000 | 2044 | 9930 | 5420 | 7998 | 35,333 | 17,886 | 78,611 |
| 2005 | 3110 | 28,853 | 8460 | 15,809 | 105,592 | 38,400 | 200,224 |
| 2010 | 4783 | 41,918 | 14,333 | 48,352 | 87,628 | 41,054 | 238,069 |
| 2015 | 4024 | 10,397 | 2500 | 33,020 | - 35,544 | 93,451 | 107,848 |
| 2016 | 2694 | 3608 | 1544 | 8922 | 1000 | 88,795 | 106,563 |
| | Share of exports in GDP (%) | | | | | | |
| 1995 | 70.7 | 53.6 | 34.9 | 44.9 | 37.3 | 45.9 | 42.1 |
| 2000 | 79.2 | 56.5 | 53.9 | 68.2 | 43.4 | 51.0 | 49.4 |
| 2005 | 83.9 | 64.0 | 63.2 | 66.0 | 57.1 | 66.7 | 61.9 |
| 2010 | 69.5 | 66.7 | 65.7 | 63.1 | 49.6 | 77.7 | 61.2 |
| 2015 | 84.6 | 54.4 | 56.1 | 56.1 | 33.4 | 100.4 | 57.4 |
| 2016 | 73.9 | 50.3 | 48.8 | 47.5 | 30.5 | 103.8 | 54.9 |
| | Increase of share of exports in GDP (%) | | | | | | |
| 95–16 | 3.2 | −3.3 | 14.0 | 2.6 | −6.8 | 57.9 | 12.9 |

**Table 6.2** (continued)

| | *Bahrain* | *Kuwait* | *Oman* | *Qatar* | *Saudi Arabia* | *United Arab Emirates* | GCC |
|---|---|---|---|---|---|---|---|
| | Share of imports in GDP (%) | | | | | | |
| 1995 | 60.7 | 43.0 | 29.4 | 43.9 | 27.7 | 37.0 | 33.0 |
| 2000 | 56.6 | 30.1 | 26.0 | 22.6 | 24.8 | 34.1 | 28.6 |
| 2005 | 64.4 | 28.3 | 35.9 | 30.0 | 24.9 | 45.7 | 32.6 |
| 2010 | 50.9 | 30.4 | 41.2 | 24.0 | 33.0 | 63.6 | 40.3 |
| 2015 | 71.7 | 45.3 | 52.5 | 36.0 | 38.9 | 74.3 | 49.6 |
| 2016 | 65.5 | 47.1 | 46.4 | 41.6 | 30.4 | 78.4 | 47.0 |
| | Increase of share of imports in GDP (%) | | | | | | |
| 95–16 | 4.8 | 4.1 | 17.0 | −2.2 | 2.7 | 41.3 | 14.0 |

Source: UNdata—National Accounts Estimates by Main Aggregates 1995–2016

process of the 10 largest economies became more complex and more interdependent. It is worth noticing that the share of intermediates in total output in China increased by 6.7 per cent, while the share of intermediates in the USA slightly declined by −0.4 per cent.

For the GCC countries combined, the share of intermediates and gross value added in total output was more or less constant during the last 20 years (see Table 6.4). However, among individual countries, the performance has been mixed. While large increases of intermediates were reported for Bahrain (+10.6 per cent), Oman (+7.7 per cent), Kuwait (+5.4 per cent) and the UAE (+4.5 per cent), these were offset by the decline of intermediates in Saudi Arabia (−4.1 per cent).

## 3 Economic Diversification of the GCC Countries in International Comparison

The sustainable development of nations involves economic, social and environmental changes. Within this process, diversification and structural change of production and demand are closely related to many areas of the economy and society.

**Table 6.3** Intermediates, value added and output of the world's ten largest economies

| | *USA* | *China* | *Japan* | *Germany* | *France* | *Italy* | *United Kingdom* | *Brazil* | *India* | *Canada* | *G10* |
|---|---|---|---|---|---|---|---|---|---|---|---|
| | Intermediates at purchasers' prices (million US dollars) | | | | | | | | | | |
| 1995 | 5,710,633 | 1,153,865 | 4,457,767 | 2,005,032 | 1,298,807 | 1,094,780 | 1,096,869 | 604,655 | 343,683 | 466,288 | 18,232,377 |
| 2000 | 7,993,838 | 2,072,385 | 3,945,944 | 1,668,162 | 1,213,918 | 1,161,085 | 1,393,193 | 550,544 | 426,770 | 645,970 | 21,071,807 |
| 2005 | 10,421,164 | 4,365,839 | 4,051,216 | 2,593,739 | 1,901,837 | 1,936,759 | 1,938,854 | 878,865 | 921,350 | 1,002,072 | 29,339,760 |
| 2010 | 11,129,144 | 12,754,844 | 5,044,160 | 3,250,704 | 2,311,177 | 2,218,062 | 2,038,790 | 1,873,727 | 1,805,076 | 1,329,212 | 43,009,619 |
| 2015 | 13,270,599 | 15,605,704 | 3,905,847 | 3,058,368 | 2,085,240 | 1,825,720 | 2,335,764 | 1,552,677 | 2,124,098 | 1,287,691 | 48,728,413 |
| | Gross value added at basic price (million US dollars) | | | | | | | | | | |
| 1995 | 6,759,151 | 708,115 | 5,287,733 | 2,280,483 | 1,404,508 | 1,019,575 | 1,051,892 | 654,215 | 335,020 | 553,449 | 20,054,141 |
| 2000 | 9,214,378 | 1,164,362 | 4,718,141 | 1,695,955 | 1,185,991 | 985,034 | 1,328,514 | 544,016 | 433,419 | 679,326 | 21,949,136 |
| 2005 | 12,660,441 | 2,192,815 | 4,769,184 | 2,589,273 | 1,972,448 | 1,670,923 | 2,274,762 | 756,995 | 781,680 | 1,088,760 | 29,068,685 |
| 2010 | 14,509,603 | 5,763,455 | 5,686,225 | 3,074,910 | 2,381,037 | 1,911,471 | 2,209,003 | 1,877,439 | 1,531,440 | 1,509,599 | 38,698,664 |
| 2015 | 17,550,250 | 7,117,538 | 4,343,472 | 3,039,094 | 2,182,051 | 1,647,215 | 2,585,059 | 1,549,670 | 1,914,110 | 1,455,566 | 42,360,316 |
| | Output at basic prices (million US dollars) | | | | | | | | | | |
| 1995 | 12,469,784 | 1,861,980 | 9,745,500 | 4,285,514 | 2,703,315 | 2,114,355 | 2,148,760 | 1,258,870 | 678,703 | 1,019,737 | 38,286,518 |
| 2000 | 17,208,216 | 3,236,745 | 8,664,085 | 3,364,117 | 2,399,909 | 2,146,119 | 2,721,707 | 1,094,559 | 860,189 | 1,325,296 | 43,020,941 |
| 2005 | 23,081,605 | 6,558,654 | 8,820,399 | 5,183,011 | 3,874,285 | 3,607,682 | 4,213,615 | 1,635,859 | 1,611,770 | 2,090,832 | 58,408,447 |
| 2010 | 25,638,747 | 18,518,299 | 10,730,385 | 6,325,614 | 4,692,214 | 4,129,533 | 4,247,792 | 3,751,166 | 3,322,592 | 2,838,811 | 81,708,283 |
| 2015 | 30,820,849 | 22,723,243 | 8,249,319 | 6,097,462 | 4,267,291 | 3,472,935 | 4,920,823 | 3,102,346 | 3,643,666 | 2,743,257 | 91,088,731 |
| | Shares of intermediates in total output (%) | | | | | | | | | | |
| 1995 | 45.8 | 62.0 | 45.7 | 46.8 | 48.0 | 51.8 | 51.0 | 48.0 | 50.6 | 45.7 | 47.6 |
| 2000 | 46.5 | 64.0 | 45.5 | 49.6 | 50.6 | 54.1 | 51.2 | 50.3 | 49.6 | 48.7 | 49.0 |
| 2005 | 45.1 | 66.6 | 45.9 | 50.0 | 49.1 | 53.7 | 46.0 | 53.7 | 57.2 | 47.9 | 50.2 |
| 2010 | 43.4 | 68.9 | 47.0 | 51.4 | 49.3 | 53.7 | 48.0 | 50.0 | 54.3 | 46.8 | 52.6 |
| 2015 | 43.1 | 68.7 | 47.3 | 50.2 | 48.9 | 52.6 | 47.5 | 50.0 | 58.3 | 46.9 | 53.5 |
| 2015–1995 | −2.7 | 6.7 | 1.6 | 3.4 | 0.8 | 0.8 | −3.6 | 2.0 | 7.7 | 1.2 | 5.9 |
| | Shares of gross value added in total output (%) | | | | | | | | | | |
| 1995 | 54.2 | 38.0 | 54.3 | 53.2 | 52.0 | 48.2 | 49.0 | 52.0 | 49.4 | 54.3 | 52.4 |
| 2000 | 53.5 | 36.0 | 54.5 | 50.4 | 49.4 | 45.9 | 48.8 | 49.7 | 50.4 | 51.3 | 51.0 |
| 2005 | 54.9 | 33.4 | 54.1 | 50.0 | 50.9 | 46.3 | 54.0 | 46.3 | 48.5 | 52.1 | 49.8 |
| 2010 | 56.6 | 31.1 | 53.0 | 48.6 | 50.7 | 46.3 | 52.0 | 50.0 | 46.1 | 53.2 | 47.4 |
| 2015 | 56.9 | 31.3 | 52.7 | 49.8 | 51.1 | 47.4 | 52.5 | 50.0 | 52.5 | 53.1 | 46.5 |
| 2015–1995 | 2.7 | −6.7 | −1.6 | −3.4 | −0.8 | −0.8 | 3.6 | −2.0 | 3.2 | −1.2 | −5.9 |

**Table 6.4** Intermediates, value added and output at current prices of GCC countries

| | *Bahrain* | *Kuwait* | *Oman* | *Qatar* | *Saudi Arabia* | *United Arab Emirates* | *GCC* |
|---|---|---|---|---|---|---|---|
| | Intermediates at purchasers' prices (million US dollar) | | | | | | |
| 1995 | 3737 | 11,878 | 5131 | 3332 | 79,090 | 38,371 | 141,539 |
| 2000 | 7101 | 16,019 | 7440 | 5351 | 99,219 | 64,268 | 199,398 |
| 2005 | 14,009 | 43,685 | 13,197 | 14,670 | 164,159 | 117,210 | 366,929 |
| 2010 | 24,374 | 79,339 | 29,639 | 55,329 | 262,678 | 194,448 | 645,807 |
| 2015 | 26,396 | 74,149 | 39,613 | 68,906 | 307,508 | 252,409 | 768,981 |
| | Gross value added at basic prices (million US dollar) | | | | | | |
| 1995 | 6782 | 26,879 | 13,685 | 8311 | 139,225 | 65,744 | 260,626 |
| 2000 | 9063 | 39,015 | 19,326 | 18,052 | 184,162 | 104,337 | 373,955 |
| 2005 | 15,969 | 83,808 | 31,784 | 44,053 | 319,210 | 180,617 | 675,441 |
| 2010 | 25,442 | 131,065 | 59,876 | 124,080 | 512,817 | 289,880 | 1,143,160 |
| 2015 | 30,835 | 131,530 | 73,675 | 165,245 | 651,097 | 357,949 | 1,410,331 |
| Output at basic prices (million US dollar) | | | | | | | |
| 1995 | 10,519 | 38,757 | 18,816 | 11,643 | 218,315 | 104,114 | 402,165 |
| 2000 | 16,163 | 55,035 | 26,766 | 23,402 | 283,381 | 168,605 | 573,353 |
| 2005 | 29,978 | 127,493 | 44,981 | 58,723 | 483,369 | 297,827 | 1,042,370 |
| 2010 | 49,816 | 210,403 | 89,515 | 179,410 | 775,494 | 484,328 | 1,788,966 |
| 2015 | 57,231 | 205,679 | 113,288 | 234,151 | 958,604 | 610,358 | 2,179,312 |
| | Share of intermediates in output (%) | | | | | | |
| 1995 | 35.5 | 30.6 | 27.3 | 28.6 | 36.2 | 36.9 | 35.2 |
| 2000 | 43.9 | 29.1 | 27.8 | 22.9 | 35.0 | 38.1 | 34.8 |
| 2005 | 46.7 | 34.3 | 29.3 | 25.0 | 34.0 | 39.4 | 35.2 |
| 2010 | 48.9 | 37.7 | 33.1 | 30.8 | 33.9 | 40.1 | 36.1 |
| 2015 | 46.1 | 36.1 | 35.0 | 29.4 | 32.1 | 41.4 | 35.3 |
| | Increase in share of intermediates in output (%) | | | | | | |
| 95–15 | 10.6 | 5.4 | 7.7 | 0.8 | −4.1 | 4.5 | 0.1 |
| | Share of gross value added in output (%) | | | | | | |
| 1995 | 64.5 | 69.4 | 72.7 | 71.4 | 63.8 | 63.1 | 64.8 |
| 2000 | 56.1 | 70.9 | 72.2 | 77.1 | 65.0 | 61.9 | 65.2 |
| 2005 | 53.3 | 65.7 | 70.7 | 75.0 | 66.0 | 60.6 | 64.8 |
| 2010 | 51.1 | 62.3 | 66.9 | 69.2 | 66.1 | 59.9 | 63.9 |
| 2015 | 53.9 | 63.9 | 65.0 | 70.6 | 67.9 | 58.6 | 64.7 |
| | Increase in share of gross value added in output (%) | | | | | | |
| 95–15 | −10.6 | −5.4 | −7.7 | −0.8 | 4.1 | −4.5 | −0.1 |

Source: UNdata—National Accounts Estimates by Main Aggregates 1995–2016; OECD

For income per capita to converge, countries must move towards more diversified and complex production structures incorporating more advanced technology and knowledge. Economic diversification means the diversification of exports and domestic production away from extreme dependence on a single dominant industry or a few natural-resource-based products, as well as towards increased complexity and quality of output.

In the following two tables, data on gross value added by industry and on GDP by type of expenditure are presented for the GCC countries in 2016 to discuss their different economic structure. In this comparison, Norway will act as the reference, as this country is highly developed in many ways, generating substantial oil and gas revenues and benefitting from a successful sovereign wealth fund. The Government Pension Fund Global, also known as Oil Fund, was established in 1990 to invest the surplus revenue of the Norwegian petroleum sector for the benefit of future generations. It has over US$1 trillion in assets, including 1.3 per cent of global stocks and shares, making it the world's largest sovereign wealth fund.

Table 6.5 reveals that the industry structure of the UAE is similar to that of Norway. It is also striking to observe that the industry structure of Iran is similar to the structure of the GCC countries combined, except for agriculture.

If we compare the industry structure of the individual GCC countries, we observe that their structure is quite diverse. The share of mining and utilities in gross value added is much higher in Kuwait than in Bahrain. Manufacturing in Kuwait has a very low share in gross value added, but in Bahrain this share is more than three times larger.

The oil and gas dependency of Kuwait and Oman was still high in 2016. The sector 'Mining and utilities' accounted for 51.4 per cent of gross value added in Kuwait, 42.6 per cent in Oman and 26.9 per cent in the GCC countries combined. The oil and gas dependency of Norway is considerably lower (19.0 per cent). And yet, if we compare Norway and the GCC in terms of the expenditure side of GDP, we observe a similar structure. But it is obvious that Norway is more developed concerning 'Other activities', which include private business services and government services.

**Table 6.5** Value added by industries 2016 at current prices in international comparison

| *Year* | *Bahrain* | *Kuwait* | *Oman* | *Qatar* | *Saudi Arabia* | *United Arab Emirates* | *GCC* | *Iran* | *Malaysia* | *Norway* | *Canada* |
|---|---|---|---|---|---|---|---|---|---|---|---|
| | Gross value added (million US dollar) | | | | | | | | | | |
| Agriculture, forestry, fishing | 107 | 550 | 883 | 279 | 17,301 | 2795 | 21,916 | 40,998 | 26,133 | 7916 | 24,252 |
| Mining, utilities | 4318 | 64,319 | 28,545 | 47,219 | 152,570 | 72,170 | 369,141 | 72,040 | 36,755 | 57,472 | 146,618 |
| Manufacturing | 5835 | 6679 | 6355 | 13,792 | 80,444 | 32,230 | 145,336 | 50,604 | 67,735 | 25,009 | 147,467 |
| Construction | 2544 | 2504 | 4158 | 18,105 | 42,571 | 35,652 | 105,535 | 21,398 | 13,114 | 22,503 | 114,853 |
| Trade, restaurants, hotels | 2262 | 5383 | 5339 | 17,124 | 73,637 | 53,157 | 156,903 | 54,679 | 53,123 | 30,016 | 177,325 |
| Transport, communication | 2402 | 6508 | 3270 | 7710 | 42,774 | 37,305 | 99,970 | 43,281 | 25,403 | 32,437 | 107,603 |
| Other activities | 14,422 | 39,146 | 18,491 | 55,014 | 230,764 | 115,435 | 473,273 | 137,031 | 70,611 | 127,898 | 716,476 |
| Value added | 31,892 | 125,091 | 67,042 | 159,243 | 640,061 | 348,744 | 1,372,073 | 420,030 | 292,874 | 303,251 | 1,434,594 |
| | Shares in gross value added (%) | | | | | | | | | | |
| Agriculture, forestry, fishing | 0.3 | 0.4 | 1.3 | 0.2 | 2.7 | 0.8 | 1.6 | 9.8 | 8.9 | 2.6 | 1.7 |
| Mining, utilities | 13.5 | 51.4 | 42.6 | 29.7 | 23.8 | 20.7 | 26.9 | 17.2 | 12.5 | 19.0 | 10.2 |
| Manufacturing | 18.3 | 5.3 | 9.5 | 8.7 | 12.6 | 9.2 | 10.6 | 12.0 | 23.1 | 8.2 | 10.3 |
| Construction | 8.0 | 2.0 | 6.2 | 11.4 | 6.7 | 10.2 | 7.7 | 5.1 | 4.5 | 7.4 | 8.0 |
| Trade, restaurants, hotels | 7.1 | 4.3 | 8.0 | 10.8 | 11.5 | 15.2 | 11.4 | 13.0 | 18.1 | 9.9 | 12.4 |
| Transport, communication | 7.5 | 5.2 | 4.9 | 4.8 | 6.7 | 10.7 | 7.3 | 10.3 | 8.7 | 10.7 | 7.5 |
| Other activities | 45.2 | 31.3 | 27.6 | 34.5 | 36.1 | 33.1 | 34.5 | 32.6 | 24.1 | 42.2 | 49.9 |
| Value added | 100.0 | 100.0 | 100.0 | 100.0 | 100.0 | 100.0 | 100.0 | 100.0 | 100.0 | 100.0 | 100.0 |

Source: UNdata—National Accounts Estimates of Main Aggregates

In Table 6.6, we present similar information for the expenditure side of GDP in 2016 for consumption, capital formation and foreign trade. Here, we observe significant differences between the GCC countries. The share of household consumption in GDP is similar in Norway (45.5 per cent), Bahrain (45.0 per cent), Kuwait (42.9 per cent) and Saudi Arabia (42.8 per cent). The share of government consumption is in the same range in Norway (24.3 per cent), Oman (28.2 per cent), Kuwait (27.7 per cent) and Saudi Arabia (25.8 per cent). By far the highest investment ratio is reported for Qatar (45.2 per cent) and for Oman (38.0 per cent). The shares of gross fixed capital formation in GDP in Norway (24.1 per cent) and in the GCC combined (23.6 per cent) are also similar.

In the bottom part of Table 6.6, we relate the results on the expenditure side of GDP to the population. The highest GDP per capita is reported for Norway (70,617 Mio $/person), followed by the GCC countries Qatar (59.324 Mio $/person), UAE (37,622 Mio $/person) and Kuwait (27,229 Mio $/person). But what matters for the well-being of the people is less GDP per capita than final consumption per capita. In this respect, among the selected countries, the gold medal goes to Norway (32,140 Mio $/person), the silver medal to Canada (24,578 Mio $/person) and the bronze medal to Qatar (15,312 Mio $/person).

### 3.1 *Product Concentration and Product Diversification*

The most widely quoted product concentration and diversification indices are published by the United Nations Conference on Trade and Development (UNCTAD).

Figure 6.1 presents these indices for the GCC for the period 1995–2016. The product concentration index measures whether the exports and imports of the GCC are concentrated on a few products or distributed in a more homogeneous manner among a series of products. The diversification index indicates whether the structure of exports or imports by product differs in the GCC from the world pattern.

Both top two indices for GCC exports show a falling trend. The concentration index for exports indicates that today a broader range of products is exported than just oil and gas. The export diversification index shows that the divergence from the world pattern was significantly reduced.

The bottom two indices for imports are relatively stable. The concentration index shows that imports are rather homogenously distributed among many of products. The diversification index for imports shows very little divergence from the pattern of world trade (Box 6.1).

**Table 6.6** GDP by type of expenditure 2016 at current prices in international comparison

| *Category* | *Bahrain* | *Kuwait* | *Oman* | *Qatar* | *Saudi Arabia* | *United Arab Emirates* | *GCC* | *Iran* | *Malaysia* | *Norway* | *Canada* |
|---|---|---|---|---|---|---|---|---|---|---|---|
| | Million US dollar | | | | | | | | | | |
| Final consumption expenditure | 19,976 | 77,912 | 37,622 | 74,565 | 439,001 | 171,895 | 820,973 | 266,759 | 200,163 | 259,072 | 1,216,211 |
| Household consumption expenditure | 14,491 | 47,342 | 19,830 | 39,348 | 276,329 | 122,964 | 520,303 | 210,086 | 162,873 | 168,887 | 891,923 |
| General government final consumption | 5485 | 30,570 | 17,792 | 35,218 | 162,673 | 48,931 | 300,669 | 56,674 | 37,290 | 90,185 | 324,288 |
| Gross capital formation | 9509 | 28,826 | 24,005 | 68,965 | 199,616 | 88,053 | 418,973 | 151,864 | 77,412 | 108,793 | 349,801 |
| Gross fixed capital formation | 8328 | 28,826 | 24,005 | 68,965 | 161,290 | 82,459 | 373,872 | 86,164 | 76,437 | 89,342 | 351,798 |
| Changes in inventories | 1181 | 0 | 0 | 0 | 38,326 | 5594 | 45,102 | 65,700 | 974 | 19,450 | - 1997 |
| Net exports of goods and services | 2694 | 3608 | 1544 | 8922 | 1000 | 88,795 | 106,563 | 6779 | 18,957 | 3204 | - 36,252 |
| Exports of goods and services | 23,782 | 55,530 | 30,845 | 72,397 | 195,169 | 362,069 | 739,793 | 95,307 | 199,271 | 126,670 | 474,344 |
| Imports of goods and services | 21,088 | 51,922 | 29,301 | 63,475 | 194,169 | 273,274 | 633,230 | 88,528 | 180,315 | 123,466 | 510,595 |
| Gross domestic product | 32,179 | 110,346 | 63,171 | 152,452 | 639,617 | 348,744 | 1,346,509 | 425,403 | 296,531 | 371,069 | 1,529,760 |

(*continued*)

Table 6.6 (continued)

| *Category* | *Bahrain* | *Kuwait* | *Oman* | *Qatar* | *Saudi Arabia* | *United Arab Emirates* | *GCC* | *Iran* | *Malaysia* | *Norway* | *Canada* |
|---|---|---|---|---|---|---|---|---|---|---|---|
| | Shares in GDP (%) | | | | | | | | | | |
| Final consumption expenditure | 62.1 | 70.6 | 59.6 | 48.9 | 68.6 | 49.3 | 61.0 | 62.7 | 67.5 | 69.8 | 79.5 |
| Household consumption expenditure | 45.0 | 42.9 | 31.4 | 25.8 | 42.8 | 35.3 | 38.6 | 49.4 | 54.9 | 45.5 | 58.3 |
| General government final consumption | 17.0 | 27.7 | 28.2 | 23.1 | 25.8 | 14.0 | 22.3 | 13.3 | 12.6 | 24.3 | 21.2 |
| Gross capital formation | 29.6 | 26.1 | 38.0 | 45.2 | 30.9 | 25.2 | 31.1 | 35.7 | 26.1 | 29.3 | 22.9 |
| Gross fixed capital formation | 25.9 | 26.1 | 38.0 | 45.2 | 26.0 | 23.6 | 27.8 | 20.3 | 25.8 | 24.1 | 23.0 |
| Changes in inventories | 3.7 | 0.0 | 0.0 | 0.0 | 4.9 | 1.6 | 3.3 | 15.4 | 0.3 | 5.2 | −0.1 |
| Net exports of goods and services | 8.4 | 3.3 | 2.4 | 5.9 | 0.4 | 25.5 | 7.9 | 1.6 | 6.4 | 0.9 | −2.4 |
| Exports of goods and services | 73.9 | 50.3 | 48.8 | 47.5 | 31.1 | 103.8 | 54.9 | 22.4 | 67.2 | 34.1 | 31.0 |
| Imports of goods and services | 65.5 | 47.1 | 46.4 | 41.6 | 30.7 | 78.4 | 47.0 | 20.8 | 60.8 | 33.3 | 33.4 |
| Gross domestic product | 100.0 | 100.0 | 100.0 | 100.0 | 100.0 | 100.0 | 100.0 | 100.0 | 100.0 | 100.0 | 100.0 |
| | Population | | | | | | | | | | |
| Memo: Population (millions) | 1.4 | 4.1 | 4.4 | 2.6 | 32.3 | 9.3 | 54.0 | 80.3 | 31.2 | 5.3 | 36.3 |
| GDP per capita ($/person) | 22,579 | 27,229 | 14,277 | 59,324 | 19,817 | 37,622 | 24,927 | 5299 | 9508 | 70,617 | 42,154 |
| Private consumption per capita ($/person) | 10,168 | 11,682 | 4482 | 15,312 | 8562 | 13,265 | 9632 | 2617 | 5222 | 32,140 | 24,578 |

Source: UNdata—National Accounts Estimates of Main Aggregates

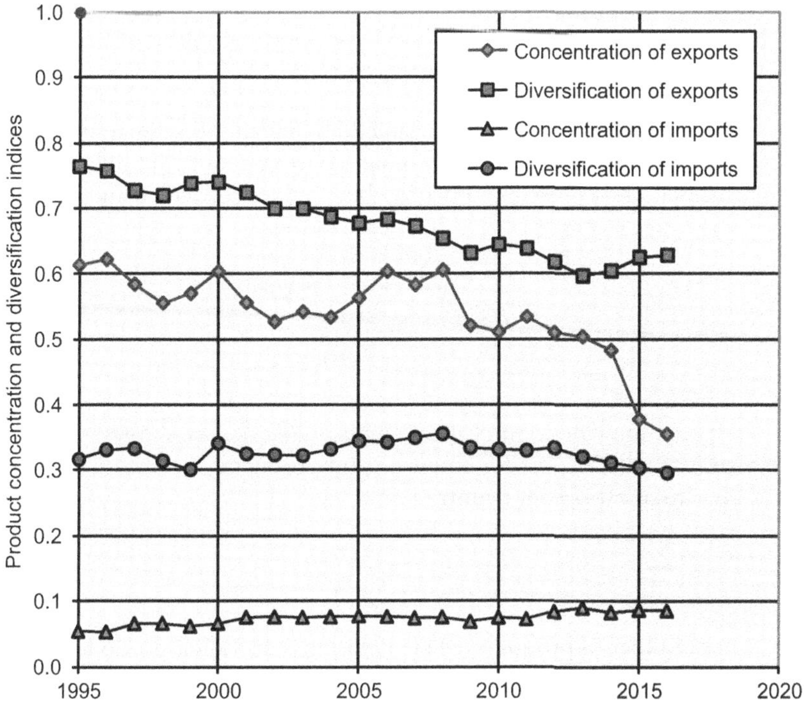


**Fig. 6.1** Product concentration and diversification of exports and imports for GCC. (Source: UNCTADStat)

### *3.2 Economic Diversification of the Industrial Sector*

The challenge is now to study the diversification of the industrial sector in the GCC countries. The key question is whether diversification has increased during the last 50 years.

The Herfindahl-Hirschman Index is the most widely used measure to evaluate market concentration and industry concentration of an economy. This time, the index is the sum of squared shares of the various industries in total gross value added. In the normalised form, the index varies from 0 to 1.

In the case of a low value, the economy has a large number of industries with similar shares in gross value added, indicating high diversity. If the

**Box 6.1 Product Concentration and Diversification Indices of Exports and Imports**

**Concentration Index:**

The concentration index, also named Herfindahl-Hirschman Index, is a measure of the degree of product concentration. The following normalised HHI is used in order to obtain values between 0 and 1:

$$H_j = \frac{\sqrt{\sum_{i=1}^{n}\left(\frac{x_{ij}}{X_j}\right)} - \sqrt{\frac{1}{n}}}{1-\sqrt{1/n}}$$

Where,

Hj = country or country group index

xij = value of export for country j and product i

Xj = total exports of country j

$$X_j = \sum_{i=1}^{n} x_{ij}$$

and

n = number of products (SITC Revision 3 at 3-digit group level).

An index value closer to 1 indicates a country's exports or imports are highly concentrated on a few products. On the contrary, values closer to 0 reflect exports or imports are more homogeneously distributed among a series of products.

**Diversification Index:**

The diversification index is computed by measuring the absolute deviation of the trade structure of a country from world structure:

$$S_j = \frac{\sum_i \left|h_{ij} - h_i\right|}{2}$$

Where,

hij = share of product i in total exports or imports of country or country group j

hi = share of product i in total world exports or imports.

The diversification index takes values between 0 and 1. A value closer to 1 indicates greater divergence from the world pattern.

Source: UNCTAD Merchandise: Product concentration and diversification indices of exports and imports, annual, 1995–2016

index reaches 1, only one industry accounts for all gross value added, and a high concentration of economic activity is given. Thus, a decline in the index signifies less concentration in the dominant industry or greater diversification. If more concentration in the dominant sector is observed, the indicator will be higher.

Figure 6.2 plots the Herfindahl-Hirschman Index for all GCC countries (Bahrain, Qatar, Kuwait, Oman, Saudi Arabia and UAE). The calculation is based on the UN Statistics Division's long time series for value added by economic activity for the period 1970–2016, covering the following seven industries: (1) agriculture, hunting, forestry and fishing, (2) mining and utilities, (3) manufacturing, (4) construction, (5) wholesale, retail trade, restaurants and hotels, (6) transport, storage and communication and (7) other activities.

Starting from the first oil boom in 1974 and the second oil boom in 1978, the index steadily fell until 1998, indicating a successful process of diversification. However, at the end of the last century, the recovery of oil

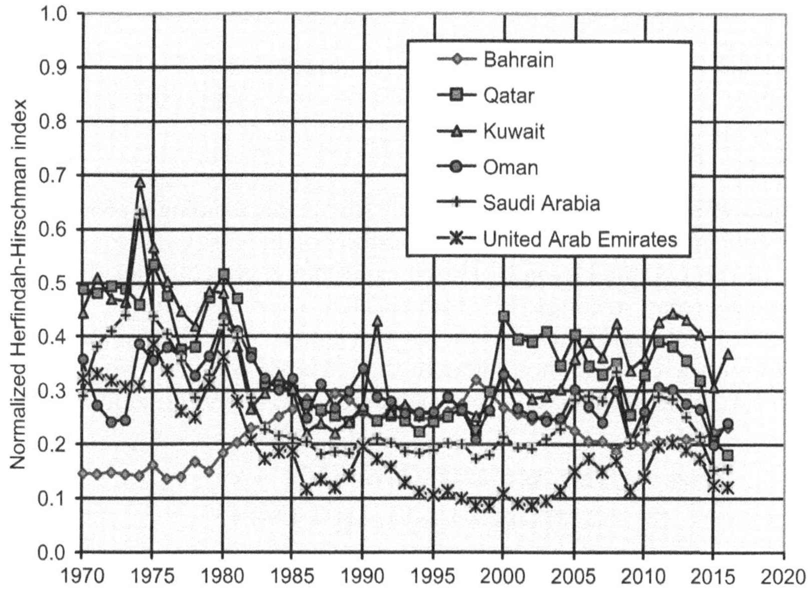


**Fig. 6.2** Industry concentration index for the GCC countries. (Source: National Accounts Estimates of Main Aggregates—United Nations Statistics Division)

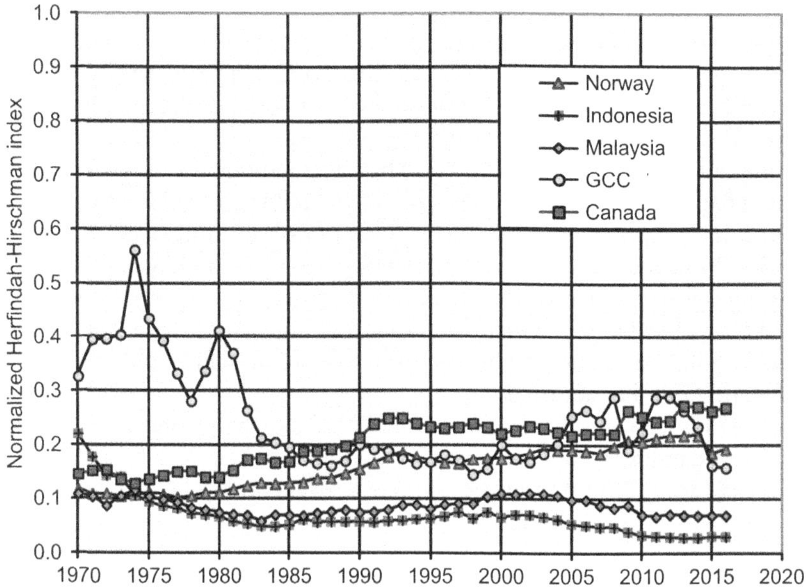


**Fig. 6.3** Industry concentration index in international comparison. (Source: National Accounts Estimates of Main Aggregates—United Nations Statistics Division)

prices after a long period of relatively low prices seemingly provoked a reversal in the trend of industrial diversification in most GCC countries.

In Fig. 6.3, the Herfindal-Hirschman Index is presented for the GCC as an aggregate in comparison with selected reference countries. This figure clearly demonstrates that the GCC countries combined moved since 1995 in the concentration range of the reference countries Norway and Canada, while Malaysia and Iran manifested lower concentration.

## 4 New Assessment for GCC Countries with Input-Output Data

The best way of measuring the relationship between intermediate consumption, gross value added and final demand is through the use of input-output tables, which are derived from supply and use tables that are an integral part of National Accounts (Beutel 2017).

The input-output table is a matrix containing detailed information on the production of goods and services in an economy. It details the intermediate and final uses of domestic and imported goods and services, and also covers net taxes on products and the gross value added of industries.

The required inputs and corresponding cost structures of industries and final demand categories (consumption, investment, exports) can be studied in the columns of the table, while the sales or output structure for goods and services and components of value added (compensation of employees, net taxes on production, consumption of fixed capital, net operating surplus) can be derived from the rows.

Extended input-output tables comprise other useful information derived from satellite systems which are integrated into the national accounts. The additional data include information in matrix form on investment and capital in values, while matrices on employment, energy, emissions, natural resources, waste, sewage and water show the corresponding quantities.

The extended input-output table of Saudi Arabia for 2010 in Table 6.7 has the following seven extensions with information in values and quantities:

1. Gross fixed capital formation (million Saudi riyals)
2. Capital stock (million Saudi riyals)
3. Employment (1000 persons)
4. Energy use (1000 tons of oil equivalent)
5. Air emissions (1000 tons)
6. Global warming, acid deposition, tropospheric ozone formation (1000 tons)
7. Water use (million cubic metres)

The first part of the extended input-output table in rows 1–18 comprises the traditional input-output table. In rows 1–8, the use of domestic products in industries and final uses is shown. Imports and net taxes on products are presented in rows 10–11, followed by the various components of gross value added in rows 13–16. Detailed information on imported products is given in a separate import matrix below the input-output table in rows 19–27.

Environmentally extended input-output tables and models have become a powerful tool in supporting environmental and economic analyses and policies. They play an important role in providing the database for studying sustainable development and the impact of

**Table 6.7** Extended input-output table with satellite systems for Saudi Arabia

| | INDUSTRIES | | | | | | | | FINAL USES | | | | | |
|---|---|---|---|---|---|---|---|---|---|---|---|---|---|---|
| | Agriculture, forestry and fishing | Mining | Manufacturing | Electricity and water | Construction | Trade, transport, communication | Financial and business services | Public and other services | Private consumption | Government consumption | Gross fixed capital formation | Changes in inventories | Exports | Output at basic prices |
| | 1 | 2 | 3 | 4 | 5 | 6 | 7 | 8 | 9 | 10 | 11 | 12 | 13 | 14 |
| **Input-output table (bn SR)** | | | | | | | | | | | | | | |
| 1 Prod. of agriculture, forestry, fi | 2 | 0 | 31 | | 2 | 1 | 0 | 1 | 15 | 1 | | 17 | 1 | 72 |
| 2 Products of mining | 0 | 2 | 79 | 6 | 8 | 5 | 0 | 0 | 2 | 3 | | 2 | 731 | 838 |
| 3 Products of manufacturing | 5 | 3 | 28 | 4 | 35 | 24 | 7 | 22 | 113 | 0 | 146 | 40 | 91 | 519 |
| 4 Electricity and water | 1 | 0 | 9 | 1 | 5 | 10 | 2 | 5 | 13 | 5 | | 2 | 0 | 53 |
| 5 Construction work | 2 | 1 | 13 | 2 | 16 | 11 | 5 | 18 | 13 | 0 | 149 | 1 | 2 | 231 |
| 6 Trade, transport and communi | 7 | 4 | 64 | 4 | 30 | 67 | 9 | 31 | 127 | 5 | 58 | 22 | 47 | 475 |
| 7 Financial and business service | 1 | 1 | 18 | 1 | 7 | 27 | 26 | 20 | 120 | 2 | | 0 | 4 | 228 |
| 8 Public and other services | 2 | 2 | 12 | 2 | 3 | 13 | 3 | 12 | 51 | 344 | | 0 | 0 | 445 |
| 9 Domestic products at basic pri | 18 | 14 | 255 | 20 | 106 | 168 | 52 | 110 | 453 | 360 | 353 | 85 | 876 | 2 860 |
| 10 Imported products | 5 | 3 | 44 | 4 | 33 | 34 | 11 | 27 | 129 | 40 | 126 | 37 | 79 | 574 |
| 11 Taxes less subsidies on produ | 0 | 0 | 1 | 0 | 1 | 0 | 0 | 1 | 4 | 0 | 4 | 1 | 2 | 15 |
| 12 Products at purchasers' prices | 24 | 17 | 300 | 24 | 141 | 193 | 64 | 137 | 585 | 400 | 484 | 123 | 957 | 3 449 |
| 13 Compensation of employees | 6 | 25 | 37 | 8 | 26 | 64 | 26 | 268 | | | | | | 459 |
| 14 Other net taxes on production | - 4 | 1 | 1 | 0 | 1 | 3 | 1 | 0 | | | | | | 4 |
| 15 Consumption of fixed capital | | | | | | | | | | | | | | |
| 16 Net operating surplus | 46 | 795 | 182 | 21 | 64 | 214 | 137 | 39 | | | | | | 1 498 |
| 17 Value added at basic prices | 48 | 821 | 219 | 29 | 91 | 281 | 164 | 308 | | | | | | 1 961 |
| 18 Input at basic prices | 72 | 838 | 519 | 53 | 231 | 475 | 228 | 445 | | | | | | 2 860 |
| **Input-output table of imports (bn SR)** | | | | | | | | | | | | | | |
| 19 Prod. of agriculture, forestry, fi | 0 | 0 | 8 | | 0 | 0 | 0 | 0 | 4 | 0 | | 5 | 0 | 19 |
| 20 Products of mining | 0 | 0 | 1 | 0 | 0 | 0 | 0 | 0 | 0 | 0 | | 0 | 6 | 7 |
| 21 Products of manufacturing | 3 | 2 | 20 | 3 | 25 | 17 | 5 | 16 | 80 | 0 | 104 | 29 | 64 | 369 |
| 22 Electricity and water | 0 | 0 | 0 | 0 | 0 | 0 | 0 | 0 | 0 | 0 | | 0 | 0 | 0 |
| 23 Construction work | 0 | 0 | 1 | 0 | 1 | 1 | 0 | 2 | 1 | 0 | 13 | 0 | 0 | 20 |
| 24 Trade, transport and communi | 1 | 1 | 10 | 1 | 5 | 11 | 1 | 5 | 20 | 1 | 9 | 4 | 7 | 75 |
| 25 Financial and business service | 0 | 0 | 3 | 0 | 1 | 4 | 4 | 3 | 18 | 0 | | 0 | 1 | 34 |
| 26 Public and other services | 0 | 0 | 1 | 0 | 0 | 2 | 0 | 1 | 6 | 39 | | 0 | 0 | 50 |
| 27 Total | 5 | 3 | 44 | 4 | 33 | 34 | 11 | 27 | 129 | 40 | 126 | 37 | 79 | 574 |
| **EXTENSIONS** | | | | | | | | | | | | | | |
| **Gross fixed capital formation (bn SR)** | | | | | | | | | | | | | | |
| 28 Buildings | 2 | 3 | 13 | 7 | 5 | 18 | 87 | 130 | | | | | | 264 |
| 29 Transport equipment | 2 | 0 | 13 | 8 | 5 | 19 | 4 | 5 | | | | | | 57 |
| 30 Machinery and other equipme | 3 | 51 | 24 | 10 | 7 | 26 | 5 | 37 | | | | | | 163 |
| 31 Total | 7 | 54 | 49 | 26 | 17 | 64 | 95 | 172 | | | | | | 484 |
| **Capital stock (bn SR)** | | | | | | | | | | | | | | |
| 32 Buildings | 47 | 99 | 126 | 57 | 64 | 221 | 1 227 | 1 272 | | | | | | 3 114 |
| 33 Transport equipment | 6 | 6 | 52 | 13 | 90 | 140 | 91 | 36 | | | | | | 434 |
| 34 Machinery and other equipme | 8 | 325 | 319 | 126 | 27 | 194 | 89 | 181 | | | | | | 1 271 |
| 35 Total | 61 | 430 | 498 | 197 | 181 | 556 | 1 408 | 1 489 | | | | | | 4 819 |
| **Employment (1.000 persons)** | | | | | | | | | | | | | | |
| 36 Saudi | 93 | 91 | 148 | 70 | 91 | 562 | 207 | 2 693 | | | | | | 3 965 |
| 37 Non-Saudi | 199 | 13 | 509 | 27 | 1 006 | 1 679 | 221 | 1 225 | | | | | | 4 880 |
| 38 Total | 293 | 104 | 657 | 97 | 1 097 | 2 241 | 428 | 3 918 | | | | | | 8 835 |
| **Energy (1.000 tons oil equivalent)** | | | | | | | | | | | | | | |
| 39 Crude oil, NGL and feedstock | | 19 | 161 112 | 20 596 | | | | | | | | - 5 046 | 294 881 | 471 562 |
| 40 Oil products | 2 736 | 7 792 | 27 393 | 14 156 | 192 | 25 563 | 37 | 122 | 10 840 | | | 2 774 | 63 134 | 154 739 |
| 41 Natural gas | | 15 138 | 21 193 | 30 161 | | | | | | | | | | 66 492 |
| 42 Electricity | 316 | 767 | 1 687 | 3 395 | 290 | 3 276 | 272 | 1 300 | 9 342 | | | | | 20 645 |
| 43 Total | 3 052 | 23 716 | 211 385 | 68 309 | 482 | 28 838 | 309 | 1 422 | 20 182 | | | - 2 272 | 358 015 | 713 438 |
| **Air emissions (1.000 tons)** | | | | | | | | | | | | | | |
| 44 Carbon dioxide (CO2) | 5 931 | 9 591 | 100 146 | 177 291 | 818 | 87 197 | 800 | 2 372 | 61 806 | | | | | 445 950 |
| 45 Methane (CH4) | 1 903 | 51 110 | 575 | 336 | 2 | 519 | 2 | 5 637 | 227 | | | | | 60 310 |
| 46 Nitrous oxide (N2O) | 2 613 | 927 | 2 426 | 197 | 1 | 29 | 0 | 25 | 32 | | | | | 6 250 |
| 47 Sulfur dioxide (SO2) | 58 | 568 | 786 | 702 | 1 | 16 | 1 | 7 | 46 | | | | | 2 185 |
| 48 Ammonia (NH3) | 33 | 7 | 23 | 2 | 10 | 21 | 10 | 19 | 0 | | | | | 126 |
| 49 Nitrogen oxides (NOx) | 39 | 7 | 40 | 594 | 8 | 766 | 4 | 7 | 91 | | | | | 1 556 |
| 50 Carbon monoxide (CO) | 33 | 29 | 163 | 35 | 28 | 1 454 | 5 | 27 | 396 | | | | | 2 170 |
| 51 Organic compounds (NMVOC | 10 | 1 552 | 200 | 37 | 21 | 330 | 128 | 59 | 57 | | | | | 2 394 |
| 52 Hydroflurocarbons (HFC) | 6 | 21 | 44 | 5 | 20 | 40 | 19 | 38 | 100 | | | | | 291 |
| 53 Sulfur hexafluoride (SF6) | 0 | 0 | 0 | 0 | 0 | 0 | 0 | 0 | | | | | | 0 |
| 54 Total | 10 625 | 63 811 | 104 403 | 179 200 | 907 | 90 371 | 971 | 8 191 | 62 753 | | | | | 521 232 |
| **Global warming, acid deposition and tropospheric ozone formation (1.000 tons)** | | | | | | | | | | | | | | |
| 55 Greenhouse gases 1) | 855 913 | 1 370 291 | 864 363 | 245 552 | 1 032 | 106 942 | 975 | 128 473 | 76 419 | | | | | 3 649 960 |
| 56 Acid deposition 2) | 85 | 573 | 814 | 1 118 | 7 | 552 | 4 | 11 | 109 | | | | | 3 274 |
| 57 Tropospheric ozone 3) | 1 984 | 52 697 | 979 | 1 003 | 58 | 3 069 | 140 | 5 730 | 770 | | | | | 66 430 |
| **Water use (Mio. cubic meter)** | | | | | | | | | | | | | | |
| 58 Desalinated water | 50 | 58 | 108 | 104 | 1 | 281 | 97 | 207 | 378 | | | | | 1 284 |
| 59 Renewable groundwater | 2 260 | 16 | 31 | 30 | 0 | 68 | 23 | 50 | 90 | | | | | 2 568 |
| 60 Non-renewable groundwater | 15 298 | 54 | 100 | 97 | 1 | 205 | 70 | 150 | 273 | | | | | 16 247 |
| 61 Total | 17 608 | 128 | 239 | 231 | 3 | 553 | 189 | 407 | 742 | | | | | 20 099 |

Saudi Arabia 2010    [white box] = Values    [shaded box] = Quantities

*(continued)*

**Table 6.7** (continued)

Source: Central Department of Statistics and Information (CDSI), Ministry of Economy and Planning (MOEP), Ministry of Water and Electricity (MOWE), International Energy Agency (IEA), World Bank (WB), World Input-Output Database (WIOD), European Commission (EDGAR), own estimates

Notes: (1) Carbon dioxide ($CO_2 = 1$), methane ($CH_4 = 21$) and nitrous oxide ($N_2O = 310$) were transformed with the documented factors to greenhouse gases in $CO_2$-equivalents; (2) sulphur dioxide ($SO_2 = 1$) and nitrogen oxides ($NO_x = 0.7$) were transformed with the documented factors to acid depositions in $SO_2$-equivalents; (3) substances causing tropospheric ozone formation: carbon monoxide (CO), non-methane volatile organic compounds (NMVOC), methane ($CH_4$), nitrogen oxides ($NO_x$)

environmental policies. If, for example, input-output tables are extended to include environmental information, a solid foundation for environmental policy analysis is provided. Life cycle analysis of products and their impact on the environment and sustainable use of natural resources are two prominent applications. The United Nations recently published the *Handbook on Supply, Use and Input-Output Tables with Extensions and Applications* (United Nations 2018, p. 518) in which the compilation of physical supply and use tables and extended input-output tables is promoted (Box 6.2). I was a member of the Editorial Board and heavily engaged in drafting the new *Handbook* for the national accounts.

### 4.1 *Traditional Input-Output Indicators for GCC Countries*

Only a few GCC countries compile and publish supply, use and input-output tables for their national accounts. At the time of writing, national tables were available only for Saudi Arabia and Kuwait.

A comparison of input-output data of Saudi Arabia and Norway reveals that the shares of intermediate consumption of products and gross value added in total output did not change much in both countries between 2005 and 2011. However, the shares of intermediates in total output were significantly lower in Saudi Arabia than in Norway (31 vs. 46 per cent of output), as were the shares of imported intermediates (5 vs. 10 per cent).

The first part of Table 6.8 is devoted to the analysis of direct input coefficients, while in the second part of the table the cumulative input coefficients are explored. The input coefficients reflect the direct input requirements of products for a specific industry, while the cumulative input coefficients represent the direct and indirect input requirements of

**Box 6.2 UN Handbook of Supply, Use and Input-Output Tables**
The United Nations Statistics Division (UNSD) announced in May 2018 that the final draft of the *Handbook on Supply, Use and Input-Output Tables with Extensions and Applications* is now available on the website of UNSD at:

https://unstats.un.org/unsd/nationalaccount/docs/SUT_IOT_HB_wc.pdf

From the preface of the UN Handbook:

*The Handbook on Supply, Use and Input-Output Tables with Extensions and Applications has been prepared as part of a series of handbooks on national accounting in support of the implementation of the System of National Accounts 2008 (2008 SNA). The objective of this Handbook is to provide a step-by-step guidance for the compilation of Supply and Use Tables (SUTs) and Input-Output Tables (IOTs) and an overview of the possible extensions of SUTs and IOTs which increase their analytical usefulness.*

Supply and use tables and the institutional sector accounts constitute the core of the national accounts.

The Institutional Sector Accounts provide detailed information on institutions (Non-financial corporations, Financial corporations, General Government, Households, Non-profit institutions serving households) in the production, the income and the capital account.

Supply and use tables show the relationship between input and output of industries in great detail including the components of value added, intermediate use of products in industries and final demand of products.

Supply and use tables are transformed into symmetric input-output tables on the basis of analytical assumption on technology and sales structure. Input-output models are often used to study the impact of exogenous changes of final demand or primary inputs on the rest of the economy. Input-output tables also provide the database for different macroeconomic models.

A social accounting matrix (SAM) comprises all information of the supply and use tables and the institutional sector accounts in one matrix. The SAM is often used as the database of Computable General Equilibrium Model (CGE).

**Box 6.2 (continued)**

The UN Handbook promotes the compilation of physical supply and use tables and extended input-output tables as an integral part of the national accounts. The additional tables offer new opportunities for the analysis of economic diversification and sustainable development models.

products at all stages of production. The cumulative input coefficients are often used to identify the backward linkages of an industry.

In its simplest form, the strength of the backward linkage of an industry is given by the column sum of the direct input coefficients. A more useful and comprehensive measure is provided by the column sum of the Leontief Inverse, which reflects the direct and indirect effects on other industries.

Backward linkages are input-oriented. The industry 'Construction' requires inputs from many other industries and therefore will have strong backward linkages. Forward linkages are output-oriented. The industry 'Electricity' supplies electricity to all other industries, and therefore, this industry is expected to have strong forward linkages (many clients) but weak backward linkages (few inputs).

The column totals of the direct input coefficients and the Leontief Inverse input coefficients reflect the intensity of backward linkages. The row totals of the direct output coefficients and the Ghosh inverse output coefficients show the intensity of forward linkages.

In Table 6.8, the cumulative input coefficients per industry are reported for domestic products (column 2) and total (domestic and imported) products (column 4). The sum of cumulative input coefficients was divided by the number of industries, as the input-output tables comprise a different number of industries in successive years. While the input coefficients for domestic products allow a high degree of substitution between domestic and imported products, the input coefficients for total products reflect technical requirements.

On average, Saudi Arabia reached 95 per cent of the Leontief Inverse for domestic products of Norway. In other words, it can be said that by 2011, Saudi Arabia has reached an international level of industrial

**Table 6.8** Input-output data of Saudi Arabia and Norway

| | *Number of industries* | *Intermediate consumption of domestic products at basic prices* | *Intermediate consumption of imported products at basic prices* | *Intermediate consumption of products at basic prices* | *Taxes less subsidies on products* | *Intermediate consumption of products at purchasers' prices* | *Gross value added at basic prices* | *Output at basic prices* |
|---|---|---|---|---|---|---|---|---|
| | *(1)* | *(2)* | *(3)* | *(4)* | *(5)* | *(6)* | *(7)* | *(8)* |
| | Saudi Arabia (Mio Saudi riyal) | | | | | | | |
| 2005 | 59 | 444,900 | 99,327 | 544,227 | 3342 | 547,569 | 1,172,399 | 1,719,968 |
| 2010 | 59 | 728,715 | 168,027 | 896,742 | 2830 | 899,572 | 1,960,873 | 2860 445 |
| 2011 | 59 | 931,300 | 191,442 | 1,122,743 | 4062 | 1,126,805 | 2,493,366 | 3,620,170 |
| | Norway (Mio Norwegian krona) | | | | | | | |
| 2005 | 59 | 1,079,314 | 310,162 | 1,389,476 | 44,267 | 1,433,743 | 1,731,949 | 3,165,692 |
| 2010 | 65 | 1,468,471 | 448,813 | 1,917,284 | 63,684 | 1,980,968 | 2260 620 | 4,241,588 |
| 2011 | 65 | 1,579,262 | 481,601 | 2,060,863 | 66,761 | 2,127,624 | 2,455,726 | 4,583,350 |
| | Saudi Arabia (input coefficients %) | | | | | | | |
| 2005 | 59 | 25.9 | 5.8 | 31.6 | 0.2 | 31.8 | 68.2 | 100.0 |
| 2010 | 59 | 25.5 | 5.9 | 31.3 | 0.1 | 31.4 | 68.6 | 100.0 |
| 2011 | 59 | 25.7 | 5.3 | 31.0 | 0.1 | 31.1 | 68.9 | 100.0 |
| | Norway (input coefficients %) | | | | | | | |
| 2005 | 59 | 34.1 | 9.8 | 43.9 | 1.4 | 45.3 | 54.7 | 100.0 |
| 2010 | 65 | 34.6 | 10.6 | 45.2 | 1.5 | 46.7 | 53.3 | 100.0 |
| 2011 | 65 | 34.5 | 10.5 | 45.0 | 1.5 | 46.4 | 53.6 | 100.0 |

| | | | | | | | | |
|---|---|---|---|---|---|---|---|---|
| | Saudi Arabia (sum of cumulative input coefficients/number of industries) | | | | | | | |
| 2005 | 59 | 1.5363 | – | 2.0067 | – | – | – | – |
| 2010 | 59 | 1.4965 | – | 1.7733 | – | – | – | – |
| 2011 | 59 | 1.5225 | – | 1.8118 | – | – | – | – |
| | Norway (sum of cumulative input coefficients/number of industries) | | | | | | | |
| 2005 | 59 | 1.6195 | – | 2.0902 | – | – | – | – |
| 2010 | 65 | 1.6004 | – | 2.0807 | – | – | – | – |
| 2011 | 65 | 1.6130 | – | 2.1135 | – | – | – | – |
| | Diversity index 'backward linkages' for Saudi Arabia (% of reference country Norway) | | | | | | | |
| 2005 | – | 94.9 | – | 96.0 | – | – | – | – |
| 2010 | – | 93.5 | – | 85.2 | – | – | – | – |
| 2011 | – | 94.4 | – | 85.7 | – | – | – | – |

Source: Statistics Norway, CDSI Saudi Arabia, author estimates

diversification. If imported intermediate inputs are included, the level of diversification appears to have reached 86 per cent of that of Norway.

In Table 6.9, a similar comparison was made for the input-output data of Kuwait and Norway. Kuwait has a long history of compiling input-output data. We used the national input-output tables of Kuwait for 2005, 2010 and 2013. We refrained from using input-output tables which were compiled for the Global Trade Analysis Project (Green 2011) as these tables have been generated in a mechanical process with only a few sources. The results for 2005–2013 show a clear trend for the use of domestic intermediates in Kuwait. In 2013, Kuwait, with 32.8 per cent of output, has almost reached the level of Norway (35.2 per cent). However, the share of imported intermediate inputs (5.5 per cent) is much lower than the share in Norway (11.1 per cent). The share of gross value added in output of Kuwait (64.8 per cent) is about 10 per cent higher than in Norway (54.6 per cent), indicating the potential for more diversification.

The backward linkages for domestic inputs in Kuwait reached 95.8 per cent that of Norway. They even exceeded the level of Norway (108.8 per cent) if imported inputs are included. Thus, in Kuwait, the potential to induce more diversification should not be based on a general policy of promoting more import substitution but rather on a specific policy of encouraging more imports of intermediates.

### 4.2 Primary Diversity Measure

Economic diversity has often been promoted as a means to achieve the economic goals of stability and growth. Empirical studies have been able to relate higher levels of diversity to both economic stability and overall levels of economic activity. Diversity measures, as used in these studies, have tended to be narrowly defined, usually emphasising the distribution of employment across industries. Such measures are unsatisfactory, because they do not capture inter-industrial linkages.

An alternative approach to measuring diversity, based on the technical coefficients matrix of an input-output model, was developed by Wagner and Deller (1998) who show that higher levels of diversification within the theoretical construct of input-output are associated with higher levels of stability. Ahmed Al-Kawaz (2008) has successfully implemented this

**Table 6.9** Input-output data of Kuwait and Norway

| | *Number of industries* | *Intermediate consumption of domestic products at basic prices* | *Intermediate consumption of imported products at basic prices* | *Intermediate consumption of products at basic prices* | *Taxes less subsidies on products* | *Intermediate consumption of products at purchasers' prices* | *Gross value added at basic prices* | *Output at basic prices* |
|---|---|---|---|---|---|---|---|---|
| | *(1)* | *(2)* | *(3)* | *(4)* | *(5)* | *(6)* | *(7)* | *(8)* |
| | Kuwait (Mio Kuwaiti dinar) | | | | | | | |
| 2005 | 80 | 10,716 | 1450 | 12,166 | –752 | 11,414 | 23,900 | 35,314 |
| 2010 | 44 | 20,584 | 2536 | 23,120 | –1429 | 21,691 | 36,182 | 57,873 |
| 2013 | 50 | 25,776 | 4518 | 30,294 | –1398 | 28,896 | 53,162 | 82,058 |
| | Norway (Mio Norwegian krona) | | | | | | | |
| 2005 | 59 | 1,079,314 | 310,162 | 1,389,476 | 44,267 | 1,433,743 | 1,731,949 | 3,165,692 |
| 2010 | 59 | 1,468,471 | 448,813 | 1,917,284 | 63,684 | 1,980,968 | 2260 620 | 4,241,588 |
| 2013 | 59 | 1,655,109 | 558,949 | 2,214,058 | 75,509 | 2,289,567 | 2,749,018 | 5,038,585 |
| | Kuwait (input coefficients %) | | | | | | | |
| 2005 | 80 | 30.3 | 4.1 | 34.5 | –2.1 | 32.3 | 67.7 | 100.0 |
| 2010 | 44 | 35.6 | 4.4 | 39.9 | –2.5 | 37.5 | 62.5 | 100.0 |
| 2013 | 50 | 31.4 | 5.5 | 36.9 | –1.7 | 35.2 | 64.8 | 100.0 |
| | Norway (input coefficients %) | | | | | | | |
| 2005 | 59 | 34.1 | 9.8 | 43.9 | 1.4 | 45.3 | 54.7 | 100.0 |
| 2010 | 59 | 34.6 | 10.6 | 45.2 | 1.5 | 46.7 | 53.3 | 100.0 |
| 2013 | 59 | 32.8 | 11.1 | 43.9 | 1.5 | 45.4 | 54.6 | 100.0 |
| | Kuwait (sum of cumulative input coefficients/number of industries) | | | | | | | |
| 2005 | 80 | 1.3990 | – | 2.2638 | – | – | – | – |
| 2010 | 44 | 1.5783 | – | 2.2671 | – | – | – | – |
| 2013 | 50 | 1.5007 | – | 2.2412 | – | – | – | – |

(*continued*)

**Table 6.9** (continued)

| | *Number of industries* | *Intermediate consumption of domestic products at basic prices* | *Intermediate consumption of imported products at basic prices* | *Intermediate consumption of products at basic prices* | *Taxes less subsidies on products* | *Intermediate consumption of products at purchasers' prices* | *Gross value added at basic prices* | *Output at basic prices* |
|---|---|---|---|---|---|---|---|---|
| | *(1)* | *(2)* | *(3)* | *(4)* | *(5)* | *(6)* | *(7)* | *(8)* |
| | Norway (sum of cumulative input coefficients/number of industries) | | | | | | | |
| 2005 | 59 | 1.6195 | – | 2.0902 | – | – | – | – |
| 2010 | 65 | 1.6004 | – | 2.0807 | – | – | – | – |
| 2013 | 65 | 1.5667 | – | 2.0591 | – | – | – | – |
| | Diversity index 'backward linkages' for Kuwait (% of reference country Norway) | | | | | | | |
| 2005 | – | 86.4 | – | 108.3 | – | – | – | – |
| 2010 | – | 98.6 | – | 109.0 | – | – | – | – |
| 2013 | – | 95.8 | – | 108.8 | – | – | – | – |

Source: Statistics Norway, CSB Kuwait, author estimates

approach for Kuwait using input-output data for Kuwait and Norway for 2000.

The primary diversity measure of Wagner and Deller puts special emphasis on inter-industry relations and provides the best way to evaluate the diversification of economies. The index is a multiplicative combination of three components:

- The relative size of an economy (SI)—number of indigenous industries
- The density of the economy (DEN)—number of non-zero elements in the Leontief Matrix, indicating the diversity of transactions
- The condition number of the Leontief Matrix (CN)—indicator of inter-industry linkages

The Primary Diversity Measure (PDM) is defined as the simple multiplicative combination of these three characteristics, PDM = SI * DEN * CN.

The relative size of the economy is defined as $SI = N/N_r$, where N is the number of indigenous industries identified in the nation, and $N_r$ is the number of indigenous industries in the reference country. This measure implies the larger the economy as compared to the base economy, the better. The larger the regional economy, in terms of the number of industries contained within the economy, the greater the ability of the economy to absorb shocks. This is a measure of relative size, but does not contain any information on inter-industry linkages.

The density of the (I - A) matrix is defined as DEN = non-zero/N * N where non-zero is the number of transactions (non-zero elements) in the Leontief Matrix, and N is again the number of indigenous industries. The greater the number of non-zero elements contained in the table, the greater the degree of possible inter-industry linkages. This measure does not capture the relative magnitudes of the elements, nor does it capture the positions of these elements within the Leontief Matrix.

The third component measures the degree of inter-industry linkages. The condition number of the Leontief Matrix defines a scalar reflecting the inter-industry linkages.

The condition number is a measure of linear independence. Most commonly, it is used to test for the uniqueness of a solution to a set of linear equations. By definition, an identity matrix, of any size, has a condition number equal to one. Any divergence from an identity matrix will cause the condition number to increase. In terms of regional economics, divergence from the identity matrix in terms of the (I - A) matrix implies more purchases from indigenous industries or a greater degree of inter-industry linkages. Thus, the higher the condition number, the more diverse the economy.

Common diversification measures focus on value added of industries and exports and imports of products. The main shortcoming of the traditional measures is the non-consideration of inter-industry nature of economic activities. Input-output indicators for diversification comprise detailed information on supply and demand. They cover intermediate consumption of domestic and imported products, value added by industry and final uses of products (consumption, investment, exports). The Primary Diversity Index puts special emphasis on inter-industry relations. In our opinion, this indicator is the best choice to evaluate the diversification of economies (Box 6.3).

**Box 6.3 Input-Output-Based Primary Diversity Index**

An alternative approach to measuring diversity based on the technical coefficients matrix of an input-output model was developed by Wagner and Deller and applied to the 50 US states. Empirical results suggest that higher levels of diversification are associated with higher levels of stability.

Ahmed Al-Kawaz implemented this approach also for Kuwait.

The diversity measure of Wagner and Deller is a multiplicative combination of three components:

Relative size of an economy (SI)—Number of indigenous industries

Density of the economy (DEN)—Number of non-zero elements in the Leontief Matrix indicating the diversity of transactions

The condition number of the Leontief Matrix (CN)—Scalar measure of the inter-industry linkages with the economy as captured in the Leontief matrix

The Primary Diversity Measure ($PDM_i$) for a country (i) is defined as the simple multiplicative combination of these three characteristics $PDM_i = SI_i * DEN_i * CN_i$.

The relative size of the economy is defined as $S_i = N_i / N_r$, where N is the number of indigenous industries in country (i) and $N_r$ is the number of indigenous industries in the reference country.

**Box 6.3 (continued)**

The density of the economy is defined as $DEN_i = \text{non-zero}_i / N_i * N_i$, where non-zero is the number of transactions in the Leontief Matrix (I-A), and $N_i$ is again the number of indigenous industries in country (i).

The third component measures the degree of inter-industry linkages. The condition number of the Leontief matrix defines a scalar reflecting the inter-industry linkages. The condition number is defined as

$CN_i = \|(I - A)\| \, \|(I - A)^{-1}\| = \delta_l (I - A) / \delta_s (I - A)$ where

$\|(I - A)\|$ is the 2-norm of the (I - A) matrix,

$\|(I - A)^{-1}\|$ is the 2-norm of the $(I - A)^{-1}$ (the Leontief inverse matrix),

$\delta_l(I - A)$ is the largest singular value of the (I-A) matrix, and

$\delta_s(I - A)$ is the smallest singular value of the (I - A) matrix, respectively.

The Primary Diversity Index is defined as $PDI = PDM_i/PDM_r$.

i = Country i

r = Reference country r

Sources:

John E. Wagner and Steven C. Deller (1998): *Measuring the Effects of Economic Diversity on Growth and Stability, in: Land Economics*, Vol. 74, No. 4. pp. 541–556; Ahmed Al-Kawaz (2008): Economic Diversification: The Case of Kuwait with Reference to Oil Producing Countries, in *Journal of Economic Cooperation*, 29, 3, pp. 23–48.

The results of the input-output-based diversity index for Saudi Arabia compared to Norway are presented in Table 6.10.

We used the input-output tables of the OECD (2018) for this analysis. All input-output tables of the database show 34 industries. In Norway and Saudi Arabia, only 33 indigenous industries were reported. This is why the size (SI) in Table 6.13 is always reported with 1.0 for all years. The number of non-zero elements in the table of Saudi Arabia was somewhat smaller than in Norway. This is why the density (DEN) of Saudi Arabia is a bit lower. The condition number (CN) reflects inter-industry linkages: in all years, it is lower in Saudi Arabia than in Norway, but the difference is not strikingly high. Consequently, the Primary Diversity Index of Saudi Arabia was lower than that of Norway for all years, but the distance has

**Table 6.10** Input-output diversity index

| | | *1995* | *2000* | *2005* | *2010* | *2011* |
|---|---|---|---|---|---|---|
| | | Norway | | | | |
| Number of indigenous industries | N | 33 | 33 | 33 | 33 | 33 |
| Size | SI | 1.0000 | 1.0000 | 1.0000 | 1.0000 | 1.0000 |
| Number of non-zero elements | Non-zero | 1090 | 1089 | 1090 | 1090 | 1090 |
| Number of elements | N*N | 1156 | 1156 | 1156 | 1156 | 1156 |
| Density | DEN | 0.9429 | 0.9420 | 0.9429 | 0.9429 | 0.9429 |
| Condition number | CN | 3.4960 | 3.7054 | 3.6340 | 3.9076 | 4.0122 |
| Primary Diversity Measure | PDM | 3.2964 | 3.4906 | 3.4265 | 3.6845 | 3.7832 |
| | | Saudi Arabia | | | | |
| Number of indigenous industries | N | 33 | 33 | 33 | 33 | 33 |
| Size | SI | 1.0000 | 1.0000 | 1.0000 | 1.0000 | 1.0000 |
| Number of non-zero elements | Non-zero | 1052 | 1053 | 1066 | 1077 | 1077 |
| Number of elements | N*N | 1156 | 1156 | 1156 | 1156 | 1156 |
| Density | DEN | 0.910 | 0.911 | 0.922 | 0.932 | 0.932 |
| Condition number C | CN | 3.2999 | 3.0747 | 3.3256 | 3.5792 | 3.9049 |
| Primary Diversity Measure | PDM | 3.0030 | 2.8008 | 3.0666 | 3.3346 | 3.6380 |
| Diversity index | PDM/ PDM | 0.9110 | 0.8024 | 0.8950 | 0.9051 | 0.9616 |

Source: OECD Input-Output Tables

shrunk. In 1995, Saudi Arabia was at 91.1 per cent of the corresponding level of Norway, but by 2011, it had increased to 96.2 per cent. Therefore, by 2011, Saudi Arabia almost reached the diversity level of Norway. This is a considerable achievement of the Saudi development policy.

## 5 Sustainable Development of Nations: Adjusted Net National Income and Savings

Since a long time, the World Bank is engaged in measuring sustainable development of nations (World Bank 2001). Given the exhaustible nature of oil resources, the long-term strategy for economic sustainability of the oil-producing countries must be to transform the non-renewable natural capital into other forms of capital like machinery, buildings and human capital, and so increase the gross national income per capita (Beutel et al. 2013). In the World Bank's World Development Indicators (World Bank 2018), we find two prominent indicators for sustainable economic development:

- Adjusted net national income, which is gross national income (GNI) minus consumption of fixed capital and natural resources depletion.
- Adjusted net savings, which are equal to net national savings plus education expenditure and minus energy depletion, mineral depletion, net forest depletion, and carbon dioxide and particulate emissions damage.

### 5.1 *Adjusted Net National Income*

Adjusted net national income is calculated by subtracting consumption of fixed capital and the depletion of natural resources from gross national income (GNI). Consumption of fixed capital reflects the decline of man-made capital (buildings, machinery, transport equipment), while the depletion of natural resources measures the decline in non-renewable natural resources through extraction.

Gross domestic product (GDP)

+ Net income from abroad<br>
= Gross national income (GNI)<br>
– Consumption of fixed capital<br>
= Net national income (NNI)<br>
– Natural resources depletion<br>
= Adjusted net national income

Where:

*Gross domestic product* at purchaser's prices is the sum of gross value added by all resident producers in the economy plus any product taxes and minus any subsidies not included in the value of the products.

*Net income from abroad* includes the net labour income and net property and entrepreneurial income components of the SNA.

*Gross national income* (GNI) is the sum of value added by all resident producers plus any product taxes (less subsidies) not included in the valuation of output plus net receipts of primary income (compensation of employees and property income) from abroad.

*Consumption of fixed capital* represents the replacement value of capital used up in the process of production.

*Net national income* is gross national income (GNI) less consumption of fixed capital.

*Natural resource depletion* is the ratio of the value of the stock of natural resources to the remaining reserve lifetime (capped at 25 years). It covers coal, crude oil and natural gas. The World Bank provides valuable information for 10 minerals, 4 energy sources and the net forest depletion.

*Adjusted net national income* is GNI minus consumption of fixed capital and natural resources depletion.

### 5.2 Adjusted Net Saving

Adjusted net saving is a national accounting aggregate designed to measure the net change in assets in a national balance sheet that includes natural and human capital. The gross stock of natural capital, produced capital and human capital is growing if the adjusted net savings of a nation are positive.

There is an intrinsic link between change in the wealth of a nation and the sustainability of a development path. If genuine (adjusted) savings are negative at any point in time, then welfare in the future will be less than current welfare. Therefore, adjusted net saving can be regarded as a sustainability indicator.

Specifically, the World Bank's definition of adjusted net savings is as follows:

Gross national savings
Consumption of fixed capital
= Net savings
+ Education expenditure
– Energy depletion
– Mineral depletion
– Net forest depletion
– Carbon dioxide emissions damage
– Particulate emissions damage
= Adjusted net savings (genuine savings)

Where:

*Net national savings* are equal to gross national savings less the value of consumption of fixed capital.

*Education expenditure* refers to the current operating expenditures in education, including wages and salaries and excluding capital investments in buildings and equipment.

*Energy depletion* is the ratio of the value of the stock of energy resources to the remaining reserve lifetime (capped at 25 years). It covers coal, crude oil and natural gas.

*Mineral depletion* is the ratio of the value of the stock of mineral resources to the remaining reserve lifetime (capped at 25 years). It covers tin, gold, lead, zinc, iron, copper, nickel, silver, bauxite and phosphate.

*Net forest depletion* is calculated as the product of unit resource rents and the excess of wood harvest over natural growth.

*Carbon dioxide damage* is estimated to be $20 per ton of carbon (the unit damage in 1995 US dollars) times the number of tons of carbon emitted.

*Particulate emissions damage* is the damage due to exposure of a country's population to ambient concentrations of particulates measuring less than 2.5 microns in diameter (PM2.5), ambient ozone pollution and indoor concentrations of PM2.5 in households cooking with solid fuels. Damages are calculated as foregone labour income due to premature death. Estimates of health impacts from the Global Burden of Disease Study 2015 are for 1990, 1995, 2000.

### 5.3 *Test for Sustainable Development*

An economy is sustainable if it saves more than the depreciation on its man-made and natural capital. In Table 6.11, an assessment has been made for all GCC countries and Norway.

The highest adjusted national income per capita (78.515 $/person) and adjusted net national saving per capita (22.363 $/person) were achieved in Norway. Among GCC countries, Qatar had the highest adjusted national income per capita (67.443 $/person), followed by UAE (38.670 $/person) and Kuwait (37,781 $/person). Qatar had also the highest adjusted net national saving per capita (34.570 $/person) followed by Kuwait (13.421 $/person).

Among the GCC countries, only Oman (−1.052 $/person) recorded negative adjusted savings, and its combined capital stock of man-made capital and natural capital declined in 2014. This is well in line with the lowest level of the adjusted net national income per capita (10,556 $/person). The net income from abroad was negative throughout the years.

Tables 6.12–6.17 contain complete information on a country by country basis, showing the evolution of sustainable development of the six GCC countries for the 20 years 1995–2015.

Earlier in the text in Table 6.5, it was reported that Bahrain had in 2016 the lowest oil dependency (13.5% of gross value added) and Kuwait (51.4% of gross value added) the highest oil dependency among the GCC countries. The test for sustainable development of Bahrain in Table 6.12 reveals that the adjusted net national income and the adjusted net saving are low compared to other GCC countries. In 2014, final consumption expenditure is high (68.2% of GNI) and education expenditure is low (2.9% of GNI), both resulting in a low adjusted savings ratio (10.3% of GNI).

The prospects for Kuwait in Table 6.13 are brighter. Despite substantial allocations for the depletion of natural resources, the savings ratios throughout the period 2000–2014 are impressive. The positive performance of the savings ratio is certainly supported by relatively low

**Table 6.11** Test for sustainable development of GCC countries in 2014

| | *Bahrain* | *Kuwait* | *Oman* | *Qatar* | *Saudi Arabia* | *United Arab Emirates* | *Norway* |
|---|---|---|---|---|---|---|---|
| | Million US $ at current prices | | | | | | |
| Gross domestic product | 33,388 | 162,632 | 81,034 | 206,225 | 756,350 | 403,198 | 499,339 |
| + Net income from abroad | −5664 | 15,641 | −4346 | −9301 | 13,553 | 659 | 18,349 |
| = Gross national income (GNI) | 27,724 | 178,272 | 76,688 | 196,924 | 769,903 | 403,857 | 517,687 |
| − Consumption of fixed capital | −1750 | −12,340 | −10,904 | −28,642 | −67,161 | −29,013 | −82,454 |
| = Net national income | 25,973 | 165,932 | 65,785 | 168,282 | 702,742 | 374,844 | 435,233 |
| − Natural resources depletion | −2343 | −23,026 | −23,975 | −8145 | −98,603 | −24,075 | −31,885 |
| = Adjusted net national income | 23,630 | 142,906 | 41,810 | 160,138 | 604,139 | 350,768 | 403,348 |
| Net national income | 25,973 | 165,932 | 65,785 | 168,282 | 702,742 | 374,844 | 435,233 |
| + Net transfers | −2364 | −20,674 | −10,301 | −17,514 | −38,734 | – | −8693 |
| − Final consumption | −18,904 | −76,005 | −38,351 | −64,441 | −439,737 | −278,449 | −314,257 |
| = Net national savings | 4705 | 69,253 | 17,133 | 86,327 | 224,272 | – | 112,282 |
| + Education expenditure | 807 | 5658 | 3537 | 4986 | 55,359 | – | 35,186 |
| − Energy depletion | −2343 | −23,025 | −23,968 | −8144 | −98,452 | −24,075 | −31,848 |
| − Mineral depletion | 0 | 0 | −5 | 0 | −151 | 0 | −36 |
| − Net forest depletion | 0 | −1 | −2 | 0 | 0 | 0 | 0 |
| − Carbon dioxide damage | −280 | −1013 | −778 | −1003 | −6116 | −2182 | −505 |
| − Particulate emission damage | −24 | −111 | 0 | 0 | 0 | 0 | −195 |
| = Adjusted net national saving | 2865 | 50,763 | −4083 | 82,165 | 174,911 | – | 114,884 |
| | Population (Mio Persons) | | | | | | |
| Population | 1.336 | 3.782 | 3.961 | 2.374 | 30.777 | 9.071 | 5.137 |
| | US $ per person | | | | | | |
| Gross domestic product | 24,983 | 42,996 | 20,458 | 86,853 | 24,575 | 44,450 | 97,200 |
| + Net income from abroad | −4238 | 4135 | −1097 | −3917 | 440 | 73 | 3572 |
| = Gross national income (GNI) | 20,745 | 47,131 | 19,361 | 82,936 | 25,016 | 44,522 | 100,772 |
| − Consumption of fixed capital | −1310 | −3262 | −2753 | −12,063 | −2182 | −3198 | −16,050 |

(*continued*)

**Table 6.11** (continued)

| | *Bahrain* | *Kuwait* | *Oman* | *Qatar* | *Saudi Arabia* | *United Arab Emirates* | *Norway* |
|---|---|---|---|---|---|---|---|
| = Net national income | 19,435 | 43,869 | 16,608 | 70,873 | 22,834 | 41,324 | 84,721 |
| − Natural resources depletion | −1753 | −6088 | −6053 | −3430 | −3204 | −2654 | −6207 |
| = Adjusted net national income | 17,682 | 37,781 | 10,556 | 67,443 | 19,630 | 38,670 | 78,515 |
| Net national income | 19,435 | 43,869 | 16,608 | 70,873 | 22,834 | 41,324 | 84,721 |
| + Net transfers | −1769 | −5466 | −2601 | −7376 | −1259 | – | −1692 |
| − Final consumption | −14,145 | −20,094 | −9682 | −27,140 | −14,288 | −30,697 | −61,173 |
| = Net national savings | 3521 | 18,309 | 4325 | 36,357 | 7287 | – | 21,857 |
| + Education expenditure | 604 | 1496 | 893 | 2100 | 1799 | – | 6849 |
| − Energy depletion | −1753 | −6087 | −6051 | −3430 | −3199 | −2654 | −6200 |
| − Mineral depletion | 0 | 0 | −1 | 0 | −5 | 0 | −7 |
| − Net forest depletion | 0 | 0 | 0 | 0 | 0 | 0 | 0 |
| − Carbon dioxide damage | −210 | −268 | −197 | −422 | −199 | −241 | −98 |
| − Particulate emission damage | −18 | −29 | 0 | 0 | 0 | 0 | −38 |
| = Adjusted net national saving | 2144 | 13,421 | −1031 | 34,604 | 5683 | – | 22,363 |

Source: The World Bank—World Development Indicators, March 2018

**Table 6.12** Test for sustainable development of Bahrain

| | 1995 | 2000 | 2005 | 2010 | 2011 | 2012 | 2013 | 2014 | 2015 |
|---|---|---|---|---|---|---|---|---|---|
| | Million US $ at current prices | | | | | | | | |
| Gross domestic product | 5849 | 9063 | 15,969 | 25,713 | 28,777 | 30,749 | 32,540 | 33,388 | 31,126 |
| + Net income from abroad | −59 | −224 | −413 | −2373 | −3384 | −3837 | −3926 | −5664 | −1736 |
| = Gross national income (GNI) | 5790 | 8839 | 15,555 | 23,340 | 25,393 | 26,912 | 28,614 | 27,724 | 29,390 |
| − Consumption of fixed capital | −588 | −834 | −1023 | −1457 | −1658 | −1585 | −1659 | −1750 | −1871 |
| = Net national income | 5203 | 8005 | 33 | 21,883 | 23,735 | 25,327 | 26,955 | 25,973 | 27,519 |
| − Natural resources depletion | −171 | −320 | −772 | −994 | −1789 | −2213 | −2347 | −2343 | −1332 |
| = Adjusted net national income | 5032 | 7685 | 13,761 | 20,889 | 21,946 | 23,114 | 24,608 | 23,630 | 26,187 |
| Net national income | 5203 | 8005 | 14,533 | 21,883 | 23,735 | 25,327 | 26,955 | 25,973 | 27,519 |
| + Net transfers | −379 | −990 | −1223 | −1642 | −2050 | −2074 | −2166 | −2364 | – |
| − Final consumption | −4321 | −5517 | −8590 | −13,914 | −15,229 | −16,435 | −18,327 | −18,904 | −19,519 |
| = Net national savings | 502 | 1498 | 4720 | 6328 | 6456 | 6817 | 6462 | 4705 | – |
| + Education expenditure | 208 | 351 | 467 | 638 | 706 | 783 | 822 | 807 | 752 |
| − Energy depletion | −171 | −320 | −772 | −994 | −1788 | −2212 | −2347 | −2343 | −1332 |
| − *Mineral depletion* | 0 | 0 | 0 | 0 | 0 | 0 | 0 | 0 | 0 |
| − Net forest depletion | 0 | 0 | 0 | 0 | 0 | 0 | 0 | 0 | 0 |
| − Carbon dioxide damage | −95 | −127 | −156 | −219 | −227 | −242 | −262 | −280 | – |
| − Particulate emission damage | −14 | −17 | −18 | −21 | −24 | −24 | −25 | −24 | −22 |
| = Adjusted net national saving | 431 | 1386 | 4241 | 5732 | 5122 | 5121 | 4651 | 2865 | – |

(*continued*)

**Table 6.12** (continued)

| | 1995 | 2000 | 2005 | 2010 | 2011 | 2012 | 2013 | 2014 | 2015 |
|---|---|---|---|---|---|---|---|---|---|
| | % of gross national income (GNI) | | | | | | | | |
| Gross domestic product | 101.0 | 102.5 | 102.7 | 110.2 | 113.3 | 114.3 | 113.7 | 120.4 | 105.9 |
| + Net income from abroad | −1.0 | −2.5 | −2.7 | −10.2 | −13.3 | −14.3 | −13.7 | −20.4 | −5.9 |
| = Gross national income (GNI) | 100.0 | 100.0 | 100.0 | 100.0 | 100.0 | 100.0 | 100.0 | 100.0 | 100.0 |
| − Consumption of fixed capital | −10.2 | −9.4 | −6.6 | −6.2 | −6.5 | −5.9 | −5.8 | −6.3 | −6.4 |
| = Net national income | 89.8 | 90.6 | 93.4 | 93.8 | 93.5 | 94.1 | 94.2 | 93.7 | 93.6 |
| − Natural resources depletion | −2.9 | −3.6 | −5.0 | −4.3 | −7.0 | −8.2 | −8.2 | −8.5 | −4.5 |
| = Adjusted net national income | 86.9 | 86.9 | 88.5 | 89.5 | 86.4 | 85.9 | 86.0 | 85.2 | 89.1 |
| Net national income | 89.8 | 90.6 | 93.4 | 93.8 | 93.5 | 94.1 | 94.2 | 93.7 | 93.6 |
| + Net transfers | −6.5 | −11.2 | −7.9 | −7.0 | −8.1 | −7.7 | −7.6 | −8.5 | – |
| − Final consumption | −74.6 | −62.4 | −55.2 | −59.6 | −60.0 | −61.1 | −64.0 | −68.2 | −66.4 |
| = Net national savings | 8.7 | 16.9 | 30.3 | 27.1 | 25.4 | 25.3 | 22.6 | 17.0 | – |
| + Education expenditure | 3.6 | 4.0 | 3.0 | 2.7 | 2.8 | 2.9 | 2.9 | 2.9 | 2.6 |
| − Energy depletion | −2.9 | −3.6 | −5.0 | −4.3 | −7.0 | −8.2 | −8.2 | −8.5 | −4.5 |
| − Mineral depletion | 0.0 | 0.0 | 0.0 | 0.0 | 0.0 | 0.0 | 0.0 | 0.0 | 0.0 |
| − Net forest depletion | 0.0 | 0.0 | 0.0 | 0.0 | 0.0 | 0.0 | 0.0 | 0.0 | 0.0 |
| − Carbon dioxide damage | −1.6 | −1.4 | −1.0 | −0.9 | −0.9 | −0.9 | −0.9 | −1.0 | – |
| − Particulate emission damage | −0.2 | −0.2 | −0.1 | −0.1 | −0.1 | −0.1 | −0.1 | −0.1 | −0.1 |
| = Adjusted net national saving | 7.4 | 15.7 | 27.3 | 24.6 | 20.2 | 19.0 | 16.3 | 10.3 | – |

Source: The World Bank—World Development Indicators, March 2018

**Table 6.13** Test for sustainable development of Kuwait

| | *1995* | *2000* | *2005* | *2010* | *2011* | *2012* | *2013* | *2014* | *2015* |
|---|---|---|---|---|---|---|---|---|---|
| | Million US $ at current prices | | | | | | | | |
| Gross domestic product | 27,192 | 37,712 | 80,798 | 115,419 | 154,028 | 174,070 | 174,161 | 162,632 | 114,567 |
| + Net income from abroad | 4883 | 6698 | 7182 | 8462 | 9180 | 12,697 | 13,254 | 15,641 | 12,692 |
| = Gross national income (GNI) | 32,074 | 44,410 | 87,979 | 123,881 | 163,208 | 186,767 | 187,416 | 178,272 | 127,259 |
| − Consumption of fixed capital | −2440 | −2774 | −5483 | −8741 | −9776 | −10,532 | −10,582 | −12,340 | −9428 |
| = Net national income | 29,635 | 41,636 | 82,497 | 115,140 | 153,432 | 176,235 | 176,834 | 165,932 | 117,831 |
| − Natural resources depletion | −1700 | −3321 | −10,466 | −11,950 | −22,863 | −28,625 | −26,399 | −23,026 | −12,215 |
| = Adjusted net national income | 27,935 | 38,315 | 72,030 | 103,189 | 130,569 | 147,609 | 150,436 | 142,906 | 105,616 |
| Net national income | 29,635 | 41,636 | 82,497 | 115,140 | 153,432 | 176,235 | 176,834 | 165,932 | 117,831 |
| + Net transfers | −1465 | −1956 | −3421 | −12,258 | −14,376 | −16,714 | −19,107 | −20,674 | −16,522 |
| − Final consumption | −20,355 | −23,761 | −38,671 | −53,119 | −60,276 | −67,387 | −72,333 | −76,005 | −75,318 |
| = Net national savings | 7814 | 15,919 | 40,405 | 49,763 | 78,781 | 92,133 | 85,394 | 69,253 | 25,991 |
| + Education expenditure | 1656 | 2304 | 3537 | 3952 | 5206 | 5958 | 5979 | 5658 | 4111 |
| − energy depletion | −1700 | −3321 | −10,466 | −11,950 | −22,863 | −28,625 | −26,398 | −23,025 | −12,214 |
| − Mineral depletion | 0 | 0 | 0 | 0 | 0 | 0 | 0 | 0 | 0 |
| − Net forest depletion | 0 | 0 | 0 | 0 | 0 | 0 | 0 | −1 | −1 |
| − Carbon dioxide damage | −324 | −365 | −580 | −854 | −882 | −968 | −1004 | −1013 | – |
| − Particulate emission damage | −25 | −41 | −74 | −109 | −173 | −183 | −182 | −166 | −111 |
| = Adjusted net national saving | 7421 | 14,495 | 32,821 | 40,802 | 60,069 | 68,315 | 63,788 | 50,707 | – |

(*continued*)

**Table 6.13** (continued)

| | 1995 | 2000 | 2005 | 2010 | 2011 | 2012 | 2013 | 2014 | 2015 |
|---|---|---|---|---|---|---|---|---|---|
| | % of gross national income (GNI) | | | | | | | | |
| Gross domestic product | 84.8 | 84.9 | 91.8 | 93.2 | 94.4 | 93.2 | 92.9 | 91.2 | 90.0 |
| + Net income from abroad | 15.2 | 15.1 | 8.2 | 6.8 | 5.6 | 6.8 | 7.1 | 8.8 | 10.0 |
| = Gross national income (GNI) | 100.0 | 100.0 | 100.0 | 100.0 | 100.0 | 100.0 | 100.0 | 100.0 | 100.0 |
| – Consumption of fixed capital | −7.6 | −6.2 | −6.2 | −7.1 | −6.0 | −5.6 | −5.6 | −6.9 | −7.4 |
| = Net national income | 92.4 | 93.8 | 93.8 | 92.9 | 94.0 | 94.4 | 94.4 | 93.1 | 92.6 |
| – Natural resources depletion | −5.3 | −7.5 | −11.9 | −9.6 | −14.0 | −15.3 | −14.1 | −12.9 | −9.6 |
| = Adjusted net national income | 87.1 | 86.3 | 81.9 | 83.3 | 80.0 | 79.0 | 80.3 | 80.2 | 83.0 |
| Net national income | 92.4 | 93.8 | 93.8 | 92.9 | 94.0 | 94.4 | 94.4 | 93.1 | 92.6 |
| + Net transfers | −4.6 | −4.4 | −3.9 | −9.9 | −8.8 | −8.9 | −10.2 | −11.6 | – |
| – Final consumption | −63.5 | −53.5 | −44.0 | −42.9 | −36.9 | −36.1 | −38.6 | −42.6 | −59.2 |
| = Net national savings | 24.4 | 35.8 | 45.9 | 40.2 | 48.3 | 49.3 | 45.6 | 38.8 | – |
| + Education expenditure | 5.2 | 5.2 | 4.0 | 3.2 | 3.2 | 3.2 | 3.2 | 3.2 | 3.2 |
| – Energy depletion | −5.3 | −7.5 | −11.9 | −9.6 | −14.0 | −15.3 | −14.1 | −12.9 | −9.6 |
| – Mineral depletion | 0.0 | 0.0 | 0.0 | 0.0 | 0.0 | 0.0 | 0.0 | 0.0 | 0.0 |
| – Net forest depletion | 0.0 | 0.0 | 0.0 | 0.0 | 0.0 | 0.0 | 0.0 | 0.0 | 0.0 |
| – Carbon dioxide damage | −1.0 | −0.8 | −0.7 | −0.7 | −0.5 | −0.5 | −0.5 | −0.6 | – |
| – Particulate emission damage | −0.1 | −0.1 | −0.1 | −0.1 | −0.1 | −0.1 | −0.1 | −0.1 | −0.1 |
| = Adjusted net national saving | 23.1 | 32.6 | 37.3 | 32.9 | 36.8 | 36.6 | 34.0 | 28.4 | – |

Source: The World Bank—World Development Indicators, March 2018

**Table 6.14** Test for sustainable development of Oman

| | *1995* | *2000* | *2005* | *2010* | *2011* | *2012* | *2013* | *2014* | *2015* |
|---|---|---|---|---|---|---|---|---|---|
| | Million US $ at current prices | | | | | | | | |
| Gross domestic product | 13,803 | 19,507 | 31,082 | 58,642 | 67,937 | 76,690 | 78,939 | 81,034 | 69,832 |
| + Net income from abroad | −397 | −753 | −1199 | −3581 | −4138 | −4463 | −3235 | −4346 | −2315 |
| = Gross national income (GNI) | 13,406 | 18,754 | 29,883 | 55,060 | 63,799 | 72,227 | 75,703 | 76,688 | 67,517 |
| − Consumption of fixed capital | −1508 | −1846 | −3110 | −6568 | −7199 | −7709 | −8678 | −10,904 | −10,504 |
| = Net national income | 11,898 | 16,909 | 26,773 | 48,492 | 56,601 | 64,517 | 67,025 | 65,785 | 57,013 |
| − Natural resources depletion | −2767 | −6149 | −9707 | −16,106 | −24,557 | −26,440 | −26,770 | −23,975 | −12,277 |
| = Adjusted net national income | 9131 | 10,759 | 17,066 | 32,386 | 32,044 | 38,078 | 40,255 | 41,810 | 44,735 |
| Net national income | 11,898 | 16,909 | 26,773 | 48,492 | 56,601 | 64,517 | 67,025 | 65,785 | 57,013 |
| + Net transfers | −1469 | −1451 | −2257 | −5704 | −7215 | −8087 | −9104 | −10,301 | −10,991 |
| − Final consumption | −10,565 | −11,089 | −15,962 | −29,403 | −31,584 | −31,904 | −34,550 | −38,351 | −38,771 |
| = Net national savings | −136 | 4368 | 8553 | 13,386 | 17,802 | 24,526 | 23,371 | 17,133 | 7250 |
| + Education expenditure | 429 | 696 | 974 | 2559 | 2959 | 3338 | 3492 | 3537 | 3114 |
| − Energy depletion | −2766 | −6149 | −9706 | −16,098 | −24,548 | −26,432 | −26,764 | −23,968 | −12,272 |
| − *Mineral depletion* | −1 | 0 | 0 | −6 | −7 | −6 | −5 | −5 | −4 |
| − Net forest depletion | 0 | 0 | 0 | −1 | −1 | −1 | −1 | −2 | −2 |
| − Carbon dioxide damage | −94 | −149 | −242 | −529 | −629 | −692 | −738 | −778 | – |
| − Particulate emission damage | −28 | −27 | −40 | −57 | −72 | −79 | −85 | −85 | −68 |
| = Adjusted net national saving | −2596 | −1261 | −461 | −747 | −4497 | 654 | −730 | −4168 | – |

(*continued*)

**Table 6.14** (continued)

| | 1995 | 2000 | 2005 | 2010 | 2011 | 2012 | 2013 | 2014 | 2015 |
|---|---|---|---|---|---|---|---|---|---|
| | % of gross national income (GNI) | | | | | | | | |
| Gross domestic product | 103.0 | 104.0 | 104.0 | 106.5 | 106.5 | 106.2 | 104.3 | 105.7 | 103.4 |
| + Net income from abroad | −3.0 | −4.0 | −4.0 | −6.5 | −6.5 | −6.2 | −4.3 | −5.7 | −3.4 |
| = Gross national income (GNI) | 100.0 | 100.0 | 100.0 | 100.0 | 100.0 | 100.0 | 100.0 | 100.0 | 100.0 |
| − Consumption of fixed capital | −11.2 | −9.8 | −10.4 | −11.9 | −11.3 | −10.7 | −11.5 | −14.2 | −15.6 |
| = Net national income | 88.8 | 90.2 | 89.6 | 88.1 | 88.7 | 89.3 | 88.5 | 85.8 | 84.4 |
| − Natural resources depletion | −20.6 | −32.8 | −32.5 | −29.3 | −38.5 | −36.6 | −35.4 | −31.3 | −18.2 |
| = Adjusted net national income | 68.1 | 57.4 | 57.1 | 58.8 | 50.2 | 52.7 | 53.2 | 54.5 | 66.3 |
| Net national income | 88.8 | 90.2 | 89.6 | 88.1 | 88.7 | 89.3 | 88.5 | 85.8 | 84.4 |
| + Net transfers | −11.0 | −7.7 | −7.6 | −10.4 | −11.3 | −11.2 | −12.0 | −13.4 | −16.3 |
| − Final consumption | −78.8 | −59.1 | −53.4 | −53.4 | −49.5 | −44.2 | −45.6 | −50.0 | −57.4 |
| = Net national savings | −1.0 | 23.3 | 28.6 | 24.3 | 27.9 | 34.0 | 30.9 | 22.3 | 10.7 |
| + Education expenditure | 3.2 | 3.7 | 3.3 | 4.6 | 4.6 | 4.6 | 4.6 | 4.6 | 4.6 |
| − Energy depletion | −20.6 | −32.8 | −32.5 | −29.2 | −38.5 | −36.6 | −35.4 | −31.3 | −18.2 |
| − Mineral depletion | 0.0 | 0.0 | 0.0 | 0.0 | 0.0 | 0.0 | 0.0 | 0.0 | 0.0 |
| − Net forest depletion | 0.0 | 0.0 | 0.0 | 0.0 | 0.0 | 0.0 | 0.0 | 0.0 | 0.0 |
| − Carbon dioxide damage | −0.7 | −0.8 | −0.8 | −1.0 | −1.0 | −1.0 | −1.0 | −1.0 | – |
| − Particulate emission damage | −0.2 | −0.1 | −0.1 | −0.1 | −0.1 | −0.1 | −0.1 | −0.1 | −0.1 |
| = Adjusted net national saving | −19.4 | −6.7 | −1.5 | −1.4 | −7.0 | 0.9 | −1.0 | −5.4 | – |

Source: The World Bank—World Development Indicators, March 2018

**Table 6.15** Test for sustainable development of Qatar

| | *1995* | *2000* | *2005* | *2010* | *2011* | *2012* | *2013* | *2014* | *2015* |
|---|---|---|---|---|---|---|---|---|---|
| | Million US $ at current prices | | | | | | | | |
| Gross domestic product | 8138 | 17,760 | 44,530 | 125,122 | 167,775 | 186,834 | 198,728 | 206,225 | 164,641 |
| + Net income from abroad | 237 | – | −5716 | −12,944 | −13,271 | −12,125 | −10,364 | −9301 | −3565 |
| = Gross national income (GNI) | 8375 | – | 38,815 | 112,179 | 154,504 | 174,709 | 188,364 | 196,924 | 161,076 |
| − Consumption of fixed capital | −1048 | −2008 | −3862 | −17,396 | −22,925 | −25,547 | −27,238 | −28,642 | −22,990 |
| = Net national income | 7327 | – | 34,952 | 94,783 | 131,579 | 149,162 | 161,126 | 168,282 | 138,087 |
| − Natural resources depletion | −1031 | – | −3382 | −5477 | −8718 | −9280 | −9524 | −8145 | −4035 |
| = Adjusted net national income | 6296 | – | 31,571 | 89,306 | 122,860 | 139,881 | 151,602 | 160,138 | 134,051 |
| Net national income | 7327 | – | 34,952 | 94,783 | 131,579 | 149,162 | 161,126 | 168,282 | 138,087 |
| + Net transfers | – | – | – | – | −12,651 | −14,058 | −14,732 | −17,514 | −15,704 |
| − Final consumption | −5201 | −6197 | −13,478 | −37,729 | −41,022 | −47,936 | −57,869 | −64,441 | −68,723 |
| = Net national savings | – | – | – | – | 77,907 | 87,168 | 88,524 | 86,327 | 53,660 |
| + Education expenditure | 175 | – | 768 | 2293 | 3346 | 3997 | 4539 | 4986 | 4078 |
| − Energy depletion | −1031 | −1827 | −3381 | −5477 | −8718 | −9280 | −9524 | −8144 | −4035 |
| − Mineral depletion | 0 | 0 | 0 | 0 | 0 | 0 | 0 | 0 | 0 |
| − Net forest depletion | 0 | – | 0 | 0 | 0 | 0 | 0 | 0 | 0 |
| − Carbon dioxide damage | −182 | −237 | −421 | −707 | −813 | −877 | −941 | −1003 | – |
| − Particulate emission damage | −17 | −20 | −47 | −57 | −72 | −77 | −82 | −82 | −64 |
| = Adjusted net national saving | – | – | – | – | 71,650 | 80,931 | 82,517 | 82,083 | – |

(*continued*)

**Table 6.15** *(continued)*

| | *1995* | *2000* | *2005* | *2010* | *2011* | *2012* | *2013* | *2014* | *2015* |
|---|---|---|---|---|---|---|---|---|---|
| | % of gross national income (GNI) | | | | | | | | |
| Gross domestic product | 97.2 | – | 114.7 | 111.5 | 108.6 | 106.9 | 105.5 | 104.7 | 102.2 |
| + Net income from abroad | 2.8 | – | −14.7 | −11.5 | −8.6 | −6.9 | −5.5 | −4.7 | −2.2 |
| = Gross national income (GNI) | 100.0 | – | 100.0 | 100.0 | 100.0 | 100.0 | 100.0 | 100.0 | 100.0 |
| − Consumption of fixed capital | −12.5 | – | −10.0 | −15.5 | −14.8 | −14.6 | −14.5 | −14.5 | −14.3 |
| = Net national income | 87.5 | – | 90.0 | 84.5 | 85.2 | 85.4 | 85.5 | 85.5 | 85.7 |
| − Natural resources depletion | −12.3 | – | −8.7 | −4.9 | −5.6 | −5.3 | −5.1 | −4.1 | −2.5 |
| = Adjusted net national income | 75.2 | – | 81.3 | 79.6 | 79.5 | 80.1 | 80.5 | 81.3 | 83.2 |
| Net national income | 87.5 | – | 90.0 | 84.5 | 85.2 | 85.4 | 85.5 | 85.5 | 85.7 |
| + Net transfers | – | – | – | – | −8.2 | −8.0 | −7.8 | −8.9 | −9.7 |
| − Final consumption | −62.1 | – | −34.7 | – | −26.6 | −27.4 | −30.7 | −32.7 | −42.7 |
| = Net national savings | – | – | – | – | 50.4 | 49.9 | 47.0 | 43.8 | 33.3 |
| + Education expenditure | 2.1 | – | 2.0 | 2.0 | 2.2 | 2.3 | 2.4 | 2.5 | 2.5 |
| − Energy depletion | −12.3 | – | −8.7 | −4.9 | −5.6 | −5.3 | −5.1 | −4.1 | −2.5 |
| − Mineral depletion | 0.0 | – | 0.0 | 0.0 | 0.0 | 0.0 | 0.0 | 0.0 | 0.0 |
| − Net forest depletion | 0.0 | – | 0.0 | 0.0 | 0.0 | 0.0 | 0.0 | 0.0 | 0.0 |
| − Carbon dioxide damage | −2.2 | – | −1.1 | −0.6 | −0.5 | −0.5 | −0.5 | −0.5 | – |
| − Particulate emission damage | −0.2 | – | −0.1 | −0.1 | 0.0 | 0.0 | 0.0 | 0.0 | 0.0 |
| = Adjusted net national saving | – | – | – | – | 46.4 | 46.3 | 43.8 | 41.7 | – |

Source: The World Bank—World Development Indicators, March 2018

**Table 6.16** Test for sustainable development of Saudi Arabia

| | *1995* | *2000* | *2005* | *2010* | *2011* | *2012* | *2013* | *2014* | *2015* |
|---|---|---|---|---|---|---|---|---|---|
| | Million US $ at current prices | | | | | | | | |
| Gross domestic product | 143,343 | 189,515 | 328,460 | 528,207 | 671,239 | 735,975 | 746,647 | 756,350 | 651,757 |
| + Net income from abroad | 2191 | 892 | 2715 | 7044 | 9684 | 9695 | 10,969 | 13,553 | 19,006 |
| = Gross national income (GNI) | 145,534 | 190,407 | 331,174 | 535,251 | 680,923 | 745,670 | 757,616 | 769,903 | 670,763 |
| − Consumption of fixed capital | −15,423 | −16,430 | −27,218 | −44,410 | −48,555 | −54,449 | −64,048 | −67,161 | −61,024 |
| = Net national income | 130,111 | 173,977 | 303,957 | 490,842 | 632,368 | 691,221 | 693,567 | 702,742 | 609,739 |
| − Natural resources depletion | −10,208 | −22,540 | −56,660 | −69,682 | −106,930 | −114,896 | −106,585 | −98,603 | −50,602 |
| = Adjusted net national income | 119,904 | 151,438 | 247,297 | 421,160 | 525,438 | 576,325 | 586,982 | 604,139 | 559,137 |
| Net national income | 130,111 | 173,977 | 303,957 | 490,842 | 632,368 | 691,221 | 693,567 | 702,742 | 609,739 |
| + Net transfers | −16,694 | −15,490 | −14,778 | −27,921 | −29,386 | −30,438 | −35,869 | −38,734 | −44,707 |
| − Final consumption | −100,455 | −117,573 | −156,602 | −277,224 | −311,953 | −356,422 | −391,268 | −439,737 | −457,473 |
| = Net national savings | 12,963 | 40,914 | 132,577 | 185,697 | 291,029 | 304,360 | 266,430 | 224,272 | 107,559 |
| + Education expenditure | 7844 | 13,576 | 24,025 | 38,364 | 48,808 | 53,533 | 54,464 | 55,359 | 47,549 |
| − Energy depletion | −10,207 | −22,539 | −56,658 | −69,577 | −106,781 | −114,608 | −106,367 | −98,452 | −50,462 |
| − Mineral depletion | −1 | −1 | −2 | −105 | −149 | −288 | −218 | −151 | −140 |
| − Net forest depletion | 0 | 0 | 0 | 0 | 0 | 0 | 0 | 0 | 0 |
| − Carbon dioxide damage | −1396 | −2022 | −3223 | −5010 | −5043 | −5467 | −5756 | −6116 | – |
| − Particulate emission damage | −358 | −468 | −515 | −801 | −889 | −1029 | −1141 | −1141 | −977 |
| = Adjusted net national saving | 8845 | 29,461 | 96,203 | 148,568 | 226,975 | 236,502 | 207,411 | 173,770 | – |

(*continued*)

**Table 6.16** (continued)

| | 1995 | 2000 | 2005 | 2010 | 2011 | 2012 | 2013 | 2014 | 2015 |
|---|---|---|---|---|---|---|---|---|---|
| | % of gross national income (GNI) | | | | | | | | |
| Gross domestic product | 98.5 | 99.5 | 99.2 | 98.7 | 98.6 | 98.7 | 98.6 | 98.2 | 97.2 |
| + Net income from abroad | 1.5 | 0.5 | 0.8 | 1.3 | 1.4 | 1.3 | 1.4 | 1.8 | 2.8 |
| = Gross national income (GNI) | 100.0 | 100.0 | 100.0 | 100.0 | 100.0 | 100.0 | 100.0 | 100.0 | 100.0 |
| – Consumption of fixed capital | −10.6 | −8.6 | −8.2 | −8.3 | −7.1 | −7.3 | −8.5 | −8.7 | −9.1 |
| = Net national income | 89.4 | 91.4 | 91.8 | 91.7 | 92.9 | 92.7 | 91.5 | 91.3 | 90.9 |
| – Natural resources depletion | −7.0 | −11.8 | −17.1 | −13.0 | −15.7 | −15.4 | −14.1 | −12.8 | −7.5 |
| = Adjusted net national income | 82.4 | 79.5 | 74.7 | 78.7 | 77.2 | 77.3 | 77.5 | 78.5 | 83.4 |
| Net national income | 89.4 | 91.4 | 91.8 | 91.7 | 92.9 | 92.7 | 91.5 | 91.3 | 90.9 |
| + Net transfers | −11.5 | −8.1 | −4.5 | −5.2 | −4.3 | −4.1 | −4.7 | −5.0 | – |
| – Final consumption | −69.0 | −61.7 | −47.3 | −51.8 | −45.8 | −47.8 | −51.6 | −57.1 | −68.2 |
| = Net national savings | 8.9 | 21.5 | 40.0 | 34.7 | 42.7 | 40.8 | 35.2 | 29.1 | – |
| + Education expenditure | 5.4 | 7.1 | 7.3 | 7.2 | 7.2 | 7.2 | 7.2 | 7.2 | 7.1 |
| – Energy depletion | −7.0 | −11.8 | −17.1 | −13.0 | −15.7 | −15.4 | −14.0 | −12.8 | −7.5 |
| – Mineral depletion | 0.0 | 0.0 | 0.0 | 0.0 | 0.0 | 0.0 | 0.0 | 0.0 | 0.0 |
| – Net forest depletion | 0.0 | 0.0 | 0.0 | 0.0 | 0.0 | 0.0 | 0.0 | 0.0 | 0.0 |
| – Carbon dioxide damage | −1.0 | −1.1 | −1.0 | −0.9 | −0.7 | −0.7 | −0.8 | −0.8 | – |
| – Particulate emission damage | −0.2 | −0.2 | −0.2 | −0.1 | −0.1 | −0.1 | −0.2 | −0.1 | −0.1 |
| = Adjusted net national saving | 6.1 | 15.5 | 29.0 | 27.8 | 33.3 | 31.7 | 27.4 | 22.6 | – |

Source: The World Bank—World Development Indicators, March 2018

**Table 6.17** Test for sustainable development of United Arab Emirates

| | *1995* | *2000* | *2005* | *2010* | *2011* | *2012* | *2013* | *2014* | *2015* |
|---|---|---|---|---|---|---|---|---|---|
| | Million US $ at current prices | | | | | | | | |
| Gross domestic product | 65,744 | 104,337 | 180,617 | 289,880 | 350,908 | 374,818 | 390,427 | 403,198 | 357,949 |
| + Net income from abroad | – | – | 2886 | −100 | 110 | 297 | 1579 | 659 | 1743 |
| = Gross national income (GNI) | – | – | 183,503 | 289,781 | 351,018 | 375,115 | 392,007 | 403,857 | 359,692 |
| − Consumption of fixed capital | −7963 | −7725 | −19,463 | −21,449 | −22,681 | −25,262 | −26,363 | −29,013 | −30,462 |
| = Net national income | – | – | 164,040 | 268,332 | 328,337 | 349,854 | 365,643 | 374,844 | 329,230 |
| − Natural resources depletion | – | – | −8512 | −13,719 | −21,316 | −24,609 | −25,906 | −24,075 | −11,851 |
| = Adjusted net national income | – | – | 155,529 | 254,612 | 307,022 | 325,245 | 339,738 | 350,768 | 317,379 |
| Net national income | – | – | 164,040 | 268,332 | 328,337 | 349,854 | 365,643 | 374,844 | 329,230 |
| + Net transfers | – | – | – | – | – | – | – | – | – |
| − Final consumption | – | – | −115,984 | −215,229 | −240,373 | −236,083 | −254,816 | −278,449 | −253,189 |
| = Net national savings | – | – | – | – | – | – | – | – | – |
| + Education expenditure | – | – | – | – | – | – | – | – | – |
| − Energy depletion | −1841 | −4097 | −8512 | −13,719 | −21,316 | −24,609 | −25,906 | −24,075 | −11,851 |
| − Mineral depletion | 0 | 0 | 0 | 0 | 0 | 0 | 0 | 0 | 0 |
| − Net forest depletion | – | – | 0 | 0 | 0 | 0 | 0 | 0 | 0 |
| − Carbon dioxide damage | −417 | −767 | −941 | −1575 | −1730 | −1902 | −2035 | −2182 | – |
| − Particulate emission damage | −213 | −294 | −411 | −687 | −926 | −944 | −966 | −1019 | −984 |
| = Adjusted net national saving | – | – | – | – | – | – | – | – | – |

(*continued*)

**Table 6.17** (continued)

| | 1995 | 2000 | 2005 | 2010 | 2011 | 2012 | 2013 | 2014 | 2015 |
|---|---|---|---|---|---|---|---|---|---|
| | % of gross national income (GNI) | | | | | | | | |
| Gross domestic product | – | – | 98.4 | 100.0 | 100.0 | 99.9 | 99.6 | 99.8 | 99.5 |
| + Net income from abroad | – | – | 1.6 | 0.0 | 0.0 | 0.1 | 0.4 | 0.2 | 0.5 |
| = Gross national income (GNI) | – | – | 100.0 | 100.0 | 100.0 | 100.0 | 100.0 | 100.0 | 100.0 |
| – Consumption of fixed capital | – | – | –10.6 | –7.4 | –6.5 | –6.7 | –6.7 | –7.2 | –8.5 |
| = Net national income | – | – | 89.4 | 92.6 | 93.5 | 93.3 | 93.3 | 92.8 | 91.5 |
| – Natural resources depletion | – | – | –4.6 | –4.7 | –6.1 | –6.6 | –6.6 | –6.0 | –3.3 |
| = Adjusted net national income | – | – | 84.8 | 87.9 | 87.5 | 86.7 | 86.7 | 86.9 | 88.2 |
| Net national income | – | – | 89.4 | 92.6 | 93.5 | 93.3 | 93.3 | 92.8 | 91.5 |
| + Net transfers | – | – | 0.0 | 0.0 | 0.0 | 0.0 | 0.0 | 0.0 | 0.0 |
| – Final consumption | – | – | –63.2 | –74.3 | –68.5 | –62.9 | –65.0 | –68.9 | –70.4 |
| = Net national savings | – | – | – | – | – | – | – | – | – |
| + Education expenditure | – | – | – | – | – | – | – | – | – |
| – Energy depletion | – | – | –4.6 | –4.7 | –6.1 | –6.6 | –6.6 | –6.0 | –3.3 |
| – Mineral depletion | – | – | 0.0 | 0.0 | 0.0 | 0.0 | 0.0 | 0.0 | 0.0 |
| – Net forest depletion | – | – | 0.0 | 0.0 | 0.0 | 0.0 | 0.0 | 0.0 | 0.0 |
| – Carbon dioxide damage | – | – | –0.5 | –0.5 | –0.5 | –0.5 | –0.5 | –0.5 | – |
| – Particulate emission damage | – | – | –0.2 | –0.2 | –0.3 | –0.3 | –0.2 | –0.3 | –0.3 |
| = Adjusted net national saving | – | – | – | – | – | – | – | – | – |

Source: The World Bank—World Development Indicators, March 2018

consumption ratios in recent years. Due to substantial net income from abroad, GNI is larger than GDP throughout the whole period.

Is Oman the problem-child of the GCC at present? Oman is the only GCC country for which the test for sustainable development failed. The adjusted net savings ratio in the last row of Table 6.14 is negative throughout the period due to heavy allocation for energy depletion and negative income from abroad. As a consequence, it must be expected that the combined capital stock of man-made capital and natural capital declined.

The test for sustainable development of Qatar in Table 6.15 shows favourable results. In 2014, the share of adjusted net savings in GNI is 41.7% despite negative income from abroad. The allocations for natural resources depletion are relatively modest and the share of consumption in GNI very low.

In Table 6.16, the test of sustainable development is applied to Saudi Arabia. Saudi Arabia has significant net incomes from abroad, low allocations for consumption of fixed capital, moderate allocation for natural resource depletion and the highest allocations for education expenditures of the GCC countries. Since 2005, adjusted net savings are in the range of 20–30 per cent of GNI. This level is an achievement as it even exceeded the corresponding level of Norway.

Table 6.17 includes the test for sustainable development of the United Arab Emirates. Unfortunately, no information is available on net transfers and education expenditure. Therefore, no estimate on net savings and net adjusted savings is available. However, the information on energy depletion allows assessing the adjusted net national income. The reported shares of energy depletion in GNI are low compared to other GCC countries.

In 2006, the World Bank published the *Little Green Data Book 2006* (World Bank 2006). In this assessment, energy depletion is equal to the product of unit resource rents and the physical quantities of energy extracted. It should be noted that a recent change in methodology in the new time series of the World Bank's World Development Indicators now estimates energy depletion as the ratio of the value of the stock of energy resources to the remaining reserve lifetime (capped at 25 years), covering coal, crude oil and natural gas. The new results for 2005 are quite different from the previous ones in the *Little Green Data Book 2006* for the GCC countries. The significant revision of the methodology to estimate the depletion of natural resources brought about a sudden, and debatable, improvement in the estimate of adjusted savings for the GCC.

## 6 Conclusion

This article has argued that, perhaps contrary to established opinion, the economic diversification of GCC countries is well underway. In many ways, the GCC countries are approaching the diversification levels of the reference country Norway. The test for sustainable development of GCC countries for the period 1995–2015 showed positive results for most years. In 2014, Qatar and Kuwait achieved high rates of adjusted net national saving per person; Qatar even surpassed the level of Norway. Bahrain and Saudi Arabia realised small positive rates. Oman was the only GCC country for which the test yielded a negative result. During the last 20 years (1995–2015), Bahrain, Kuwait, Qatar and Saudi Arabia had positive rates of adjusted net saving throughout the whole period, whereas for Oman the rate was only positive in 2012.

A full implementation of the input-output approach will only be possible if supply and use tables become available for all GCC countries, which are comparable, have the same number of products and industries and use the same classification of the System of National Accounts 2008—SNA 2008 (United Nations 2009). At the moment, only selected input-output tables are available for Kuwait and Saudi Arabia. The National Statistical Offices of the GCC countries should be encouraged to compile annual supply and use tables as an integral part of their national accounts, which are in line with the SNA 2008.

## References

Al-Kawaz, Ahmed (2008): Economic Diversification: The Case of Kuwait with Reference to Oil Producing Countries, *Journal of Economic Cooperation*, 29, pp. 23–48.

Beutel, Joerg (2012): Conceptual Problems of Measuring Economic Diversification as Applied to the GCC Countries, in: Giacomo Luciani (ed.), *Resources Blessed: Diversification and the Gulf Development Model*, Gulf Research Centre, Gerlach Press, pp. 29–70.

Beutel, Joerg, Isabelle Rémond-Tiedrez, José M. Rueda Cantuche (2013): The Importance of Input-Output Data for the Regional Integration and Sustainable Development of the European Union, in Joy Murray and Manfred Lenzen (eds.), *The Sustainability Practitioner's Guide to Multi-Regional Input-Output Analysis*, Champaign, IL, USA, pp. 220–239.

Beutel, Joerg (2017): The supply and use framework of national accounts, in: Thijs ten Raa (ed.): *Handbook of Input-Output Analysis*, Cheltenham, UK, pp. 41–129.

Green, David (2011). *Input-Output Table for Kuwait 2005, GTAP-Global Trade Analysis Project*, Center for Global Trade Analysis, Purdue University, Indiana.
OECD (2018): Harmonised National Input-Output Tables.
United Nations (2009): System of National Accounts 2008.
United Nations (2018): *Handbook on Supply, Use and Input-Output Tables with Extensions and Applications.*
Wagner, John E.; Steven C. Deller (1998): Measuring the Effects of Economic Diversity on Growth and Stability, *Land Economics*, Vol. 74, No. 4, pp. 541–556.
World Bank (2001): *The Changing Wealth of Nations: Measuring Sustainable Development in the New Millennium.*
World Bank (2006): *Little Green Data Book 2006*, Washington DC.
World Bank (2018): *World Development Indicators*, Washington, DC.

CHAPTER 7

# Redefining Economic Sustainability in Resource-rich States: Comparative Lessons

*Manal Shehabi*

## 1 Introduction

In economic development discussions in both policy and academic milieus alike, the term "economic sustainability" has been repeatedly used, but its definition remains elusive. At times, it is implicitly defined, while at others it is defined in terms of economic indicators (such as gross domestic product (GDP) per capita or share of export to GDP). The way the term is used can have various economic foci—such as fiscal, debt, export, and growth sustainability. In the context of "resource-dependent economies"—defined herein as resource-rich states that depend on the exportation of natural resources such as hydrocarbons as a primary source of economic revenue—the issue of "economic sustainability" is also variant and has come to mean different things over time. Initially, economic sustainability was linked to trade theory, and thought to be dependent on the export of natural resources, but that theory failed to achieve the promised growth and development in many resource-rich states—many of which were the world's poorest and most troubled economies. With time,

M. Shehabi (✉)
Oxford Institute for Energy Studies, Oxford, UK
e-mail: manal.shehabi@oxfordenergy.org

 
G. Luciani, T. Moerenhout (eds.), *When Can Oil Economies Be Deemed Sustainable?*, The Political Economy of the Middle East,
https://doi.org/10.1007/978-981-15-5728-6_7

"sustainability" has changed to reflect the political and economic problems in resource economies. In recent and contemporary discussions, economic diversification has mostly become the dominant recipe for economic sustainability in resource-dependent economies. While diversification is important, adopting policies that focus solely on it fails to reflect the economic performance of these states and the full scope of economic challenges they face. Defining economic sustainability is important to the extent that it affects the design and implementation of economic policy meant to achieve long-term growth and development.

Against this background, this chapter redefines economic sustainability in resource-dependent states. The definition reflects the economic challenges they face due to resource dependence with reference to economic growth theory and by way of a comparative assessment of successful states. First, the chapter summarizes the changing definition of economic sustainability in this context. The chapter then redefines economic sustainability as an issue that needs to be considered from two main perspectives: a volatility perspective and a depletion perspective. The definition then reverts to economic growth theory as a basis for designing sustainability-promoting economic policies that take into account both perspectives. A comparative assessment follows, drawing on examples from three main cases: Botswana, Chile, and Kuwait (the primary case study herein).

## 2 The Changing Understanding of Economic Sustainability in Resource States

In its basic form, "economic sustainability" entails continuation of the economy; in the context of resource-dependent states, its definition has substantially changed over time. In the wake of the expansion of the export-led growth economic theory, the dominant driver for economic sustainability in the 1970s (especially in poor countries) was believed to be economic growth achieved by specialized exportations of abundant natural resources. Support for this strategy drew from international trade theories, which proved important in explaining economic performances and business motives.

Export-led growth was based on the concept of comparative advantage, a key driver in economic trade theory. The Ricardian model articulated patterns of and gains from trade in terms of comparative advantage: countries specialize in the activities in which they are relatively more productive.

The relationship between factor endowment and patterns of trade, known as the Heckscher–Ohlin Theorem, states that a country will have a comparative advantage in, and therefore export, the goods that use relatively intensively that country's abundant factor (Heckscher 1919; Ohlin 1933). Hecksher–Ohlin models are driven by differences in factor intensities across countries, entailing that sustainability in resource-rich states would require exporting natural resources.

Export-led growth shaped development policies of resource states, supported by evidence of a positive relationship between exports and economic growth (such as Michaely 1977; Bhagwati 1978; Balassa 1978; Krueger 1984). Yet such models had shortcomings in explaining actual trade trends, first noted in the US in the 1950s.[1] Further, export-led growth theory neither predicted weak economic performance in resource-dependent economies (like Bolivia, Nigeria, and Venezuela), nor explained problems they commonly faced (such as lower growth or price and exchange rate volatility). Thus, the validity of the export-promotion hypothesis was challenged (by studies like that of Jung and Marshall (1985)).

Achieving sustainability by exporting natural resources was further challenged following the exposure of surprising political, economic, and developmental realities in resource-dependent states by the exhaustively debated "resource curse" literature in the 1990s (a term coined by Auty (1993) similar to the "paradox of plenty" by Karl (1997)). Since the 1970s, many of these states have underperformed resource-poor countries and failed to achieve expected socio-economic and political development commensurate with their resource wealth. A significant negative relationship is shown to exist between economic growth and the intensity of petroleum and other mineral resources (Sachs and Warner 1995). The *curse* and various associated elements, such as state intervention, also feature as a common denominator in many political economy models of resource-dependent economies. This counterintuitive situation has been attributed to a host of factors, such as institutional quality (Sala-i-Martin and Subramanian 2003; Bulte et al. 2005), rent-seeking behavior (Gelb 1988; Auty 2001; van der Ploeg 2011), weak public governance, and corruption. Yet no general consensus exists regarding its exact cause.

[1] This discrepancy is pointed out by Leontief (1953) in what is known as the Leontief paradox. It refers to the US's exporting labor-intensive commodities and importing capital-intensive commodities despite being the most capital-abundant nation in the 1950s.

Subsequent research emerged extending the neoclassical framework of trade patterns beyond resource endowments and examining determinants of comparative advantage other than the abundance of natural resources. Such determinants include infrastructure (such as transportation), institutions, innovation, macroeconomic volatility, and human capital and education of labor supply (Lederman and Xu 2007). Shortcomings of the resource curse literature have not passed unnoticed. Studies on the relationship between resources and growth (such as Delacroix 1977; Davis 1995; Monzano and Rigobon 2001; Maloney 2002; Wright and Czelusta 2004; Alexīv and Conrad 2009) show inconclusive or no evidence on the resource curse and suggest its literature ignores the natural cycle associated with resource exploitation. Nevertheless, the performance of resource-dependent economies continued to be explained as part of the resource curse, so the understanding of sustainability development changed to entail movement away from resource dependence.

Sustaining exportation of a finite resource is linked to optimal supply, but theoretical predications also have limitations. The leading Hotelling economic theory has influenced countries' supply. It suggests that the (real) resource price should rise over time at the rate of interest that affects producer extraction decisions (Hotelling 1931). Evidence, however, has been found that oil production in existing wells does not respond to price, unlike drilling activity and costs (Black and LaFrance 1998). Hotelling's rule is also found to be neither a sufficient condition for firm-level extraction decisions nor explanatory of entire industries' production levels by linear aggregation of firms' extraction behaviors (Brazee and Cloutier 2006). Moreover, oil exports, most notably by members of the Organization of Petroleum Exporting Countries (OPEC), are impacted by global oil market dynamics and geopolitical factors not accounted for by Hotelling's rule.

In the economics literature, too, interactions of industrialization and economic sustainability have been explained by the "Dutch disease" (Corden and Neary 1982; Corden 1984; Venables and van der Ploeg 2010). The Dutch disease attributes economic problems of resource dependence to adverse effects on non-resource tradable sectors and expansion of non-traded service sectors which result from significant appreciation of nominal (and real) exchange rates (or inflation in countries with fixed exchange rate regimes) following resource exports (Corden 1984, 2012; Corden and Neary 1982; Venables and van der Ploeg 2010; Tyers 2015; Tyers and Walker 2016). Importantly, this "disease" is an economic reality of only some sectors and occurs even in successful resource-based economies such as Norway, Canada, and Australia.

Other explanations of distortions in economic structure include poor interindustry linkages that lead to reliance on imported rather than domestic inputs to the resource sector (Tornell and Lane 1999). Another factor may be *immiserizing* growth—the possibility that a country's export growth for a product is so large relative to the world market that a decline in its terms of trade is caused. There is no obvious reason, however, as to why this phenomenon should occur more due to resource exploitation/export than the expansion of other export industries. Nonetheless, its existence instigated calls for economic diversification.

Countries that depend largely on resources appeared "unsustainable" due to lack of diversification. Thus, economic sustainability has come to entail (and heavily advocate) diversifying the industrial base and, at times, a reduction of fiscal rigidities, especially given the exhaustibility of natural resources, especially hydrocarbons. Accordingly, many countries (such as oil-dependent Gulf Cooperation Council (GCC) states) adopted developed plans and "visions" that placed diversification at the heart of its long-term goals.

Yet considering economic diversification as synonymous with economic sustainability appears inadequate and confusing at face value. First, the lack of diversification fails to explain the economic challenges faced by resource-dependent economies or the successful economic performance of states like Botswana, Chile, Australia, Malaysia, and Norway. Another example is the GCC states which, despite heavy hydrocarbon dependence, achieved enviable wealth, high measures of per capita income levels, and impressive improvements in socio-economic and human development indicators. Second, diversification is not well defined. Although considered undiversified, GCC states have a sizable share of non-energy sectors in their value added, yet they remain vulnerable to economic challenges, the depth of which has been exposed since the oil price collapse in mid-2014. Further, their export base and government revenue are dominated by the energy sector (hydrocarbons, electricity, and water desalination) with relatively low levels of industrial diversification in exports, as Table 7.1 shows.

Undoubtedly, sustaining an economy is largely facilitated by export diversification, but diversification affects more than exports. Even a closed economy is by definition self-reliant and, therefore, quite diversified; but it is not necessarily sustainable. Its resources are not optimized and its

**Table 7.1** GCC energy sectors, 2014

| *Country* | *Share of energy sectors (%)* | | |
|---|---|---|---|
| | *In value added* | *In exports* | *In government revenue* |
| Bahrain[a] | 40 | 69 | 83 |
| Kuwait[b] | 61 | 91 | 91 |
| Oman[c] | 54 | 84 | 79 |
| Qatar[d] | 32 | 85 | 90 |
| Saudi Arabia[e] | 50 | 80 | 88 |
| UAE[f] | 45 | 78 | 60 |

Notes: Unless otherwise stated, the share in government revenue was based on data reported by the government and may exclude income from oil and gas investments (which were captured in non-energy industries); thus, it may be understated. For United Nations value-added data (2018), sectoral data on the share of non-energy in value added were not available in the national accounts, so UN data for mining, manufacturing (including refining), and utilities were used as proxies. Therefore, these figures may be overestimated. For Qatar, government revenue includes income from oil and gas investments. Sources are listed below. For UAE, the share of energy exports is for 2016

[a]For Bahrain: National accounts and budget from Bahrain Open Portal Data (2018); government budget form Bahrain Ministry of Finance (2018); United Nations value added data (2018)

[b]For Kuwait: National accounts from Kuwaiti Central Statistical Bureau (2018); government budget form Kuwait Ministry of Finance (2018)

[c]For Oman: National accounts from Oman National Centre for Statistics and Information (2018); government budget from Oman Ministry of Finance (2015); United Nations value added data (2018)

[d]For Qatar: National accounts from Ministry of Development Planning and Statistics (2018), Staff Concluding Statement for the 2018 Article IV Mission (2018); International Monetary Fund (IMF) Qatar Country Report (2013)

[e]For Saudi Arabia: National accounts from Saudi General Authority for Statistics (2018); government budget from Saudi Arabia Ministry of Finance (2018)

[f]For UAE: United Nations value added data (2018); UAE Annual Economic Report (2016, 2017)

welfare is not maximized, rendering its performance possibly worse than if it were open. The point here is that "diversification" is neither a guarantee for sustainability nor assurance for solving problems arising from resource dependency, but meaningful diversification is necessary nonetheless.

A continuously changing understanding of "economic sustainability" is problematic to the extent that it impacts economic policies (locally and those promoted by international organizations, such as the IMF). It threatens the durability and consistency of policy implementation and of development plans intended to manage the resource sector and to achieve long-term economic growth. This problem demands adapting the way economic sustainability is defined and, accordingly, introducing new policies that are consistent over time. This is the focus of the next section.

# 3 Redefining Economic Sustainability

## 3.1 *Why a Redefinition Is Required*

Ensuring economic sustainability is a common goal in economic development plans of resource-dependent states. In the wake of inadequacies of existing theories in predicting or solving economic challenges common among them, how *economic sustainability* is understood needs to be adapted to form the basis of successful economic policy.

Economic sustainability is a complex concept with various elements. Of course, comparative advantage remains a key driver of economic specialization and trade. Beyond that, economic performance is affected by public debt, public investment, economies of scale, technology, transportation costs, and noncompetitive industrial structures. Economic outcomes are shaped by public policy, not only in infrastructure, public investments, industrial policy, and trade policy, but also in taxation, distribution of resource rents among the population, employment, and education. Further, in practice, sustainability assessments and subsequent policy design have tended to be discipline-driven. If economic rigor is divorced from political and/or social analyses, policy becomes myopic and counterintuitive.

While economic diversification is necessary for continuity of the economy, it is insufficient. Setting it as the main sustainability-promoting policy has inadequacies, for the following reasons.

First, diversification is at odds with the theory of comparative advantage. Second, as previously explained, the term is not well defined. Diversification can be pursued in either export, budget, or value-added each of which has separate implications on policy and sustainability. Although most definitions imply export diversification, a closed economy is, by definition, diversified.

Third, successful diversification is constrained by various factors: economic development (resource availability, capital limitations, or economic structure), political development (conflict, government structure, political economy, and oligarchs), or social development (population, health, education levels, and constituencies' requirements). Shehabi (2019) shows that the GCC economies' unique structural constraints and economic distortions, and not the Dutch disease, are the primary reasons for limited diversification of export and government revenues.

Fourth, as mentioned, the lack of diversification has in the past allowed the realization of economic advantages, at least for the GCC states. It was not until the collapse of the oil price in mid-2014 that the depth of their fiscal and economic challenges was revealed, instigating reforms. Development plans including reforms in energy, rent distribution, industrial or private sector development, and labor are intertwined but can also be contradictory and present trade-offs. Some states even adopted reforms in energy subsidies, human capital development, and resource windfall management; but implementation has been weak and resource sectors continue to dominate the economy.

Fifth, diversification is required not only due to resource depletion, but also due to the lack of sustainability of exports during the life of the resource. Export capacity depends on a petrostate's proven reserves, production capacity, local energy needs, and on world demand for conventional energy sources. Local consumption is key in determining a petrostate's ability to sustain its petroleum exports. Many states have had to shift production to satisfy local demand, ultimately becoming net-importers for a number of years; these include Syria in 2008, Indonesia in 2003, and Egypt in 2012. Others, like Oman and Brunei Darussalam, grapple with export diversification plans as their proven petroleum reserves are expected to run out in 15 and 22 years, respectively. In addition, world demand for and the affordability of conventional sources are threatened by technological advancements and investments in shale oil and gas from one side, and renewable energy from the other. The role of the energy sector in climate change has also led to pressures to reduce consumption of fossil fuels and expand reliance on renewables. These factors have contributed not only to the collapse of oil prices in mid-2014, but also to expectations that low oil prices would persist despite minor recovery. Collectively, these factors increase uncertainty in oil prices and challenge OPEC's historical ability to manipulate prices through controlling supplies.

The role of technology is not to be underestimated; a case in point is the sharp decline of the share of Chile's once-dominant copper industry from the third highest in the world to 4% of the world's production in 1911 following a failure to keep up with the technological advances of American companies (Collier and Sater 1996: 139).

The sustainability of the revenue of natural resource exports is only a part of the economic sustainability equation. Export sustainability is also affected by exchange rates and constraints on production and pricing decisions in light of domestic demand requirements and export quotas.

Moreover, resource-dependent economies, both rich and poor, face unique yet common policy challenges, chief among them being the volatility of commodity prices. This volatility harms economic performance and output growth in both petrostates and petroleum-importing countries (Ramey and Ramey 1995; van der Ploeg and Poelhekke 2009). It exposes economies to boom–bust cycles, hampering economic performance and output. These cycles are often accompanied by pro-cyclical (rather than countercyclical) fiscal policy, where governmental expenditures greatly expand during booms and contract during busts. Such tendencies are often further exacerbated by domestic macroeconomic and political instability (Frankel 2011). van der Ploeg and Poelhekke (2009) find that there is a direct effect of resource dependence on growth which is positive, but is often dominated by a negative and indirect volatility effect. Periods of oversupply in the world market that reduce petroleum prices and export revenues expose structural defects in state services, as in Nigeria and Venezuela in 1983 and in Russia in 1998.[2] More recently, since mid-2014, a 71% price collapse has caused severe fiscal deficits and real declines in exports, government revenues, and GDP for petrostates. Russia, for example, has suffered a redirection of financial flows and exchange rates instability,[3] while members of the GCC have faced unprecedented deficits, despite a history of significant fiscal surpluses.

Moreover, the problems faced by resource-rich states are often exacerbated by the political economy of resources and of rent distribution. Although heterogeneous, the experiences of many resource-rich states indicate the widespread use of politically motivated reactionary policies in response to resource price volatility, including price controls, producer subsidies, nationalization, restrictions on foreign participation, and stockpiles. These policies are focused on short-term consequences, rather than on the source of the volatility, and do not reduce the exposure of economies to commodity volatility. Resource dependence is often accompanied by overconsumption during or after booms, and buoyed by policies favoring short-term windfall over long-term benefit.

[2] In the case of OPEC, while any one country in the cartel can expand supply and raise its revenue, this is not possible for the group as a whole if the initial production level is jointly profit-maximizing.

[3] This situation has been exacerbated by the international sanctions over the Ukrainian crisis.

Therefore, resource-dependent states face considerable pressures in implementing efficient fiscal policy solutions that reduce cyclicality by harnessing oil rent windfalls during revenue booms to cushion downturns during busts, while simultaneously ensuring efficient allocation of resources and diversification of the revenue base to achieve long-term sustainable growth, development, and employment objectives. Stabilizing resource income and securing non-exhaustible sources of revenue are also important policy objectives. Striking variations in petrostates' economic and political statuses point to the conclusion that economic performance depends upon both preexisting heterogeneous conditions and the quality of policy regimes. For these reasons, redefining economic sustainability considering said challenges is crucial.

### 3.2 A Redefinition: Back to Basics

This section redefines economic sustainability by offering a broader definition launching from the linguistic meaning of "sustainability," both during resource dependence and after the depletion of resource export revenue, and anchored in economic growth theory.

The premise for the redefinition is the following. Sustainability entails continuity of economic activity to achieve continuous growth and development. While revenue from an exhaustible resource is, by definition, unsustainable, a resource industry no matter its size is only one part of an economy. Therefore, the continuity of an economy ought to be ensured and evaluated before the exhaustion of the resource as well as after it. A resource-rich economy with proven reserves lasting for a given number of decades has a "permanent" source of income for that period, which can be sufficient to ensure growth and development for that time. Yet the actual outcome depends on the way export windfalls are used and the economy is managed. As such, economic sustainability can be achieved by continuing to exploit an economy's comparative advantage, but strategically in a manner that extends the length of the resource revenue as well as better manages the resource windfall. At the same time, resource revenue is, by definition, volatile, so managing it requires reducing macroeconomic volatility. In light of resource exhaustibility, sustainability necessitates securing alternative sources of revenue generation in the long run. Both of these can be achieved by various means, one of which is diversification indeed.

As such, it is argued herein that economic specialization based on comparative advantage is not contradictory to sustainability or economic diversification, and that both concepts are relevant to, but insufficient for economic sustainability in resource-dependent states. Dependence on resources in and of itself is not unsustainable. Instead, the policies adopted to manage this dependence and the ability to maintain income generation from it are what determine the extent to which such resource dependence is sustainable. Economic sustainability must be achieved by managing resource supply, export windfalls, and price volatility, as well as by developing the resource sector and value-adding non-resource industries.

Therefore, I posit that sustainability in a resource-based economy is an issue that has two perspectives: managing the volatility of resource-driven revenue and managing the depletion of the resource and/or its export revenue after the commodity export revenue has been exhausted. Importantly, both perspectives are relevant in the short and long terms, but in the short term the volatility perspective is more relevant while in the long term, depletion is more prominent and can be offset with the accumulation of other assets.

Accordingly, this chapter proposes to consider economic sustainability by designing economic policies that address drivers of economic growth taking into account both the volatility and the depletion perspectives. Indeed, economic sustainability does not strictly entail economic growth. Moreover, the latter in reality is neither simple nor basic. Yet focusing on the drivers of economic growth offers the distinct advantage of evaluating an economy continuously over time rather than during the resource dependence episode or after it only. What follows reverts to the literature on economic growth in an elementary growth model, and then adapts it to reflect resource-export dependence.

### *3.2.1 Growth in Theory*[4]

An elementary explanation of growth in resource-dependent states is grounded in the traditional neoclassical growth theory emphasizing gradual, steady growth during which the economic structure remains stable. In the context of the vintage neoclassical growth theory of Solow (1956, 1957), as well as Swan (1956) (and subsequent models), the emphasis is on gradual, steady growth, during which the structure of the economy remains stable.

[4] Add reference to annex

In modified versions of the Solow (1956, 1957) model that include human capital,[5] the disparity of income between countries is more effectively captured. The steady growth that manifests with physical and human capital depreciation and population pressure is offset by new investments in physical and human capital, as well as by human capital through education and health, technological growth, and productivity. In this steady state, the respective economy progresses at a long-term, underlying rate of innovation. Newer growth models incorporate a more realistic assumption of economy-wide, increasing returns to scale and endogenous technological changes at a steady rate.[6] When the economy is shocked (by an event such as a war or a decline in resource price), the concavity of production in physical and human capital ensures that investment exceeds the effects of depreciation and population growth during the transition to a new steady state. This transitional growth performance can be considerably improved if higher savings rates can be mobilized.

In a resource-dependent economy, the production factors which drive economic growth are summarized as physical capital, human capital, natural resource, land and environment, and the residual institutions and technology. For illustration purposes and to account for the exhaustibility of both natural resources and land and environment, these factors will be treated as constant, thus having no impact on the growth rate. Accordingly, to demonstrate the Solow (1956, 1957) growth model, this section references an augmented, simplified version to account for both physical capital and human capital, collectively, as capital ($K$), given that both types of capital accumulate in a similar manner. The model simulates the shock of the advent of war and recovery postwar, as in the case of Kuwait. The Elementary model is described in Appendix 1. In it, the population growth rate ($n$) matches the labor force and employment growth rates. Knowledge (being labor-augmenting technology; $\theta$) grows at a constant exogenous rate. The economy grows at a steady state, in terms of per effective worker $\overline{k_e}$.

To simulate the war, a shock is first represented by a large destruction in capital and a significant reduction in population. Initially, output will fall, and the remaining capital falls short of the steady-state level, leaving investment at a level above its capital break-even point, causing growth to

[5] For further reading, see Mankiw et al. (1992).

[6] For additional reading, see Aghion and Howitt (1992), Grossman and Helpman (1993), Lucas (1988), and Romer (1986, 1990).

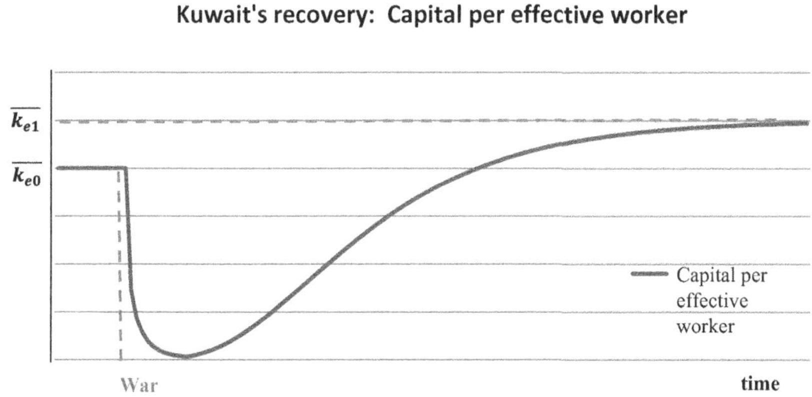


**Fig. 7.1** Rapid GDP growth rate at the start of the recovery. (Source: Author's analysis)

increase in the period following the war. This effect is enlarged if the saving rate also rises. The surplus of new investment is then over the break-even point and larger with an even faster recovery.

As the capital per worker rises temporarily, the productivity of capital increases, inducing a temporarily higher growth rate in the reconstruction (post-shock) phase. Diminished returns eventually return and the growth approaches the long-term path of growth. Additional accumulation of capital during this phase would shift the production possibility frontier curve, reaching a higher level of growth. Similarly, efficiency gains (represented normally as a movement along the production possibility or frontier curve) or higher savings produce new investment surplus over the break-even point, speeding up the recovery.

Upon the end of the shock and at the commencement of recovery efforts, the capital-to-labor ratio converges to pre-shock levels and GDP grows at diminishing rates that were larger than pre-shock levels to eventually converge with pre-shock growth levels, as in Fig. 7.1. The steady state postwar ($\overline{k_{e1}}$) exceeds that prewar ($\overline{k_{e0}}$).

### *3.2.2 An Elementary Numerical Simulation*

Notwithstanding shortcomings of the theoretical model, this section demonstrates the impact of sustainability-promoting policies on Kuwait's economic growth postwar through an elementary numerical simulation following from the aforementioned theory.[7] The simulation examines policies that favor capital accumulation (through sovereign wealth fund (SWF) assets), savings and investments, human capital, and technological advancement.

Postwar, Kuwait's economy grew at unmatched speed, owing largely to SWF investment that funded reconstruction. Although a large part of this growth was temporary (reflecting a reconstruction boom), a period of extraordinary growth was driven also by a surge in investment, growth in technology (including that employed in the rebuilding of the oil industry), and an inflow of labor (human capital), mostly of expatriates. A large part of the human capital was maintained and postwar reform focused on increasing the skill set of domestic labor.[8]

To simulate postwar recovery, the capital-to-labor ratio is first shocked. The savings rate is positively shocked to account for faster capital accumulation and further investment in human capital postwar. The rate of technical growth is also positively shocked to simulate the acquisition of knowledge and technical assistance from abroad.

Results show that the rate of output growth (dY/Y in Fig. 7.2) was larger than that of capital accumulation. They also reflect the expectation of capital and output growth rate exceeding prewar levels and growing at substantial, yet diminishing, rates and reaching a new higher level of steady-state growth, as Fig. 7.2 shows.

Although the simulation is elementary, its results mirror actual results of the Kuwaiti GDP recovery path postwar. This confirms that Kuwait's postwar growth is partially due to a reconstruction boom, but also partially due to the recovery of capital, human capital, savings, and technological advancement.

### 3.2.3 *Natural Resources and Technology*

While excluding natural resources is common practice in standard growth models (as in the previous example), examining economic growth in the

[7] The rates at which physical and human capital accumulate and depreciate are assumed to be equal for purposes of this illustration. Although not entirely realistic, this assumption is rationalized based upon the low elasticity of Kuwait's output with respect to human capital (0.10), based on 2004 data within the GTAP VIII database for Kuwait.

[8] Additionally, postwar growth in income per capita occurred due to an initially low population because Kuwait had lost close to 40% of its prewar residents. This growth slowed as the population recovered due to new labor policies until 1997 when the population reached its prewar level.

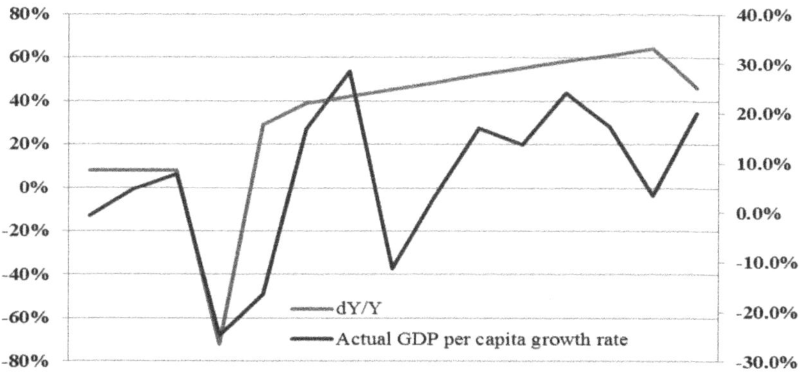


**Fig. 7.2** Simulated Kuwait recovery reaching a new steady state due to shocks in capital-to-labor ratio, savings rates, and productivity. (The impact of the war on Kuwait was simulated through the following shocks: (a) an 80% destruction in physical capital and (b) a 45% reduction in population. Postwar recovery simulation included a 30% increase in savings rates, reflecting increased investment in human capital and a doubling of productivity. Output elasticities of capital and labor were calibrated to equal the following values in the GTAP VIII database entry for Kuwait in 2004: 0.75 for capital (being the sum of elasticities of capital and skilled labor of 0.68 and 0.07, respectively. Source: Author's simulations)

presence of depleted natural resources is consistent with the goals of achieving economic sustainability. As mentioned earlier, economic sustainability entails extending the usable length of the natural resource, which requires ensuring appropriate extraction rates. For demonstration, a simple example is presented, setting the natural resource as the only input of production, and the available resource that can be extracted is a portion of the total available resource value. The mathematical representation to solve this elementary example is expressed in Appendix 1.

If the state of technology is constant over time, the output will eventually become nil when no more of the resource can be extracted. Nevertheless, if the level of technology is allowed to change over time, it counteracts the exhaustibility of the resource. Relaxing the assumption of constant technology, technology is now growing at a constant exogenous rate ($g$). This set-up enables addressing the question of sustainability: what is the trade-off between producing output today and leaving the resource in the ground for future production?

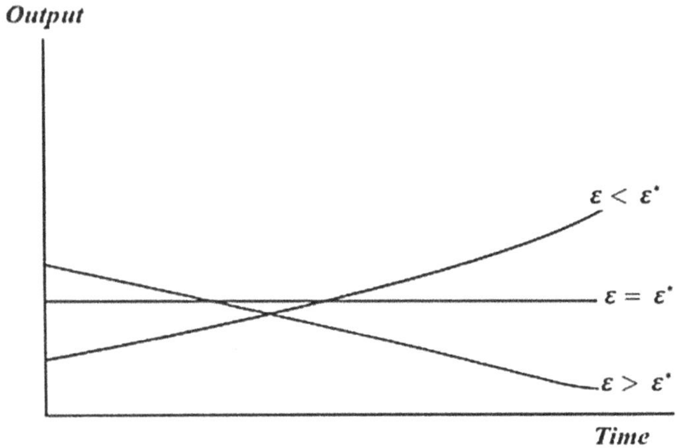


**Fig. 7.3** Relationship between resource use and growth

Assuming the resource is extracted at a constant exogenous rate ($\varepsilon$), the quantity of the remaining resource will decline exponentially over time. Solving the equation yields an extraction rate that is consistent with steady-state levels and a zero growth rate of output, which is $g = \varepsilon^*$.

It is clear that there is a positive relationship between the rate of extraction and output: a lower extraction rate would lead to a lower level of output. A higher extraction rate, on the other hand, would lead to a lower growth rate of output, which could even become negative at very high levels of extraction. Setting this value at zero yields an extraction rate that is consistent with steady-state levels and a zero growth rate of output in $\varepsilon^*$, which thus represents the sustainable level of output. Therefore, the levels of output in economy can be determined by examining the extraction rate chosen in relations to $\varepsilon^*$. If it exceeds $\varepsilon^*$, then the initial level of output will exceed the sustainable level, but will decrease overtime, as Fig. 7.3 shows.

Importantly, including the remaining factors of production would complicate the algebraic representation but yield the same general conclusion.

### 3.3 *The Takeaway*

The theoretical model shows that higher economic growth than that recorded in the previous period (reaching a new steady state) can be achieved by increased savings, increased capital per effective worker, improvements in labor productivity, and improvements in labor-augmenting technology to increase economic efficiency. This technology captures the residual of policies that impact economic growth, including but not limited to noneconomic factors—such as institutional quality, legal environments, cultural standards, industrial regulation, and the political economy. Achieving sustainability requires designing policies that address drivers of economic growth while balancing both the volatility and the depletion perspectives, in the short and long runs. This balance is key and difficult as the different terms can have competing interests and trade-offs.

As the theory confirms, there is an intuitive trade-off between producing output today and leaving the resource in the ground for future production, a trade-off that in reality is known and present in the minds of many policymakers, international aid institutions, and citizens of resource-rich states. This is difficult because competition over rent distribution favors faster production, which contributes to larger short-term growth but faster depletion. Yet the model highlights the role of technological progress and its impact on the sustainability of the resource that is already available. Advancements in technology can contribute to efficient extraction as well as extending the length of the available resource and its export revenue.

Another critical factor at the heart of this balancing act is the trade-off between local consumption and exports of a given supply. This is a serious challenge particularly for countries that depend on resources as an important input in production (such as oil being an input in manufacturing or transportation) as well as final demand. In resource-dependent states, the political economy of resources has historically resulted in large government expenditure and financial commitments to fund generous resource rent distribution and guaranteed public employment, which, in turn, have translated into excessive consumption and severe fiscal pressures. The balance between monetary, fiscal, and exchange rate policy is critical in moderating commodity-sourced volatility and in mitigating Dutch disease effects, which is necessary for long-term diversification and sustainability. Technological and efficiency advancement as well as policy reform (at the fiscal, institutional, social, legal, and even cultural levels) are key in moderating domestic consumption and extending the levels of output available for exportation. Striking a balance requires transparent use of resource rents in the short run, and the design and implementation of welfare-improving, productive long-term development. This development can be achieved through policies that focus on (a) human capital development, which boosts value added and productivity and can build future industries and knowledge economies; (b) physical capital, which can contribute to industrialization and domestic infrastructure; (c) savings, which can provide new accumulation of assets for the future; and (d) technological advancements, which can contribute to the emergence of new industries and increase efficiencies (like renewable energies, tradable services, and knowledge economies).

## 4 Lessons and Comparative Assessments

For resource-dependent states, navigating resource price volatility demands the best from their economic wealth management and economic policies. Although growth performance of many resource-dependent economies has been poor, a small group of economies have done well, such as Norway, Australia, Botswana, Malaysia, Chile, and the Gulf states. This section highlights comparative successful policy lessons drawing on successful experiences of developing states:

- Chile and Botswana, both states with exemplary performances that avoided the so-called *curse*
- Kuwait, an example of a GCC state with an enviable economic performance

The subsequent discussion highlights how policies focused on drivers of economic growth and also enabled said states to manage challenges associated with dependence on an inherently volatile and nonrenewable resource, and move toward sustainability.

### *4.1 Fiscal Rigor and Savings for Capital Accumulation*

Growth-enhancing policies that consider both perspectives entail managing resource windfalls and accumulating savings and investments for the future; SWFs are a means of achieving these two objectives. SWFs are government-owned investment funds commonly established during periods of government surplus.[9] They are typically created to serve as stabilization funds, savings funds, pension reserve funds, or reserve investment corporations. In resource-rich states, they are established to reduce the

[9] For further reading, see Collier et al. (2010) and van der Ploeg and Venables (2012).

impact of volatile petroleum windfall on exchange rates. SWFs also offer a mechanism to reinforce fiscal discipline through rules mandating recurrent contributions to the funds and withdrawal limitations from them. They also offer a mechanism to diversify government portfolios and add new dimensions to foreign assets management. Unlike reserves, SWFs buffer the real exchange rate from terms of trade shocks with fixed exchange rate regimes and in relatively closed economies, with varying results. An IMF study concludes that there is no evidence that fiscal rules have had an effect on fiscal outcomes (Ossowski et al. 2008). Yet among resource economies, the successful cases of Kuwait, Norway, Chile, and Botswana have all had successful SWF experiences.

SWFs are key in achieving sustainability in resource-rich states because they (a) impose fiscal rigor and revenue management; (b) offer savings that can be used in fund development as well as non-resource industries; and (c) save a portion of the resource income in an asset that can be used in the future after the natural resource is exhausted to produce non-resource goods. These factors are drivers of growth in the short, medium, and long terms.

Chile's fiscal constitutions for its copper-based stabilization SWF have been largely successful (Frankel 2011). Chile has had an overall successful experience of resource dependence—unlike its troubled fellow Latin American country Venezuela—which is caused by successful management of resource windfall. Figure 7.4 shows various SWF accumulations of Latin American countries, with Chile being the leader of the pack.

Botswana, a great success story, has delinked the government from current resource revenues, so that revenues that do not meet government spending and investment criteria are invested abroad through Botswana's SWF (IMF 2012).

The role of SWFs is clear in Kuwait's case, which diverts petroleum revenue to external assets invested in its SWF, managed by the Kuwait Investment Authority (KIA). Established in 1953 (eight years prior to independence), the KIA is the oldest country-owned fund in the world.[10] The KIA managed a fiscal stabilization fund (General Reserves Fund (GRF)) and a savings fund (the Future Generations Fund (FGF)).

Uses of KIA funds can address both the volatility and the depletion perspectives of sustainability, as follows. Either budget surpluses are

[10] Predated only by two state-owned funds within the US state of Texas, which were established in 1854 and 1876. It was nearly 20 years before other countries established SWFs.

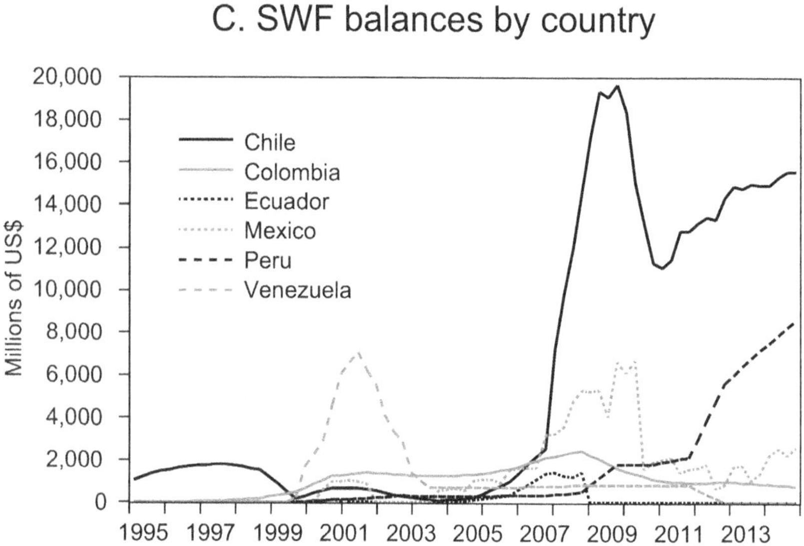


**Fig. 7.4** SWF accumulations in select Latin American economies. (Source: Aizenman and Riera-Crichton (2014))

invested in the GRF or funds are withdrawn from it to smooth out short-term governmental expenditures during deficits, thus shielding the economy from the negative impacts of petroleum price volatility. As for the FGF, Kuwaiti law requires that a minimum of 10% (increased by an additional 15% in 2012 and reverted in 2014) of the annual general revenues allocated to this fund are employed by the Ministry of Finance for various investments, with the associated returns and profits also allocated to the same fund. KIA savings created large foreign asset portfolios that acted as an insurance policy to finance capital accumulation and fiscal commitments during low oil price periods. The funds are a primary source of Kuwait's fiscal stabilization. Its funds were invested in ways that diversified revenue sources away from petroleum, hedged oil income risks, and earned interest. In an economy-wide analysis, Shehabi (2017) confirms that revenue inflow from the KIA is a key adjustment mechanism that partially absorbs negative impacts of oil price shocks.

The indispensable role of the KIA in Kuwait's postwar recovery following the Iraqi invasion in 1990–1991 demonstrates the successful role of

SWFs. The war caused massive destruction of Kuwait's petroleum assets and industry. Funds accumulated in KIA from prewar fiscal surpluses and SWF contributions (estimated at $25–$50 billion in 1989[11]) funded post-war reconstruction (from extinguishing oil-well fires to rehabilitating the petroleum industry). The ensuing petroleum production and exportation allowed the resumption of capital savings.

Despite its advantages, Kuwait's SWF accumulations are juxtaposed against substantial government expenditures and fiscal commitments. One shortcoming is that Kuwait's petroleum windfall has been invested abroad rather than domestically where it could expand industrial capacity[12] (Shehabi 2019).

The main lesson here pertains to the role of fiscal policy and savings of nonrenewable-resource rents in managing volatile petroleum windfall revenue, shaping economic performance, and providing savings for capital accumulation and development. Such savings can be used for expanding non-energy industries and their exports, and local infrastructure, as well as developing human capital and capabilities for acquiring new comparative advantages.

### 4.2 *Human Capital*

Investments in human capital are key to successful economic growth. Not only do they improve the performance of resource-dependent economies, but they are also essential for growing the non-resource sector necessary for diversification efforts and managing the depletion perspective. Many of the countries that have experienced the resource curse have missed this opportunity. Meanwhile, countries like Japan and Germany after the Second World War were able to recover despite complete destruction of their economies in part due to the preservation of prewar human capital and the subsequent investments in it.

The role of guest workers in economic development is a key unique feature of the GCC economies. Drawing one example from Kuwait, guest workers represent an important part of these economies' human capital

[11] Due to legal requirements preventing the disclosure of the funds' details, this range is estimated by calculating annual contributions and an estimated low-range return to capital of 2%. A study by the Arab Planning Institute in 1992 estimated that Kuwait spent approximately $40–$50 billion of its foreign investment portfolio on reconstruction.

[12] This is true even during low oil prices when non-energy sectors can theoretically expand (through reverse Dutch disease).

and are drivers of economic growth and development. A major feature of Kuwait's petroleum economy has been the accompanying spike in population and human capital as expatriates moved to Kuwait on a temporary basis (without a path for naturalization or permanent residency) to fill shortages in some unskilled and mostly skilled human capital given Kuwait's relatively high levels of illiteracy and small population size during the 1970s.[13] The ability to secure human capital from expatriate labor was also key to achieving postwar reconstruction, as explained earlier (Shehabi 2018a).

Notwithstanding the challenges surrounding dependence on foreign labor, the structure of their labor contracts had an unintended positive outcome on the economy. Expatriate laborers work temporarily in the GCC, with their stay linked to employer-sponsorship programs through the *kafala* system, and they have flexible labor contracts, often with lower wages than their national counterparts. In model-driven simulations, the expatriate labor exit offers another key stabilization mechanism of the Kuwaiti economy in the face of the volatility of petroleum revenue (Shehabi 2017). The dynamics of the labor market entails different impacts of this contractionary shock on the two labor segments. Typically, as the real wages of expatriate workers are assumed to be sticky (in both the short and long runs), employment levels adjust instead. Since most Kuwaitis are employed by the public sector, where contracts are rigid, their employment is unaffected, while their real wages change slightly. By contrast, the flexibility of expatriate labor contracts allows affected industries to adjust their employment levels, causing similar declines in the employment levels of skilled and unskilled expatriates. The flow of guest workers exponentially increased during oil booms, and decreased during busts. Thus, an example of successful policy includes model-driven policy solutions that focus on the management of labor contracts for expatriate labor (Shehabi 2017). While expatriate labor exit acts as a cushion absorbing oil price shocks, there is a distinct trade-off between efficiency gains from expatriate labor and the goals of Kuwaitization of the labor force

[13] The surge in liquidity in the 1970s (owing to booming oil revenues following the 1973 embargo and the nationalization of the petroleum industry) could only be initially accommodated by investments in land, construction, and real estate projects, which required importing a significant number of workers, especially as many Kuwaiti males were not interested in construction-type jobs.

(Shehabi 2018b). Managing this will be important in ensuring future economic sustainability.

Human capital is very important in the sustainability of resource-rich economies because they can contribute to acquiring new comparative advantages. They can create new industries, products, and services. Human capital is also the seed for growth-enhancing innovation and competition.

### 4.3 *Technology, Institutions, and Taxes*

Technology is undoubtedly a transformative engine and a contributor to growth. In its most traditional definition, technological advancement enhances productivity and efficiency and improves economic performance. In growth theory, as mentioned in the Sect. 3.2.3, technology captures the residual of policies that impact economic growth, including but not limited to noneconomic factors such as institutional quality, legal environments, cultural standards, industrial regulation, and the political economy.

Technology can play an unequivocal role in extending resource export revenue, transform resource industries, and even reverse the threat of depletion and overcome volatility. To demonstrate, the challenges of diminished oil export revenue with climate change mitigation can be alleviated if new technologies enable oil exporters to continue to export energy that is green and adaptive to changing demand trends in the oil market. Such technologies include investments in energy efficiency, carbon capture and storage, and hydrogen. This is an opportunity for oil exporters given their existing comparative advantage and specialization in energy.

Undoubtedly, institutions (part of the overall "technology" driver of growth) are important to securing sustainable long-term economic growth. Botswana offers a success story in striking institutional quality that is favorable to growth. It has not only achieved democratic progress with transparency in rent distribution, but has also affected the development of the resource sector. Instead of the commonly used auctions, Botswana negotiated rights to diamond extraction with the single dominant market player De Beers (Venables 2016; Besada and O'Bright 2018). In doing so, Botswana was able to avoid the lack of competitiveness and transparency, as well as a limited state's share of revenue that would have resulted from awarding the auction to a single market player. The country was also successful in attracting foreign direct investment, thus achieving economic diversification while also promoting linkages (Besada and O'Bright 2018). Further, Botswana's institutional quality is recognized by Mehlum et al. (2006), who find negative (and larger) effects of resource-richness on growth only for economies with poor institutional quality, with the break-even point being around the institutional quality of Botswana.

Taxation is another crucial factor. Norway, Australia, and Canada—resource-dependent states that are considered successful—have sophisticated tax systems (personal, corporate, indirect, and international taxes). These taxes offer a source of government revenue that serves three purposes: increasing the size of non-energy contribution to government revenue; offering additional sources of revenue that can be used to fund current or capital expenditures and public infrastructure; and reducing the need to withdraw from SWF assets during times of fiscal pressures. Taxation and low subsidies are necessary for economic sustainability—although difficult to achieve from a political economy perspective in many resource-dependent states.

### 4.4 *Industrial Reform and Regulation of Oligopolies*

While regulation is a part of institutions and, therefore, of the larger "technology" growth driver, regulation of oligopolies merits its own distinct discussion. This is due to its unequivocal role in ensuring competition, efficiency, long-term creative destruction, and economic growth. In assessing efficiency and economic policy, the role of oligopolies is critical. Oligopolies exist in almost all countries, including the most advanced ones. Some industries, such as utilities and communications, are also naturally oligopolistic because they require very large fixed costs to be established and require large scales to run them. In small resource-dependent economies, it is not surprising that the high levels of minimum efficient scale delivered by modern technology and the smallness of these economies should lead to the emergence of oligopolies or monopolized industries, particularly in protected services.

Nevertheless, these oligopolies can play a detrimental role in economic sustainability. It has been accepted in economic theory that competition induces innovation and, therefore, economic growth through the process of "creative destruction". This process, coined by Schumpeter, entails an ongoing process in which innovation is induced by competitive forces and destroys rents conferred in the short term by former innovation, maintaining efficiency (Schumpeter 1942: 82–83). Oligopolies limit innovation-inducing competition. They sustain high markups, which are pure profits above the average cost, thus distorting economic efficiency. Oligopolies also collude on prices or quantities, further distorting markets. In resource-rich states, oligopolies are pervasive to varying degrees and contribute directly to limiting competition and, consequently, sustainable economic development and growth. A large number of resource-rich states have state-owned resource industries and large public sectors that control industries which are commonly monopolies or oligopolies. They cause a big loss in economic efficiency. In an economy-wide assessment, Tyers (2015) suggests that the full exploitation of oligopoly market power in Australia would cause a reduction of real GDP by as much as a third in the long run.

Subsidizing consumption widens the gap between actual prices paid at the market and prices set by producers, thus increasing the profits earned by oligopolistic industries. Moreover, oligopolies tend to be owned by the business elite who tend to influence or depend on the government (or the ruling class), as is evident in the case of the GCC states. Through these dynamics, oligopolies can interfere with market dynamics and economic policy through their power in the market and/or the policymaking process, advancing their own interests at the expense of economic growth and development.

To limit the role of oligopolies and enhance competition, the government can use regulation to ensure that oligopolies do not collude or set their prices above the average price of production. Competition policy refers "to government policy to preserve or promote competition among market players and to promote other government policies and processes that enable a competitive environment to develop" (UNCTAD 2009: 3). Various instruments can be used to implement competition policy, such as competition law, regulation of prices, competition advocacy, consumer protection, standards, intellectual property rights, international trade, investment, and licensing (UNCTAD 2009: 3).

Policy lessons are drawn in this chapter from illustrations from Kuwait and Chile. In Kuwait, like in many resource-dependent and GCC states, oligopolies capture a large part of efficiency in the economy and are highly distortionary. Shehabi (2017, 2019) shows that oligopoly markups capture a very large size of economic efficiency in the market. The expansion of exports of non-energy sectors is very limited in Kuwait mainly because oligopolies have few incentives to export internationally where they cannot sustain large markups as they do domestically (Shehabi 2019). Even in a low oil price environment and after reducing energy subsidies, the large size of oligopolies and their pure profits continue to entail limited expansion of trade and limited economic diversification effects in the export market of non-energy sector (reverse Dutch disease) dynamics (Shehabi 2019). Achieving sustainability thus requires oligopoly reform.

Competition and productivity reforms are an important example of a policy solution that promotes economic sustainability, as shown by Shehabi (2017, 2019) in model-driven assessments. These reforms improve the overall economic performance and restore the fiscal balance, and can accompany important sustainability-promoting reforms, such as energy subsidy reform, which alone is not sufficient to restore the fiscal balance (Shehabi 2017, 2019). Further, competition reform can yield substantial improvements in performance by reducing oligopoly markups. These reductions yield effects on overall economic activity that largely exceed the neoclassical gains in allocative efficiency from removing price distortions due to taxes, subsidies, and regulation (Shehabi 2017, 2019). Therefore, it has positive impacts of economic growth and fiscal sustainability both in the short and long runs.

Another example of successful industrial policy is a real-life example from Chile's forest resources and creating a competitive advantage in them. Chile, dubbed the "tiger" of Latin America, achieved large economic growth advancements owing to economic reforms of a military dictatorship government (Clapp 1995). For more than 60 years, Chile had very oligopolistic, inefficient, highly subsidized, and corrupt plantations and forest resources. It also had a high 75% reforestation subsidy established in 1974. Chile applied forest policy, a form of industrial policy, to create one of the world's most competitive forest resources at relatively low costs (Clapp 1995) within a brief period spanning approximately one generation. Government policies included state activism to promote forest expansions; state withdrawal through reduction of subsidies; privatization of the states' pulp mills and other parts of the industry; and guarantees against expropriations (Clapp 1995). These policies attracted foreign direct investments specifically to expand the forest resources industry. Combined with the opportunity of an increased forest products trade, the implementation of these policies enabled Chile to earn a comparative advantage in a new industry and export it.

## 5 Key Messages

The chapter argues that economic sustainability in resource-rich states is not economic diversification only; rather, it is a process that pans out in the period during which revenue is dependent on an abundant yet exhaustible commodity, as well as in the subsequent period, when the commodity export revenue is depleted. The chapter offers a broader definition of economic sustainability that has two main perspectives, volatility and depletion, spanning both the short and long runs. It links economic sustainability in resource-dependent states to that of the drivers of economic growth through policies that focus on both the volatility and the depletion perspectives. Lessons are offered from successful examples of three different cases, namely Kuwait, Botswana, and Chile.

Successful growth-sustaining policies include a focus on appropriate management of resource windfall to achieve fiscal sustainability along with long-term diversification. Examples of successful policies include balancing capital investments between domestic and foreign markets, the design of countercyclical (rather than the oft-used pro-cyclical) fiscal policy, human capital investments, and avoiding boom–bust cycles by delinking spending from the volatility dynamics of resource export revenues. Successful experiences of Chile, Botswana, and Kuwait confirm the role of SWF assets in offering fiscal stabilization as well as long-term savings. Other examples include model-driven policy solutions that focus on the management of labor contracts for expatriate labor, improvements in institutional quality, and competition reform as important policy solutions for achieving economic sustainability. Notwithstanding implementation challenges, these policies were successful in achieving favorable economic performance because, as this chapter argues, they address both the volatility and the depletion challenges.

In addition to evidence from these real-case examples, achieving sustainability requires designing policies that also manage the trade-off between producing output today and leaving the resource in the ground for future production, as well as the trade-off between local consumption and exports of a given supply. Technology plays a pivotal role not only in creating new industries but also in transforming the resource industry into a more sustainable source that can overcome threats of depletion.

A key message here is that comparative advantage can be acquired, and not only restricted to the factor or product most abundantly available. This acquisition will in turn contribute to generating new sources of revenue that are diversified not only in the non-tradable but also in the exporting sectors. To that end, the growth-enhancing factors will help build new comparative advantages most importantly through technological advancements, human capital, industrial regulation, the political economy, and even the institutions, cultures, and mores needed for economic transformation.

Therefore, a key policy question for achieving sustainability in resource-dependent economies is how to best apportion investment into various components of financial, physical capital, and human capital assets. Sustainability-promoting policies are those that target enhancing the drivers of economic growth while addressing solutions for the volatility and the depletion perspectives, in both the short and long terms.

## Appendix 1: Classical Growth Model

Under the Solow (1956, 1957) assumptions of full employment and a constant rate of labor force participation, the population growth rate ($n$) matches the labor force and employment growth rates, as indicated in the following calculation:

$$\frac{N_{t+1} - N_t}{N_t} = \frac{L_{t+1} - L_t}{L_t} = n. \tag{7.1}$$

Knowledge (being labor-augmenting technology; $\theta$) also grows at a constant exogenous rate ($g$) where

$$\frac{\theta_{t+1} - \theta_t}{\theta_t} = g. \tag{7.2}$$

With a closed economy assumption absent government spending, output equals the sum of consumption and investment, and the equation of motion for collective capital stock is

$$K_{t+1} - K_t = I_t - \delta K_t, \tag{7.3}$$

where the second term is depreciation at an exogenous and constant rate.[14] Savings and investment decisions are exogenous, as are factor accumulation and technological growth, which represents one of the shortcomings of the original Solow approach. Employing a standard Cobb–Douglas production function, with both physical and human capital ($K$), labor ($L$), and labor-augmenting technology ($\theta$), and ignoring for now the production factors of land and natural resources, constant returns to scale and diminishing marginal returns are

$$Y_t = AK_t^{\alpha}\left(\theta_t L_t\right)^{\alpha-1}, \tag{7.4}$$

whereby $\theta L = L_e$, effective labor, and $0 < \alpha < 1$.

After some manipulation, the following is a corresponding equation of motion for output per effective worker with population growth:

$$k_{et+1} - k_{et} = \frac{i_{et} - (\delta + n + g)k_{et}}{1 + n + g} \tag{7.5}$$

where $i_e$ is investment per effective worker, which is the savings rate times output per effective worker, expressed as $sf(k_e)$ where $f$ is the production function or the relationship between output and capital per effective worker. It follows that, multiplying all variables in the function by $1/\theta L$, yields the following corresponding function in terms of per effective worker:

$$y_{et} = Ak_{et}^{\alpha}. \tag{7.6}$$

The steady state of capital per effective worker remains the same from one period to the next and is expressed as

[14] In the Mankiw et al. (1992) model, human capital is a separate factor of production represented by $H$ and grows in a manner whereby $H_{t+1} - H_t = I_t^H - dH_t$.

$$\frac{i_e}{1+n+g} = \frac{sy_{et}}{1+n+g} = \frac{(\delta+n+g)k_{et}}{1+n+g}$$
$$= \frac{i_{be}}{1+n+g} =, \text{or } i = sf(k_{et}) = i_{be} = (\delta+n+g)k_{et} \tag{7.7}$$

where $i_{be}$ is the break-even level of investment that maintains a constant level of capital per effective worker. Thus, in an economy growing at a steady state, in terms of per effective worker,

$$sf\left(\overline{k_{et}}\right) = (\delta+n+g)\overline{k_{et}} \tag{7.8}$$

where the steady state is

$$\overline{i_e} = sA_t k_t^{\alpha} = \overline{i_b} = (\delta+n+g)\overline{k_{et}}. \tag{7.9}$$

This equation for the steady-state capital per effective worker can be solved in the following manner:

$$\overline{k_e} = \left(\frac{sA}{\delta+n+g}\right)^{\frac{1}{1-\alpha}}. \tag{7.10}$$

From this, the following relationships emerge between the population growth rate and real output per effective worker and real consumption per effective worker:

$$\overline{y_e} = A\overline{k_{et}}^{\alpha} = \left(\frac{sA}{\delta+n+g}\right)^{\frac{\alpha}{1-\alpha}} A^{\frac{1}{1-\alpha}}. \tag{7.11}$$

$$\overline{c_e} = \overline{y_e} - \overline{i_e} = \overline{y_e} - s\overline{y_e} = \overline{y_e}(1-s) = (1-s)\left(\frac{As}{\delta+n+g}\right)^{\frac{\alpha}{1-\alpha}} A^{\frac{1}{1-\alpha}}. \tag{7.12}$$

The standard approach is to show this steady state graphically, as in Fig. 7.5, where the steady-state levels of capital, consumption, and income per effective worker are

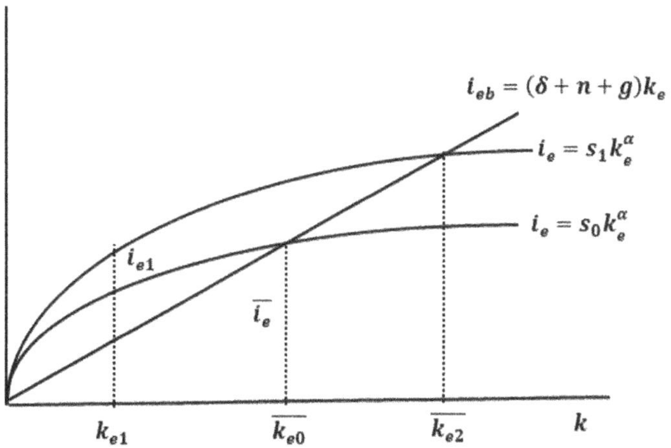


**Fig. 7.5** Steady-state growth and transition following a capital shock in the Solow model

$$\overline{k_e}, \overline{c_e}, \overline{i_e}, \overline{y_e} = \overline{c_e} + \overline{i_e}.$$

As such, aggregate variables of effective workers, capital, and output exhibit the same behavior growing at the same rate, as indicated in the following calculation:

$$\frac{\partial \theta L}{\theta L} = \frac{\partial K}{K} = \frac{\partial Y}{Y} = (n + g + ng) \approx (n + g). \tag{7.13}$$

In the context of resource-rich states, to simulate the impact of commodity price volatility or conflict, a shock in the economy is first represented by a large destruction in capital ($\gamma$) so that

$$K_{t+1} = (1 - \gamma) K_t, \text{where } 0 < \gamma < 1. \tag{7.14}$$

Initially, output will fall so that

$$\frac{\Delta Y_{t+1}}{Y_t} = A(1 - \gamma)^\alpha (n + g + ng)^{1-\alpha} < (n + g + ng). \tag{7.15}$$

The remaining capital falls short of the steady-state level, leaving investment at a level ($i_{t1}$) above its capital break-even point, causing growth to increase in the period following the war, as indicated in the following calculation:

$$\frac{\Delta Y_{t+2}}{Y_{t+1}} = \frac{\Delta K_{t+2}}{K_{t+1}} = \frac{\delta + n + g}{s} \frac{\left(n + g + ng\right)^{1-\alpha}}{\left(1-\gamma\right)^{1-\alpha}} - \delta > \left(n + g + ng\right). \quad (7.16)$$

This effect is enlarged if the saving rate also rises (see Fig. 7.5). The surplus of new investment is then over the break-even point and larger with an even faster recovery.

Thus, a temporarily contracting economy decreases the capital-to-labor ratio and raises capital per worker, increasing the productivity of capital and leading to temporarily higher growth rates in the reconstruction phase. Diminished returns eventually return as the long-term growth path is approached. If there is an accumulation of capital during the recovery (or reconstruction) phase, this recovery can be represented as a shift in the production possibility frontier curve. A speedy economic recovery post an economic shock may occur due to efficiency gains, represented normally as a movement along the production possibility or frontier curve. The shock (whether a drop in the resource price or a war) would typically have an impact on labor, increasing unemployment and decreasing the labor force and population.

Using a numerical simulation that represents a resource economy on the eve of a shock, the shock is simulated through a large negative shock to capital and a significant reduction in population, shown in Fig. 7.6.

Upon the end of the shock and at the commencement of recovery efforts, the capital-to-labor ratio converges to pre-shock levels and GDP grows at diminishing rates that are larger than pre-shocks levels to eventually converge with pre-shock growth levels, as Fig. 7.1 shows.

### *Natural Resources and Technology*

Evaluating economic growth in the presence of depleted natural resources is consistent with the goals of achieving economic sustainability in resource-dependent economies. As the standard growth model, presented earlier, does not take into account natural resources as an input, the following representation examines growth in the presence of natural resources.

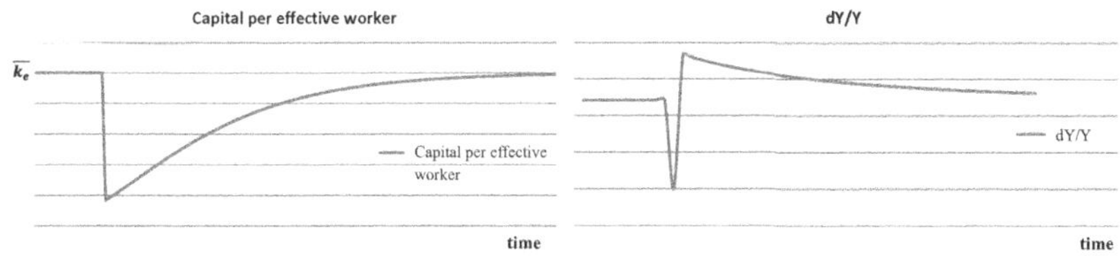


**Fig. 7.6** Transition in capital per effective worker and growth rate following a shock to capital stock in the augmented Solow model

To demonstrate this, the total available quantity of a resource ($X$) at a given time $t$ can be represented as $X_t$, while the quantity of the resource extracted can be expressed as $E_t$, so that $E_t = X_t$.

For simplification purposes, the other factors of production examined earlier will be ignored, assuming that the natural resource is the only input of production for now. The Cobb–Douglas production function can be expressed as

$$Y_t = AE_t. \tag{7.17}$$

The assumption of constant technology can be relaxed and technology is now growing at a constant exogenous rate ($g$), expressed in (7.2). This set-up enables addressing the question of sustainability: what is the trade-off between producing output today and leaving the resource in the ground for future production? Assuming the resource is extracted at a constant exogenous rate ($\varepsilon$), so that

$$Et = \varepsilon Xt,$$

the quantity of the resource remaining will decline exponentially over time, as follows:

$$X_t = X_0 e^{-\varepsilon t}. \tag{7.18}$$

The initial level of output will thus be

$$Y_0 = A_0 \varepsilon X_0. \tag{7.19}$$

It is thus clear that there is a positive relationship between the rate of extraction and output: a lower extraction rate will lead to a lower level of output. The growth rate of output can thus be expressed as the following, noting that this growth rate of extraction is the same as the growth rate of the resource stock:

$$\frac{\partial Y}{Y} = \frac{\partial A}{A} + \frac{\partial E}{E} = \left(g - \varepsilon\right). \tag{7.20}$$

This equation clearly shows that a higher extraction rate leads to a lower growth rate of output, which could even become negative at very high levels of extraction. Setting (7.20) equal to zero yields the extraction rate that is consistent with steady-state levels and a zero growth rate of output, which is

$$g = \varepsilon^{*}. \tag{7.21}$$

$\varepsilon^*$ thus represents the sustainable level of output. Therefore, examining the levels of output in the economy can be determined by examining the extraction rate chosen in relation to $\varepsilon^*$. If it exceeds $\varepsilon^*$, the initial level of output will exceed the sustainable level, but will decrease over time, as Fig. 7.4 shows.

Importantly, including the remaining factors of production would complicate the algebraic representation but yield the same general conclusion.

## References


Aghion, P. and Howitt. P. (1992). A Model of Growth Through Creative Destruction. *Econometrica 60*(2), 323–351.

Aizenman, J., & Riera-Crichton, D. (2014). Liquidity and foreign asset management challenges for Latin American countries. *Commodity Prices and Macroeconomic Policy, 22*, 91–134. Caputo, R. & Chang, R. eds, Santiago: Central Bank of Chile.

Alexīv, M., & Conrad, R. (2009). The elusive curse of oil. *Review of Economics and Statistics, 91*(3), 586–598. https://doi.org/10.1162/rest.91.3.586.

Auty, R. M. (1993). *Sustaining Development in Mineral Economies: the Resource Curse Thesis*. London: Routledge.

Auty, R. M. (2001). The political economy of resource-driven growth. *European Economic Review, 45*(4–6), 839–846. https://doi.org/10.1016/S0014-2921(01)00126-X.

Balassa, B. (1978). Exports and economic growth: Further evidence. *Journal of Development Economics, 5*(2), 181–189. https://doi.org/10.1016/0304-3878(78)90006-8.

Besada, H., & O'Bright, B. (2018). Policy impacts on Africa's extractive sector: Botswana, diamond dependence, and diversification in the post-diamond period. *Revue Gouvernance, 15*(22), 86–105.

Bhagwati, J. (1978). Foreign trade regimes and economic development. *Anatomy and Consequences of Exchange Control Regimes.* Cambridge, MA: Ballinger for the National Bureau of Economic Research.

Black, G., & LaFrance, J. (1998). Hotelling's Rule Relevant to Domestic Oil Production. *Journal of Environmental Economics and Management, 36*, 149–169.

Brazee, R. J., & Cloutier, L. M. (2006). Reconciling Gray and Hotelling: Lessons from early exhaustible resource economics. *American Journal of Economics and Sociology, 65*(3), 827–856. https://doi.org/10.1111/%28ISSN%291536-7150/issues.

Bulte, E. H., Damania, R., & Deacon, R. T. (2005). Resource intensity, institutions, and development. *World Development, 33*(7), 1029–1044. https://doi.org/10.1016/j.worlddev.2005.04.004.

Clapp, R. A. (1995). Creating competitive advantage: Forest policy as industrial policy in Chile. *Economic Geography, 71*(3), 273–296.

Collier, S., & Sater, W. F. (1996). *A History of Chile, 1808–1994.* Cambridge: Cambridge University Press.

Collier, P. et al. (2010). Managing Resource Revenues in Developing Economies. *IMF Staff Papers, 57*, 84–118.

Corden, W. M. (1984). Booming sector and Dutch disease economics: Survey and consolidation. *Oxford Economic Papers, 36*(3), 359–380.

Corden, W. M., & Neary, J. P. (1982). Booming sector and de-industrialisation in a small open economy. *Economic Journal, 92*(368), 825–848.

Corden, W. M. (2012). Dutch disease in Australia: Policy options for a three-speed economy. *Australian Economic Review, 45*(3), 290–304. https://doi.org/10.1111/j.1467-8462.2012.00685.x.

Davis, G. (1995). Learning to love the Dutch Disease: Evidence from the mineral economies. *World Development, 23*, 1765–1779.

Delacroix, J. (1977). The export of raw materials and economic growth: A cross-national study. *American Sociological Review, 42*, 795–808.

Frankel, J. A. (2011). *A solution to fiscal procyclicality: The structural budget institutions pioneered by Chile* (NBER Working Paper No. 16945). Cambridge, MA: National Bureau of Economic Research. Retrieved from https://ideas.repec.org/p/nbr/nberwo/16945.html

Gelb, A. H. (1988). *Oil windfalls: blessing or curse?* New York: Published for The World Bank [by] Oxford University Press.

Grossman, G. M. and Helpman, E. (1993). Endogenous Innovation in the theory of growth. *Journal of Economic Perspectives, 8*(1), 23–44.

Heckscher, E. (1919). The effect of foreign trade on the distribution of income. *Ekonomisk Tidskriff*, 497–512. Translated as chapter 13 in American Economic Association.

Hotelling, H. (1931). The economics of exhaustible resources. *The Journal of Political Economy, 39*, 137–175.

IMF (International Monetary Fund). (2012). Botswana: 2012 Article IV Consultation. *IMF Country Report,* no. 12/234. Washington. D.C.: Author.

Jung, W. S., & Marshall, P. J. (1985). Exports, growth and causality in developing countries. *Journal of Development Economics, 18*(1), 1–12. https://doi.org/10.1016/0304-3878(85)90002-1.

Karl, T. (1997). *The paradox of plenty: Oil booms and petro-states.* Berkeley, CA: University of California Press.

Krueger, A. O. (1984). Trade policies in developing countries. *Handbook of International Economics*, 1, 519–569. Jones, R. W. & Kenan, B. P., eds, Elsevier. Retrieved from http://www.sciencedirect.com/science/article/pii/S1573440484010145.

Lederman, D., & Xu, L. C. (2007). Comparative advantage and trade intensity: Are traditional endowments destiny?. *Natural Resources: Neither Curse nor Destiny.* Lederman D. and Maloney, M. F. eds. Stanford: Stanford University Press.

Leontief, W. (1953). *Domestic Production and Foreign Trade; The American Capital Position Re-Examined Proceedings of the American Philosophical Society*, *97*(4).

Lucas, R. E. (1988). On the Mechanics of Economic Development. *Journal of Monetary Economics, 22*, 3–42.

Maloney, W (2002). Missed opportunities: Innovation, natural resources and growth in Latin America. *Economia, 3*(1), 111–169.

Mankiw, N. G., Romer, D., & Weil, D. N. (1992). A contribution to the empirics of economic growth. *The Quarterly Journal of Economics, 107(2)*, 407–437.

Mehlum, H., Moene, K., & Torvik, R. (2006). Institutions and the resource curse. *Economic Journal, 116*(508), 1–20.

Michaely, M. (1977). Exports and growth: An empirical investigation. *Journal of Development Economics, 4*(1), 49–53. https://doi.org/10.1016/0304-3878(77)90006-2.

Monzano, O., & Rigobon, R. (2001). *Resource curse or debt overhang?* (NBER Working Paper No. 8390). Cambridge, MA: National Bureau of Economic Research. Retrieved from http://www.nber.org/papers/w8390.

Ohlin, B. (1933). *Interregional and International Trade.* Cambridge, MA: Harvard University Press, 1966.

Ossowski, R., Villafuerte, M., Medas, P. A., & Thomas, T. (2008). Managing the oil revenue book: The role of fiscal institutions. *IMF Occasional Papers*, no. 260. Washington, DC: The International Monetary Fund (IMF).

Ramey, G., & Ramey, V. (1995). Cross country evidence on the link between volatility and growth. *The American Economic Review, 85*(5), 1138–1151.

Romer, P. M. (1986). Increasing Returns and Long-Run Growth. *Journal of Political Economy, 94*(5).

Romer, P. M. (1990). *Endogenous Technological Change Journal of Political Economy, 98*(5).

Sachs, J. D., & Warner, A. M. (1995). *Natural resource abundance and economic growth* (NBER Working Paper No. 5398). Cambridge, MA: National Bureau of Economic Research. Retrieved from http://ideas.repec.org/p/nbr/nberwo/5398.html.

Sala-i-Martin, X., & Subramanian, A. (2003). *Addressing the natural resource curse: An illustration from Nigeria* (Working Paper No. 9804). Cambridge, MA: National Bureau of Economic Research. Retrieved from http://www.nber.org/papers/w9804.

Schumpeter, J. A. (1942, reprinted 1976). *Capitalism, Socialism and Democracy.* London: Routledge.

Shehabi, M. (2017). Assessing Kuwaiti energy pricing reforms. *OIES Paper No. MEP 117.* Oxford Institute for Energy Studies. Oxford, UK. DOI: 10.26889/9781784670931. https://www.oxfordenergy.org/wpcms/wp-content/uploads/2017/11/Assessing-Kuwaiti-Energy-Pricing-Reforms-MEP-17.pdf

Shehabi, M. (2018a). Maintaining the order: Contemporary Kuwaitisation dynamics and their historical perspectives. *Gulfization of the Arab World, Exeter Critical Gulf Series,* 1, 59–88. Jones M. O., Porter, R., & Valeri, M. eds, Berlin: Gerlach Press.

Shehabi, M. (2018b). *Success and failure in a petrostate: economic policy and performance in Kuwait since the Gulf War.* https://doi.org/10.26182/5b4c0d3d38b2b.

Shehabi, M. (2019). Is energy subsidy reform in an oil-exporting small economy beneficial to trade? Illustrations from Kuwait. *OIES Paper: MEP 21.* Oxford Institute for Energy Studies. Oxford: UK. https://doi.org/10.26889/9781784671273.

Shehabi, M. (2020). Diversification effects of energy subsidy reform in oil exporters: Illustrations from Kuwait. *Energy Policy, 123,* 110966. https://doi.org/10.1016/j.enpol.2019.110966.

Solow, R. A. (1956). A contribution to the theory of economic growth. *Quarterly Journal of Economics, 70*(1), 65–94.

Solow, R. A. (1957). Technical change and the aggregate production function. *Review of Economics and Statistics, 39,* 312–320.

Swan, T. (1956). Economic growth and capital accumulation. *Economic Record, 32*(63), 334–361.

Tornell, A., & Lane, P. R. (1999). The voracity effect. *American Economic Review, 89*(1), 22–46. https://doi.org/10.1257/aer.89.1.22.

Tyers, R. (2015). Service oligopolies and Australia's economy-wide performance. *Australian Economic Review, 48*(4): 333–356.

Tyers, R., & Walker A. (2016). Quantifying Australia's 'three-speed' boom. *Australian Economic Review, 49*(1), 20–43.
UNCTAD (United Nations Conference on Trade and Development). (2009). *The Relationship between Competition and Industrial Policies in Promoting Economic Development*. Geneva: UNCTAD Secretariat. https://unctad.org/en/Docs/ciclpd3_en.pdf.
van der Ploeg, F. (2011). Natural resources: curse or blessing. *Journal of Economic Literature, 49*(2), June 2011.
van der Ploeg, F., & Poelhekke, S. (2009). Volatility and the natural resource curse. *Oxford Economic Papers, 61*(4), 727–60.
van der Ploeg, F., & Venables, A. J. (2012). *Natural Resource Wealth: the Challenge of Managing a Windfall Annual Review of Economics, 4*, 315–337.
Venables, A. J. (2016). Using natural resources for development: Why has it proven so difficult? *Journal of Economic Perspectives, 30*(1), 161–184.
Venables, A. J., & van der Ploeg, F. (2010). *Absorbing a Windfall of Foreign Exchange: Dutch Disease Dynamics* (No. Discussion Paper No. 8086). London, UK: Centre for Economic Policy Research.
Wright, G., & Czelusta, J. (2004). Why economies slow: The myth of the Resource Curse, *Challenge, 47*(2), 6–38. Published online 2014. https://doi.org/10.1080/05775132.2004.11034243.

CHAPTER 8

# Fuel and Electricity Reform for Economic Sustainability in the Gulf

*Tom Moerenhout*

## 1 Introduction

After the oil price started plummeting in the summer of 2014 and remained low throughout the next years (until 2017 when the OPEC Reference Basket average price recovered to USD 52), the fuel producing countries of the Gulf Cooperation Council started implementing energy pricing reforms. For many years, such pricing reforms had been planned and envisioned but, due to their quintessential role in the domestic political economies of Gulf countries, never really implemented. This is not unlike the broad and abstract goal of diversifying the economy away from oil to a more productive economy, and of changing the domestic social contract from one that is reliant on public employment and universal price support to one with private employment, market-based pricing structures and targeted social safety nets.

In recent years, since the oil price drop, all GCC countries have taken energy pricing reform measures (Fig. 8.1). Both fuel and electricity prices

T. Moerenhout (✉)
Columbia University, School of International and Public Affairs, New York, NY, USA
e-mail: tm2794@columbia.edu



G. Luciani, T. Moerenhout (eds.), *When Can Oil Economies Be Deemed Sustainable?*, The Political Economy of the Middle East,
https://doi.org/10.1007/978-981-15-5728-6_8

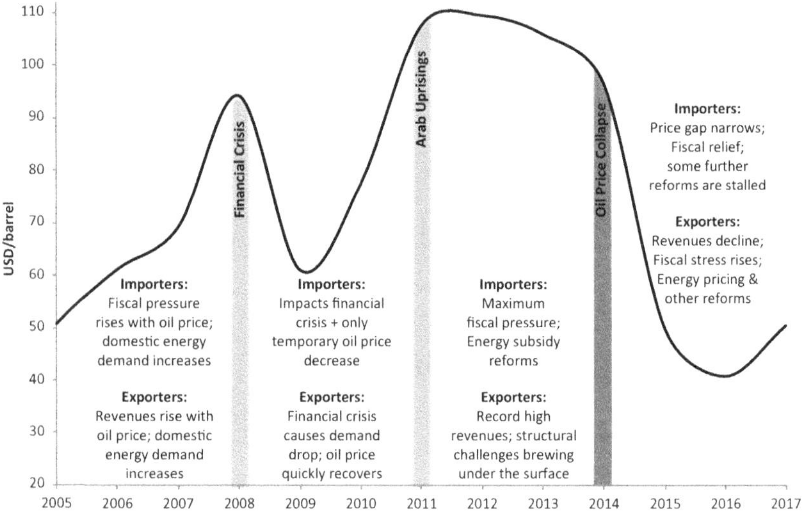


**Fig. 8.1** Oil price movements and energy price reforms. (Source: Author)

have increased substantially. This chapter takes stock of those energy price increases and aims at giving a holistic account of where countries are at toward the end of 2018. After explaining the rationale for low energy prices, this chapter takes stock of fuel pricing reform and electricity tariff policies in all GCC countries.

## 2 The Triple Rationale of Low Energy Prices

There are three compelling reasons that make energy pricing reform so difficult in GCC countries. A first reason is the centrality of low energy prices in welfare protection and distribution. Governmental involvement in domestic pricing policies is legitimized across the developing world because energy has no close substitute and provides essential functions to human life (Maxwell 2010). Low energy prices are intended to alleviate poverty (Komives et al. 2008; Commander 2012) by safeguarding commodity prices, keeping inflation in check and sheltering consumers from the volatility of international commodity markets (Commander 2012; Fattouh and El-Katiri 2013; Kojima 2013; Overland 2010). In energy producing countries in the Gulf, pricing policies are an important part of the social contract. This means low prices can be considered as both a right of the people and an obligation of the government (Hartley and Medlock 2009; Luong and Weinthal 2010; Victor et al. 2011). GCC states are rentier states, in which the government is the principal receiver of oil revenues and the key responsible one to redistribute this income (Beblawi 1990). Increasing energy prices, ceteris paribus, represents a unilateral modification of the social contract.

A second rationale for low energy prices lies in their economic function. Low energy prices have been used to promote economic development by supporting factors of production in general and competitiveness for international trade in particular (Whitley and van der Burg 2015). In other words, low prices have been used as a part of industrial policy with the explicit goal of supporting export competitiveness of domestic industries (Commander 2012). Resource-rich countries in the Gulf in particular have used their hydrocarbon endowment to incentivize energy-intensive industrialization (IEA, OPEC, OECD and World Bank 2010). Energy pricing reforms have two types of impacts on firms. First, they affect firms directly by increasing their energy input cost. Second, there are indirect effects via the rise of prices from intermediary goods or services (that also rely on energy). These indirect effects can affect the supply chains of non-energy intensive goods or services (Rentschler and Kornejew 2017). The sectors that suffer the most are logically energy-intensive industries such as heavy manufacturing, transport, petrochemicals, cement, aluminum and steel (Rentschler and Kornejew 2017; Bazilian and Onyeji 2012; Rentschler et al. 2017). These sectors are dominant industries for within-diversification in Gulf countries.

The third rationale is political. Given the potential social and economic impacts of pricing reform, implementing price increases is often politically costly and can even threaten political stability (Baig et al. 2007). It is now uniformly recognized that political economy factors are the primary barriers to reforming energy prices. Low energy prices are considered an instrument to stay in power and control political stakeholders through systems of patronage and rentierism (Cheon et al. 2014; Hartley and Medlock 2009; Overland 2010; Victor 2009). Governments use them to direct (financial) benefits to key political stakeholders, thereby consolidating power (Cheon et al. 2014). In rentier states like the ones in the GCC, low prices are used across the board to maintain the support of the public for the elite in power (Overland 2010). It should be no surprise that worldwide, fuel prices are lower in more centralized, non-democratic states (van Beers and Strand 2013).

## 3 The Sustainability Rationale for Energy Pricing Reform, and Results in the GCC

All GCC states reformed prices after the fuel price drop in 2014. Even though the nature of popular reception varies in different countries, energy price reforms seem to be one policy that authorities have recognized to be crucial for adjustment to falling oil revenues. The fiscal benefit from increasing energy prices is key to economic sustainability in two main ways: first because it gives governments more domestically sourced revenue, and second because it lowers demand of its key export commodity.

In terms of fiscal consolidation, overall pre-tax energy (gasoline, diesel and electricity) subsidies fell from $116 billion in 2014 to $47 billion in 2016 based on a price-gap model (IMF 2017). This type of subsidy quantification model reflects the opportunity costs of resources consumed domestically as opposed to being traded on the international market. That said, the fiscal capacity opened up by charging higher energy prices has been of crucial importance to GCC governments. Increasing energy prices has been considered as one of the most important measure to reduce non-wage recurrent spending (IMF 2017). The additional fiscal revenue has helped governments to prevent the international oil price collapse from threatening their domestic stability, while awaiting a recovery of oil prices that eventually started in 2017 (Fig. 8.2).

So far, it appears that fuel pricing reforms have paid off in terms of lowering demand. Average annual gasoline and diesel demand growth was around 6.2% and 4%, respectively (2010–2015) but slowed down to 0.4% and −6% in 2016. It is however difficult to attribute this slowdown to pricing reforms, as economic output also reduced over the last few years (APICORP 2017). In Saudi Arabia, gasoline demand leveled out even before the January 2018 reforms and diesel demand had already fallen 10%. The latter can be attributed to less economic activity and the sourcing of more gas for power generation. In Oman, gasoline and diesel consumption fell by 6.2% and 7.2%, respectively, from 2015 to 2016 (APICORP 2017). Unsurprisingly, given the relatively modest fuel price increases, the UAE has seen less demand reactions in recent years. On the contrary, fuel demand has actually increased in 2016.

The fiscal benefit of pricing reform and the reduction of demand are the main reasons to implement reforms for economic sustainability, but they are not the only ones (van Asselt and Skovgaard 2016). The fiscal crisis resulting from the oil price collapse has been in certain ways a

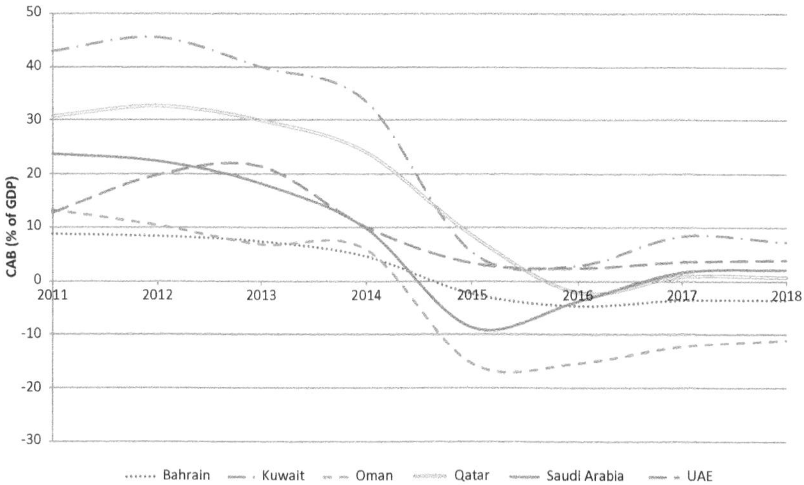


**Fig. 8.2** Current account balance (% of GDP, 2011–2018). (Source: IMF 2016 (up until 2013); IMF 2017 (as of 2014). Note: 2017 and 2018 are projections)

blessing to target excessive rentierism. Low domestic prices are correlated with higher levels of corruption through the involvement of government officials in black markets or industries that rely on cheap energy (Strand 2016).

Underpricing also leads to a misallocation of resources within the economy, often wrongly incentivizing energy intensive sectors and crowding out more productive sectors (Fattouh and El-Katiri 2013; IEA 2014; Maxwell 2010). Energy-intensive sectors tend to offer few and low-skilled jobs. In addition, in some countries with a lot of expatriate labor, it is less costly to focus on that labor as factor of production than it would be to focus on capital. Low-skilled labor is often abundant, while energy efficiency investments would be expensive and necessitate higher-skilled and therefore more expensive employees. In many countries, this has inevitably led to large shares of expensive public labor (Commander 2012), further burdening state coffers. The misallocation of energy resources has resulted in a rise in energy intensity of GDP and lowered incentives for energy efficiency, particularly in end-use sectors (Fattouh and El-Katiri 2013; IEA 2014; Anand et al. 2013).

A positive reallocation of resources within the economy does take time, and initially there may be a slowdown as sectors adapt, consolidate or close down. In an environment of low oil prices coupled with end-use energy price reforms, overall growth prospects in the medium-term have been subdued, though non-oil growth has been improving in some countries. Estimates from IMF staff show that the reforms indicated by GCC governments could lead to an increase of 1.7 to 6.6% of their non-oil GDP by 2020 based on each country's reform path, and an additional 1.5 to 3.0 percentage points of non-oil GDP would be generated with the introduction of VAT, as proposed by different countries and contained in the ratified GCC VAT agreement.

## 4 Fuel Pricing Reform in the GCC

While all GCC countries reformed energy prices, they did so in remarkably different ways. The UAE, Oman and Qatar were able to implement a periodic adjustment system that linked prices to international and regional prices. Kuwait tried to do the same but was unable to implement it. Like Saudi Arabia and Bahrain, they relied on one-off price increases. Unlike Kuwait, however, Saudi Arabia and Bahrain were able to implement more than just one ad hoc price adjustment. Below, we report price adjustments until June & July 2018. In the summer of 2018, the international oil price saw its highest price level since the 2014 plummet. In the fall of 2018, the crude oil price fell again before recovering in the spring of 2019. During this time, ad hoc reformers implemented no further reforms, whereas gradual reforms kept adjusting prices in line with international price movements (both downward and again upward). In June 2019, price levels were almost identical in all GCC countries compared to those of the year before (and reported here).

### *4.1 The 'Gradual' Fuel Price Reformers*

The UAE has been the first of GCC countries to seriously reform its fuel prices. In August 2015, transport fuel prices were liberalized and linked to international market prices using price formulae. As a result, gasoline price increased by 25% and diesel prices decreased by 29% (IMF 2015). Since, international market prices have been gradually going up and UAE prices have followed this trend. Fuel prices have been liberalized for all actors, including industry and commerce (Table 8.1).

Like many other countries, Oman first started raising energy prices for commerce and industry, before moving to residential consumers. Indicative

**Table 8.1** Fuel Prices in the UAE (July 2018)

| *Product* | *Unit* | *July 2015* | *August 2015* | *July 2018* |
|---|---|---|---|---|
| Gasoline 98 | USD/L | 0.50 | 0.61 | 0.7 |
| Gasoline 95 | USD/L | 0.47 | 0.58 | 0.67 |
| Gasoline 91 | USD/L | 0.44 | 0.56 | 0.65 |
| Diesel | USD/L | 0.79 | 0.56 | 0.72 |

Source: Collected by author

**Table 8.2** Fuel Prices in Oman (continuously adjusted since January 2016)

| *Product* | *Unit* | *2015* | *Jan 2016* | *Feb 2017* | *Jun 2018* |
|---|---|---|---|---|---|
| Gasoline 95 | USD/L | 0.31 | 0.42 | 0.47 | 0.59 |
| Gasoline 91 | USD/L | 0.31 | 0.36 | 0.45 | 0.56 |
| Diesel | USD/L | 0.38 | 0.42 | 0.53 | 0.65 |

Source: Collected by author

of the fiscal need for reform in Oman is the fact that they already increased natural gas prices for industry and power producers in January 2015, well before reforms happened in the UAE and Saudi Arabia. It reformed fuel prices in the wake of Saudi reforms in January 2016. At the same time, it introduced a new pricing formula that links Omani prices to prices on the international market and in the UAE. Ever since, Oman has stuck to the formula and increased energy prices alongside international market prices (Table 8.2).

When reforming fuel prices, Oman also experienced opposition. In exchange, they installed a cap twice, and broke it twice. The second cap was set at 0.48 USD/L for regular gasoline. The government abolished that cap after having introduced the 'National Subsidy Scheme'. This scheme continues to provide subsidized fuel at a rate of 0.47 USD/L to less wealthy families. In total, 220,000 people signed up to the scheme in January 2018 (Times of Oman 2018). Under this scheme, nationals who own a car or boat and have a monthly income of below 950 riyals can apply for 200 L of subsidized fuel per month. The scheme only applies to regular gasoline, and when the price is above 0.47 USD/L (Al Mukrashi 2017; Sultanate of Oman 2018).

Like Oman, Bahrain and Saudi Arabia, Qatar also revised gasoline prices in early 2016. In a very early move, Qatar had reformed diesel prices

Table 8.3 Qatar fuel prices (adjusted continuously since June 2016)

| *Product* | *Unit* | *2015* | *Jan 2016* | *Jun 2017* | *Jul 2018* |
|---|---|---|---|---|---|
| Gasoline 97 | USD/L | 0.23 | 0.32 (35%) | 0.43 | 0.56 |
| Gasoline 90 | USD/L | 0.28 | 0.36 (30%) | 0.44 | 0.55 |
| Diesel | USD/L | 0.28 | 0.41 (48%) | 0.45 | 0.56 |

Source: Collected by author

already in 2014 (by about 50% for local companies but 75% for joint ventures) (Walker and Kovessy 2016). In May of the same year, the Qatari government announced plans to liberalize fuel prices and adjust them alongside international market prices, regional prices and cost of production—a move similar to the one taken by Oman and the UAE. This plan was implemented as of June 2016, after which there were monthly revisions to the fuel price. This move has been rather remarkable, given Qatar's high population-reserve ratio. Much like Kuwait, it could be expected that the Qatari government had less of a need to implement deep pricing reforms. One explanation, however, may be the impacts of Qatar's political isolation from other GCC countries, and the opportunism of its Government to use geopolitical threats to advance domestic reforms (Table 8.3).

### 4.2 *The 'Ad hoc' Fuel Price Reformers*

Saudi Arabia implemented significant energy pricing reforms in two episodes in January 2016 and January 2018. In January 2016, the Government reformed prices predominantly in reaction to the fiscal crisis as a result of the oil price drop. The Saudi Government targeted a full energy and water subsidy phase out by 2020 under its Vision 2030 plan, spearheaded by the then Deputy Crown Prince Mohammed bin Salman (IMF 2016) (Table 8.4).

The first stage of reforms was successful without introducing compensation measures or communication plans (APICORP 2018). After the first price reforms, energy demand growth decreased from 3.5% to 1.7% even though the net effect of pricing reforms is difficult to estimate as overall GDP went down as well (APICORP 2018). The anti-corruption campaign of Crown Prince Mohammed bin Salman was believed to help

**Table 8.4** Energy pricing reforms in Saudi Arabia (January 2016)

| *Product* | *Unit* | *2015* | *2016* |
|---|---|---|---|
| Natural gas | USD/mmbtu | 0.75 | 1.25 (67%) |
| Ethane | USD/mmbtu | 0.75 | 1.75 (133%) |
| Gasoline 95 | USD/L | 0.16 | 0.24 (50%) |
| Gasoline 91 | USD/L | 0.12 | 0.2 (67%) |
| Diesel transport | USD/L | 0.07 | 0.12 (79%) |
| Diesel industry | USD/barrel | 9.11 | 14.1 (55%) |
| Arab light crude | USD/barrel | 4.24 | 6.35 (50%) |
| Arab heavy crude | USD/barrel | 2.67 | 4.4 (65%) |
| Kerosene | USD/barrel | 23.00 | 25.7 (12%) |

Source: Collected by author

public perception on the reforms. At the same time, reforms were unpopular, and inflation spiked right after reform.

In subsequent years, however, reform slowed down and was repeatedly postponed as the oil price recovered and the Government planned for compensation measures. There was a transparent commitment not to implement further reforms before the cash transfer scheme (citizen's income) was ready to be launched. At the same time, delays happened because the government was cautious not to slow down industrial output (Mahdi and Nereim 2017). In its latest budget, the Saudi government pushed back the plan to remove subsidies to 2025 (Gnana 2017). The January 2018 price rises mainly targeted gasoline prices, but also increased diesel prices for industry (Table 8.5).

As opposed to the first reforms, the second reforms were prepared with a large communication campaign and accompanied, as promised, by the cash transfer program. Citizen's Account started operating in January and aimed at distributing SAR 30 billion to 3.7 million households in 2018. However, there were social media reactions against the cash transfers as the first round was not enough to compensate for the distributive losses (APICORP 2018). SAR 30 billion was allocated to Citizen's Account for 2018, but SAR 50 billion was going to military and civil servants (APICORP 2018). This indicates the relative importance of these stakeholders: the people matter, but perhaps the military and civil servants are more important to retain the ruling coalition. Saudi Arabia is also careful about its energy intensive industries. While there was indeed a pricing reform in 2018 for industrial diesel but not for transport diesel, the cost

**Table 8.5** Fuel pricing reforms in Saudi Arabia (percentage changes between brackets)

| *Product* | *Unit* | *2015* | *Jan 2016* | *Jan 2018* |
|---|---|---|---|---|
| Natural gas | USD/mmbtu | 0.75 | 1.25 (67%) | Unchanged |
| Ethane | USD/mmbtu | 0.75 | 1.75 (133%) | Unchanged |
| Gasoline 95 | USD/L | 0.16 | 0.24 (50%) | 0.54 (127%) |
| Gasoline 91 | USD/L | 0.12 | 0.2 (67%) | 0.37 (83%) |
| Diesel transport | USD/L | 0.07 | 0.12 (79%) | Unchanged |
| Diesel industry | USD/barrel | 9.11 | 14.1 (55%) | 16.15 (15%) |
| Arab light crude | USD/barrel | 4.24 | 6.35 (50%) | Unchanged |
| Arab heavy crude | USD/barrel | 2.67 | 4.4 (65%) | Unchanged |
| Kerosene | USD/barrel | 23.00 | 25.7 (12%) | Unchanged |

Source: Collected by author

per liter of industrial diesel remains lower than the one for transport diesel (0.10 USD/L versus 0.12 USD/L).

Like Oman, Bahrain had its first taste of energy pricing reforms by increasing natural gas tariffs for industrial users in March 2015. An earlier attempt to reform electricity prices for residential users had failed due to the opposition from members of parliament (APICORP 2018). Again, like Oman, it followed in the footsteps of Saudi Arabia in January 2016 and reformed fuel prices. It raised gasoline prices and planned for diesel price increases of 0.05 USD/L on an annual basis. Gasoline prices were not reformed annually. They remained stable until a next reform in January 2018.

For diesel, on the other hand, Government had been planning since 2013 to raise prices to 0.37 USD/L in January 2015, 0.42 USD/L in January 2016 and 0.47 USD/L in January 2017. These price rises were delayed but since January 2015, the annual price increase of 0.05 USD/L did happen. Fishermen were given support, but it did not cover their losses (Table 8.6).

The Kuwaiti government increased diesel prices by 200% in January 2015, but immediately had to scale back reforms to 100% after parliamentary and other protests. In addition, large users (such as in industry) continued to receive diesel at the prior price. In September 2016, the Government implemented a gasoline price increase of about 70% on average and combined this with a plan to revise prices alongside international price movements. After heavy protests and a challenging of the price rise

**Table 8.6** Fuel prices in Bahrain (percentage changes between brackets)

| *Product* | *Unit* | *2015* | *Jan 2016* | *Jan 2018* |
|---|---|---|---|---|
| Gasoline 95 | USD/L | 0.27 | 0.42 (58%) | 0.53 |
| Gasoline 91 | USD/L | 0.20 | 0.33 (56%) | 0.37 |
| Diesel | USD/L | 0.26 | 0.32 | 0.42 |

Source: Collected by author

**Table 8.7** Fuel prices in Kuwait

| *Product* | *Unit* | *2014* | *Jan 2018* |
|---|---|---|---|
| Gasoline 98 | USD/L | 0.27 | 0.50 |
| Gasoline 95 | USD/L | 0.20 | 0.32 |
| Gasoline 91 | USD/L | 0.18 | 0.26 |
| Diesel | USD/L | 0.17 | 0.35 |

Source: Collected by author

before Court, the government was able to maintain its gasoline price increase, but dropped the plan to adjust prices periodically (Moerenhout 2018). It is not planning any further price increases at the moment (Kuwait Times 2018) (Table 8.7).

### 4.3 GCC Fuel Prices in Comparison

In comparison to prices in other countries in the region, such as Jordan and average world prices, GCC countries still offer fuel at some of the lowest rates worldwide. Within the GCC, the UAE, Oman and Qatar have higher prices than other countries, even though Saudi Arabia and Bahrain have also invested heavily into cutting gasoline consumption (Fig. 8.3).

It appears fashionable to conclude that countries in the Gulf still have a long way to go in terms of fuel pricing reform. Experiences in fuel price changes across the world indeed show that changes may be reversed in the wake of popular protest or in the wake of changing international oil prices. This is of course no different in the case of GCC countries. That said, progress has been remarkable on two fronts. First, three countries were able to implement periodic fuel price adjustment systems and have stuck to revising fuel prices upward so far. Especially in Qatar, this has led to strong price increases over two years' time. That aside, Saudi Arabia and

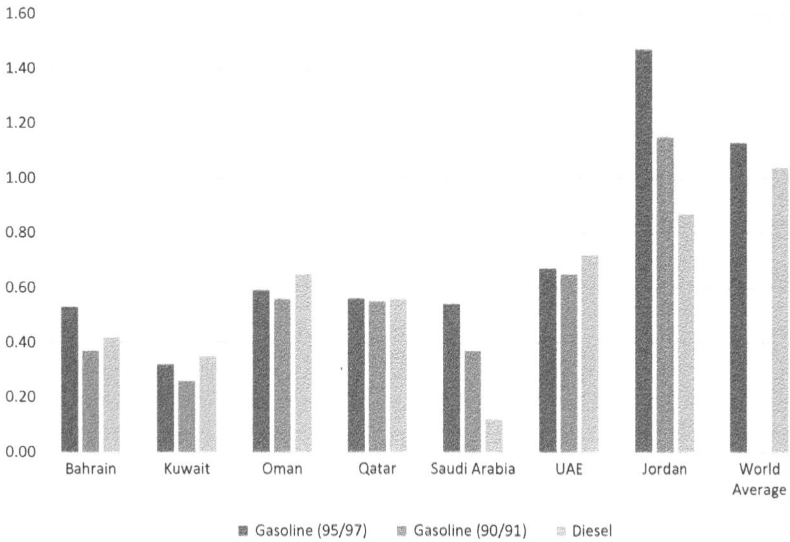


**Fig. 8.3** Transport fuel prices (USD/L, July 2018). (Source: Author)

Bahrain have proven that one-off reforms can be followed up again, especially with the introduction of mitigation measures. Of all GCC-countries, Saudi Arabia has by far increased gasoline prices the most. Even if it came from far, the fact GDP/capita is lower and a rentier mentality potentially higher makes such adjustments all the more surprising (Fig. 8.4).

## 5 GCC Electricity Pricing Reforms and Tariffs

Most GCC countries have also increased electricity prices over the past few years. This has happened in various steps, to various extents and according to various pricing policies. Who bears the burden varies across countries, but in all countries, there are blocked tariffs and some form of cross-subsidization, either between expatriates and nationals, or between various consumer groups. Many electricity prices likely do not reach cost recovery levels. In the absence of production costs, governmental prices are often an (imperfect) proxy to cost-recovery levels. Most known electricity tariffs for government hover around 0.08 USD/kWh. This is close to the official production cost figure quoted by the Bahraini Government of 0.77 USD/kWh (Fattouh et al. 2016).

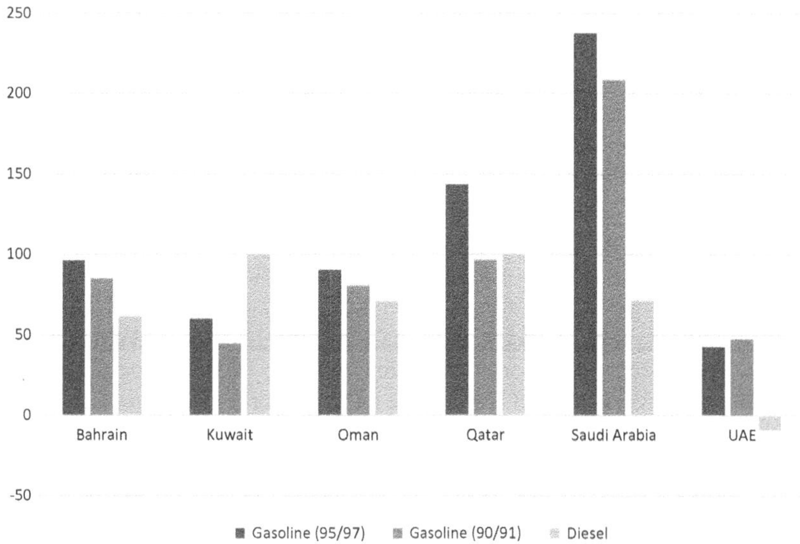


**Fig. 8.4** Price changes (%) before and after reforms (July 2018). (Source: Author)

### *5.1 Cross-Subsidies Between Consumer Groups*

Many GCC countries provide some level of cross-subsidization to industry. Between the start of reforms in 2016 and January 2018, Saudi Arabia increased electricity tariffs for residential users (< 6000 kWh/month) by on average 260%. Industrial users, however, were shielded from price rises, even if they still fall well below what the government pays. They now pay the same tariff as the lowest residential consumer group (Saudi Electric Company 2018). Similarly, in Dubai, industrial tariffs up to 10,000 kWh/month are the same as the lowest residential consumer group (up to 2000 kWh/month) (Dubai Electricity & Water Authority 2018).

Some GCC countries have flat rates for industry, disregarding however much electricity is consumed per month. Oman has such a flat rate, even if the level of the rate depends on the time of year (it is double as high in the 4 warmest months of the year) (Qatar General Electricity & Water Corporation 2018). Qatar also has a flat rate for industrial users, which is less than half the rate paid by Government and about half the rate paid by the highest residential consumer groups (Qatar General Electricity & Water Corporation 2018). Kuwait also has flat rates for industry and commerce. The latter pays the same rate as the lowest residential consumer group, while industrial users still pay less than the highest residential users. Saudi Arabia's industrial tariff is also a flat rate and the same as the lowest residential consumer bracket (Saudi Electricity Company 2018). Besides industrial and commercial users, agricultural tariffs are also often cross-subsidized (for example in Oman, Saudi Arabia and Abu Dhabi).

Another type of cross-subsidization is between expatriates and nationals. In Abu Dhabi, nationals pay more for electricity when they consume more than 30 kWh/day for apartments and 400 kWh/day for villas. For expatriates, however, this is respectively 20 kWh/day and 200 kWh/day. Tariffs also diverge strongly. Nationals pay 0.018 USD/kWh or 0.02 USD/kWh (depending on their consumption), while expatriates pay 0.073 USD/kWh or 0.083 USD/kWh (Abu Dhabi Distribution Company 2017). Bahrain also charges non-nationals and nationals with multiple accounts more than nationals with one single account (Bahrain Electricity and Water Authority 2018).

### 5.2 *Blocked Tariff Designs*

All GCC countries work with some type of blocked tariff designs to distinguish between high and low users. Blocked tariffs are particularly used in the residential sector and can vary by daily usage (Abu Dhabi) or monthly usage (others). The lowest consumption bracket in Qatar and Dubai is 2000 kWh/month, in Kuwait, Bahrain and Oman 3000 kWh/month, and in Saudi Arabia 6000 kWh/month. Some jurisdictions also have additional subsidies for low income groups. For example, in Abu Dhabi, social cardholders that are also nationals pay nothing up to 333 kWh/day and then 0.018 USD/kWh above that. Expatriates with social cards pay nothing up to 79 kWh/day, after which they pay 0.073 USD/kWh (Abu Dhabi Distribution Company 2017).

Many GCC countries have flat rates for industry. Blocked tariff designs for industrial users are however present in Abu Dhabi in the case of large industrial users with an installed capacity of more than 1 MW and only when they consume electricity during peak hours (Abu Dhabi Distribution Company 2017). This measure is to maintain the stability of the grid and adequacy of supply. As opposed to many countries, Bahrain also has blocked tariffs for Industry and Commerce that are similar to what residential consumer groups pay (Bahrain Electricity and Water Authority

2018). This shows the particular controversy around residential electricity tariffs, particularly for less wealthy households with just one account.

### 5.3 *Electricity Pricing Reforms and Tariffs in Comparison*

All GCC countries have been reforming electricity prices since the drop of fuel prices. In Saudi Arabia, the recent reform of January 2018 hikes tariffs for the lower consumption bracket. In total, consumers up to 6000 kWh/month pay 2.6 times more than in 2015 (APICORP 2018). In 2016, low-level consumers (<4000 kWh/month) were shielded from reforms when prices were raised by 2/3rd for the middle bracket (4001–6000 kWh/month) and raised and unified for any consumer above that. The Authority of Electricity Regulation of Oman has cut subsidies to large consumers (using more than 150 mWh/year) in Government, Commerce and Industry in January 2017. As a result, 10,000 of such users no longer receive subsidies and pay cost-reflective tariffs. Government hoped for savings of roughly RO 100 million annually from this measure alone (GI Consultancy 2017). Oman is currently conducting a review of how these measures impacted the specific large consumers. In the UAE, prices went up during the last years, but these mainly impacted expatriates. Also in Qatar, an unexpected tariff increase in October 2015 was followed by others in subsequent years. Kuwait also increased tariffs, starting with the commercial sector (Figs. 8.5 and 8.6).

## 6 Commerce, Industry and Energy Pricing Reforms

Commercial and industrial stakeholders have not been saved from reforms, and the question remains as to which place will be given to energy intensive industries in the various economic diversification plans of GCC states. A pertinent concern for energy-intensive industries that have been beneficiaries of the state's largesse in the form of subsidized energy input will be how to address the effect of new prices on their production activities, while governments seek to tackle possible cost-push inflation and the burden on consumers in addition to advancing their economic diversification agenda.

So far, little attention has been given to industrial users and specifically energy-intensive industries in terms of compensation measures for energy pricing reforms. These industries will have the greatest interest in preserving low energy prices not only because they have been accustomed to years

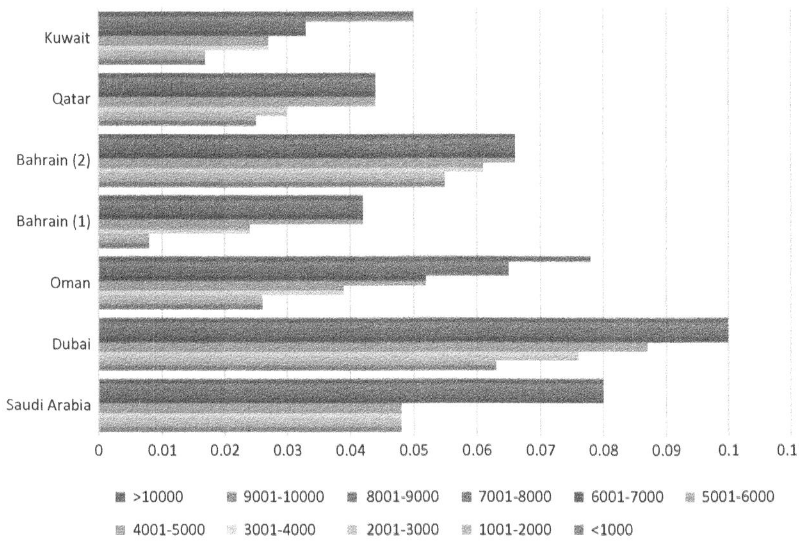


**Fig. 8.5** Residential electricity block tariffs in the GCC (July 2018). (Source: Collected by author. Note: Bahrain (1) = Bahraini nationals with only one account; Bahrain (2) = Non-Nationals and Bahraini nationals with multiple accounts)

of low prices following long-time government strategies to maximize the country's comparative advantage. Well-designed policy plans can help to consolidate the gains from the upward revision of energy prices, while also allowing enough space to producers and consumers to adjust to new economic realities.

Energy intensive industries take up a unique position in the political economies of many Gulf countries. From one side, they are used to recycle oil rents and are often closely linked to government (either informally or through the use of state-owned enterprises). From another side, they mainly benefit rich owners, while they do not offer large employment possibilities to the well-educated youth. Rather they rely on migrant labor and leave the public sector to deal with the employment of nationals. As a result, the performance (and survival) of energy intensive industries is determined by two opposing forces: one that seeks to maintain rents to keep politically connected businesses in check, and one that seeks to reform industrial prices to avoid deeper residential price increases.

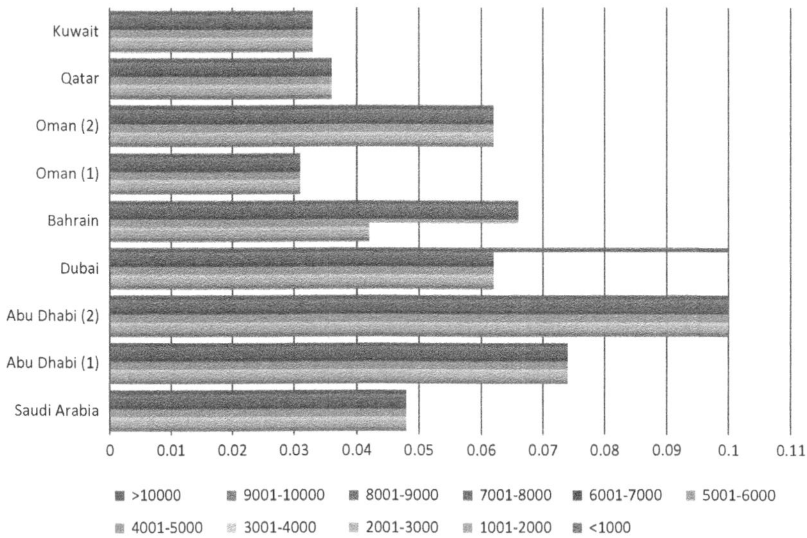


**Fig. 8.6** Industrial electricity tariffs in the GCC. (Source: Collected by author. Note: Oman (1) = 8 non-summer months; Oman (2) = 4 summer months; Abu Dhabi (1) = for industrial users with installed capacity below 1 MW and those with installed capacity above 1 MW, but outside of peak hours; Abu Dhabi (2) = for industrial users with installed capacity above 1 MW and consuming power during peak hours)

In some ways, residents benefit from low energy prices given to industries: directly, when they purchase goods from these companies, or indirectly, when other goods use energy intensive products as intermediary inputs. These inflationary effects can seep through the value chain and add to the costs of regular residents. At the same time, however, residents also compete with industries and commerce for cheap energy. Since they do not benefit from private sector employment much, the vast value of low energy prices stays with high-level managerial positions. As a result, it could be expected that when public discontent threatens political stability, commercial and industrial tariffs will be adjusted upward first (like it has happened in many other countries). On the contrary, when the wider public is able to accept reforms, residential consumers will also be targeted. This political reality does not help to design a comprehensive plan for economic diversification. It does, however and again, prove the quintessence of developing more targeted welfare protection mechanisms, such as cash transfers.

## 7 Conclusion

As a result of the oil price collapse in 2014, 2015 and 2016, all GCC countries have implemented considerable energy pricing reforms. While the level of depth of reforms has depended on various factors, such as resource reserves, population size, GDP per capita and geopolitical drivers, all countries have stuck to implemented reforms to date (June 2019). This is noteworthy, as it often happens that pricing reforms in resource-rich countries are reversed when the international price increases again, as was the case in 2017 and 2018.

GCC countries have chosen different strategies to adjust fuel prices. Saudi Arabia, Kuwait and Bahrain have opted for ad hoc pricing reforms, whereas the UAE, Qatar and Oman have implemented an automatic price setting mechanism that links domestic prices to regional and international oil price movements. The literature suggests that the implementation of an automatic price adjustment mechanism is more sustainable in the long run. However, it is important to observe the results from ad hoc price increases as well, particularly in Saudi Arabia, which has seen the largest percentage increase in domestic fuel prices. Relative to regional peers and global averages, GCC countries still have rather low fuel prices, but it is uncontestable that the price gap between domestic and international prices has been significantly reduced in all countries. Similar to fuel, GCC countries have also all implemented electricity pricing reforms since the collapse of the oil price.

The price increases have contributed to economic sustainability in two main ways. First, they have allowed governments to raise more domestic revenue and, as a result, prevent a further collapse of their already sizeable deficits in the wake of the oil price collapse. This can be reasonably linked to domestic stability more generally. Since conflict has longer-term impacts on economic sustainability, the ability to mitigate international oil price fluctuations via domestic pricing reforms is a rather essential tool to protect an economic collapse. Second, they have led to a reduction of energy demand in many (but not all) GCC countries. It is however difficult to attribute this slow down to pricing reforms, as economic output also reduced in a general economic downturn.

Despite these advantages, the impact of pricing reforms on economic sustainability suffers from a potential temporal conundrum. Besides the aforementioned short-term benefits, pricing reforms should also restore a more proper allocation of resources in the economy. Short-term effects,

**Table 8.8** Transition process from allocation state to productive economy

| *Allocation state* | *Transition process* | *Productive economy* |
|---|---|---|
| In-kind benefits with infrequently changed ad hoc pricing | Pricing reforms (energy, food, other) & introduction of value added & excise taxes | Market-based pricing that incentivizes responsible consumption |
| Universal and untargeted transfers | Ability to collect social data & design policies accordingly | Targeted welfare subsidies and social safety nets |
| Mostly public employment of nationals | Labor market, education and immigration policy reform to 'nationalize' private sector employment | Mostly private sector employment and narrowing of public-private wage gap |
| No or low taxation on business | Levy of corporate taxes and enhanced policies aimed at private sector | Interdependency between private sector, citizens and government |
| No or low taxation on income | Introduction of progressive income taxation scheme | Progressive income taxes to finance government's social responsibility |

Source: Author

however, can be negative, particularly in countries that try to use within-sector diversification to add productivity to their allocation state models. If countries adjust prices too quickly, they may undermine their own comparative advantage and they may change the social contract unilaterally, provoking public discontent. All-in-all, now the oil price has picked up, we may see a slowing down of pricing reforms among ad hoc reformers to restore stability for residential and industrial consumers. While the need for counter-cyclical economic and fiscal policies is widely recognized, this is not realistic in the GCC given the domestic political economy. This type of policy-making can only be achieved when the hydrocarbon-rich countries of the GCC simultaneously advance on other much-needed reforms, including the development of targeted welfare protection, taxation reform, education reform and, most importantly, labor market reform (Table 8.8).

## References

Abu Dhabi Distribution Company. 2017. "Slab Tariff," https://www.addc.ae/en-US/residential/Documents/02-English.pdf.

Al Mukrashi, Fahad. 2017. "Oman Introduces New Fuel Subsidy Scheme." *GulfNews*, December 16, 2017. https://gulfnews.com/news/gulf/oman/oman-introduces-new-fuel-subsidy-scheme-1.2141838.

Anand, Rahul, David Coady, Adil Mohommad, Vimal Thakoor, and James P. Walsh. 2013. "The Fiscal and Welfare Impacts of Reforming Fuel Subsidies in India." IMF Working Paper 12/128. IMF. https://www.imf.org/en/Publications/WP/Issues/2016/12/31/The-Fiscal-and-Welfare-Impacts-of-Reforming-Fuel-Subsidies-in-India-40593.

Apicorp. 2017. "Lowdown in GCC Domestic Fuel Consumption: Temporary Slump or New Reality?". *Apicorp Energy Research* 2, no. 11: 3.

———. 2018. "Saudi Energy Price Reform Getting Serious." *Apicorp Energy Research* 3, no. 5: 3.

Asselt, Harro van, and Jakob Skovgaard. 2016. "The Politics and Governance of Energy Subsidies." In *The Palgrave Handbook of the International Political Economy of Energy*, 269–88. Palgrave Handbooks in IPE. Palgrave Macmillan, London. https://doi.org/10.1057/978-1-137-55631-8_11.

Bahrain Electricity and Water Authority. 2018. "Electricity and Water Tariffs,". http://www.ewa.bh/en/Customer/BillsTariffs/electricity-water-tariffs.

Baig, Taimur, Amine Mati, David Coady, and Joseph Ntamatungiro. 2007. "Domestic Petroleum Product Prices and Subsidies: Recent Developments and Reform Strategies." IMF Working Paper 07/71. IMF. https://www.imf.org/en/Publications/WP/Issues/2016/12/31/Domestic-Petroleum-Product-Prices-and-Subsidies-Recent-Developments-and-Reform-Strategies-20608.

Bazilian, Morgan, and Ijeoma Onyeji. 2012. "Fossil Fuel Subsidy Removal and Inadequate Public Power Supply: Implications for Businesses." *Energy Policy* 45, no. Supplement C: 1–5. https://doi.org/10.1016/j.enpol.2012.02.057.

Beblawi, Hazem. 1990. "The Rentier State in the Arab World." In *The Arab State*, edited by Giacomo Luciani, 85–98. University of California Press.

Cees van Beers, and Jon Strand. 2013. "Political Determinants of Fossil Fuel Pricing." The World Bank Development Research Group.

Cheon, Andrew, Maureen Lackner, and Johannes Urpelainen. 2014. "Instruments of Political Control: National Oil Companies, Oil Prices, and Petroleum Subsidies." *Comparative Political Studies*, 1–33.

Commander, Simon. 2012. "A Guide to the Political Economy of Reforming Energy Subsidies." IZA policy paper No. 52. http://ftp.iza.org/pp52.pdf.

Dubai Electricity & Water Authority. 2018. "Slab Tariff,". https://www.dewa.gov.ae/en/customer/services/consumption-services/tariff.

Fattouh, Bassam, and Laura El-Katiri. 2013. "Energy Subsidies in the Middle East and North Africa." *Energy Strategy Reviews* 2: 108–15.

Fattouh, Bassam, Tom Moerenhout, and Anupama Sen. 2016. "Striking the Right Balance? GCC Energy Pricing Reforms in a Low Price Environment." Oxford Energy Comment. Oxford: Oxford Institute for Energy Studies. https://www.oxfordenergy.org/wpcms/wp-content/uploads/2016/04/Striking-the-Right-Balance-GCC-Energy-Pricing-Reforms-in-a-Low-Price-Environment.pdf.

GI Consultancy. 2017. "Oman Energy Master Plan: Two Years on – What Has Happened?". http://www.thegulfintelligence.com/mediafiles/cataloguedatasheet/8a911c5f-8437-4db4-bf42-9153fb8006fe.pdf.

Gnana, Jennifer. 2017. "Saudi Arabia Extends Deadline for Energy Price Reform." *The National*. https://www.thenational.ae/business/energy/saudi-arabia-extends-deadline-for-energy-price-reform-1.689473.

Hartley, Peter R., and Kenneth B. Medlock. 2009. "Potential Futures for Russian Natural Gas Exports." *The Energy Journal* 30, no. Special Issue: 73–96.

IEA. 2014. "World Energy Outlook 2014." Paris: OECD/IEA. http://www.worldenergyoutlook.org/resources/energysubsidies/.

IEA, OPEC, OECD, World Bank Joint Report. 2010. "Analysis of the Scope of Energy Subsidies and Suggestions for the G-20 Initiative." Report prepared for submission to the G-20 Summit Meeting Toronto (Canada), 26–27 June 2010. http://www.oecd.org/env/45575666.pdf.

IMF. 2016. "2016 Article IV Consultation: Saudi Arabia,". http://www.imf.org/external/pubs/ft/scr/2016/cr16326.pdf.

———. 2015. "Energy Price Reforms in the GCC – What Can Be Learned From International Experiences,". https://www.imf.org/external/np/pp/eng/2015/111015b.pdf.

———. 2017. "The Economic Outlook and Policy Challenges in the GCC Countries." GCC Surveillance Paper. IMF.

Kojima, Masami. 2013. "Petroleum Product Pricing and Complementary Policies: Experience of 65 Developing Countries since 2009." Policy research working paper 6396. The World Bank.

Komives, Kristin, Vivien Foster, Jonathan Halpern, Quentin Wodon, and Roohi Abdullah. 2008. "Water, Electricity and the Poor: Who Benefits from Utility Subsidies." The World Bank. https://openknowledge.worldbank.org/handle/10986/11745.

Kuwait Times. 2018. "No Plans for Kuwait to Raise Fuel Prices Again." *Kuwait Times*, January 9, 2018. http://news.kuwaittimes.net/website/no-plans-kuwait-raise-fuel-prices/.

Luong, Pauline Jones, and Erika Weinthal. 2010. *Oil Is Not a Curse: Ownership Structure and Institutions in Soviet Successor States*. New York: Cambridge University Press.

Mahdi, Wael, and Vivian Nereim. 2017. "Saudi Arabia Delays Energy-Subsidy Cuts." *Bloomberg.Com*. https://www.bloomberg.com/news/articles/2017-07-06/saudis-are-said-to-delay-energy-subsidy-cuts-amid-economic-pain.

Maxwell, Steve. 2010. "Market Outlook: Historical Water Price Trends." *Journal (American Water Works Association)* 102, no. 4: 24–28.

Moerenhout, Tom. 2018. "Energy Pricing Reforms in the Gulf: A Trend but Not (yet) a Norm." International Institute for Sustainable Development. https://www.iisd.org/sites/default/files/publications/energy-pricing-gulf-trend-but-not-norm.pdf.

Overland, Indra. 2010. "Subsidies for Fossil Fuels and Climate Change: A Comparative Perspective." *International Journal of Environmental Studies* 67, no. 3 (2010): 303–17. http://www.tandfonline.com/doi/abs/10.1080/00207233.2010.492143?tab=permissions&scroll=top.

Qatar General Electricity & water Corporation. 2018. "Tariff". https://www.km.com.qa/CustomerService/Pages/Tariff.aspx.

Rentschler, Jun, and Martin Kornejew. 2017. "Energy Subsidy Reforms and the Impacts on Firms: Transmission Channels and Response Measures." *Oxford Energy Forum*, no. 108: 16–19.

Rentschler, Jun, Martin Kornejew, and Morgan Bazilian. 2017. "Fossil Fuel Subsidy Reforms and Their Impacts on Firms." *Energy Policy* 108, no. Supplement C: 617–23. https://doi.org/10.1016/j.enpol.2017.06.036.

Saudi Electricity Company. 2018. "Consumption Tariff,". https://www.se.com.sa/en-us/customers/Pages/TariffRates.aspx.

Strand, Jon. 2016. *The Economics and Political Economy of Energy Subsidies*. MIT Press.

Sultanate of Oman. 2018. "National Subsidy System". https://nss.gov.om/site/about?ln=en.

Times of Oman. 2018. "Fuel Prices for February 2018 Announced in Oman." *Times of Oman*. http://timesofoman.com/article/127292.

Victor, David. 2009. "The Politics of Fossil-Fuel Subsidies." *SSRN ELibrary*. http://papers.ssrn.com/sol3/papers.cfm?abstract_id=1520984.

Victor, David G., David R. Hults, and Mark C. Thurber, eds. 2011. *Oil and Governance: State-Owned Enterprises and the World Energy Supply*. Cambridge University Press.

Walker, Lesley, and Peter Kovessy. 2016. "Qatar Increases Petrol Prices from Midnight Tonight." Doha News, January 14, 2016. https://dohanews.co/qatar-increases-petrol-prices-from-midnight-tonight/.

Whitley, Shelagh, and Laurie van der Burg. 2015. "Fossil Fuel Subsidy Reform in Sub-Saharan Africa: From Rhetoric to Reality." Overseas Development Institute. http://newclimateeconomy.report/workingpapers/wp-content/uploads/sites/5/2016/04/FFS-Reform-in-Africa_NCE-ODI_final.pdf.

CHAPTER 9

# Fiscal Sustainability and Hydrocarbon Endowment Per Capita in the GCC

*Monica Malik and Thirumalai Nagesh*

## 1 Introduction

GCC economies remain highly dependent on the hydrocarbon sector in terms of the composition of GDP and as a source of government revenue and export base. However, variations persist across the region, with the economies of Bahrain and Dubai the most diversified and Kuwait the least. The key non-oil sectors across the GCC tend to be in the following areas—construction, real estate, tourism, retail, hospitality, aviation and petrochemical and energy-intensive industries (such as aluminium). The policy focus again intensified towards economic diversification from end-2014 with the sharp fall in the oil price. New development plans have been announced to reach this objective, most notably in Saudi Arabia, with the National Transformation Plan and Vision 2030, both announced in 2016. These plans are a blueprint to diversify the Saudi economy (fiscally and in terms of composition of GDP) and highlight the government's commitment to transform the economy away from oil.

M. Malik (✉) • T. Nagesh
The Hills Compound, Abu Dhabi, United Arab Emirates
e-mail: monica.malik@adcb.com; Thirumalainagesh.Venkatesh@adcb.com



G. Luciani, T. Moerenhout (eds.), *When Can Oil Economies Be Deemed Sustainable?*, The Political Economy of the Middle East,
https://doi.org/10.1007/978-981-15-5728-6_9

In this chapter, we focus on fiscal sustainability, given the importance of fiscal policy in the GCC. Most of the hydrocarbon earnings (via nationalized companies) accrue to the regional governments who then are the main conduit for this revenue to enter the domestic economy and the central factor dominating economic activity and domestic demand. As such, governments in the GCC play a much larger role in the economy than is generally the case for other emerging market countries and are often the key employers, especially for the national population. The role of government spending is especially vital for the macro-management of the GCC economies, given their limited control of monetary policy with their currencies being pegged to the USD. We highlight that the fiscal side is just one aspect of overall economic sustainability, albeit linked to other areas. Indeed, the diversification in the economic base will support the deepening non-oil revenue. However, fiscal reforms and consolidation at times of lower oil price cycles result in weaker domestic activity, including investment (public and indirectly private) and impeding diversification plans. In this chapter, we focus on fiscal reforms between 2014 and 2018, particularly on the UAE and Saudi Arabia as they have seen the greatest fiscal reforms during this period, though it also highlights fiscal strength and developments in other GCC countries.

Despite the renewed emphasis on shifting away from the hydrocarbon sector, the GCC countries that are currently the most resilient and best able to cope with the low oil price environment are those with large hydrocarbon reserves relative to their populations (hydrocarbon rich per capita). These countries tend to use less of their resources supporting the population, resulting in stronger fiscal fundamentals that bolster their economic sustainability despite high exposure to the hydrocarbon sector. We highlight in the chapter that the economic strength of these countries are reflected in a number of economic indicators, including low debt and high foreign currency reserves, which in turn feed into their stronger sovereign rating. This reflects their stronger saving rates when oil prices are higher, alongside lower budget breakeven oil prices. This does not mean that these economies should not look to diversify. Rather, it highlights that their fiscal strength results in less pressure on the economy during times of low prices and reform. Thus, fiscal sustainability at this stage is mostly unrelated to the degree of economic diversification. Fiscal diversification has largely been weak across the region, with hydrocarbon revenue remaining the main source of government income. That some countries have a relatively higher share of non-oil revenue in total revenue, in many cases, reflect weaker hydrocarbon endowment rather than a diversified revenue base (especially tax) or lower subsidy levels.

The underlying fiscal strength and sustainability of GCC economies is especially important at a time when the region has seen relatively limited success in widening the economic base (composition of GDP) much beyond pre-2014 levels. There have been some signs of greater fiscal reforms, especially in the UAE and Saudi Arabia. However, this has placed downward pressure on their economies, including the private sector—who are often proposed key drivers of economic diversification and raising employment opportunities for nationals. Thus, government-led initiatives will be central to boosting and deepening economic activity, alongside developing a framework to access private capital and expertise—local or international. Notably, a number of GCC countries are promoting labor nationalization programs, in varying degrees, to help create job opportunities for nationals, especially the youth. Reducing the share of government spending on wages will also be critical to boosting fiscal sustainability across the GCC, especially for the hydrocarbon poorer per capita countries, alongside reforms to increase non-oil revenue.

The other area of policy adjustment and economic support has been on the oil front since 2017—the pullback in OPEC+ oil production has helped to reduce global inventories and has provided fundamental support to the oil price. The rise in global oil prices since mid-2017 reduces some of the fiscal (and external) pressure seen from end-2014, particularly in 1H2016. In turn, this has allowed an increase in government spending for 2018. GCC fiscal positions improved in 2018 despite the increase in government spending, thanks to the higher oil revenue. However, the fall in oil price at the end of 2018 and the ongoing development in hydrocarbon technology that are likely to impact future supply and demand (including shale oil, renewable energy and electric vehicles) highlight the need for further reforms to strengthen and deepen fiscal sustainability in the GCC.

## 2 Variations in GCC Hydrocarbon Endowments and Economic Sustainability

The size of hydrocarbon reserves per capita is a key factor for fiscal sustainability currently rather than just the size of the hydrocarbon reserves themselves or hydrocarbon revenue as a percentage of total government revenue. Based on per-capita hydrocarbon endowments, the GCC countries can be broadly divided into two categories (Figs. 9.1 and 9.2):

- **Hydrocarbon richer per capita:** These countries have higher hydrocarbon reserves (and production) relative to their populations. They include Kuwait, Qatar and the UAE (Abu Dhabi led).
- **Hydrocarbon poorer per capita:** These countries have relatively smaller hydrocarbon reserves (and production) relative to their populations. They include Bahrain, Oman and Saudi Arabia. Within this group, Oman and Bahrain are in relatively weaker positions versus Saudi.

The per-capita hydrocarbon ratio is particularly important in the GCC, given the relationship between the state and citizens—that is, the social contract. In simple terms, in GCC economic frameworks, hydrocarbon revenue falls to the government (via the national oil and gas companies) and is then distributed and mobilized for the well-being of the population and country. This has also meant that there is a framework to support the citizens from cradle to grave in various ways, including covering education and healthcare, highly subsidized utility prices, land and cheap financing

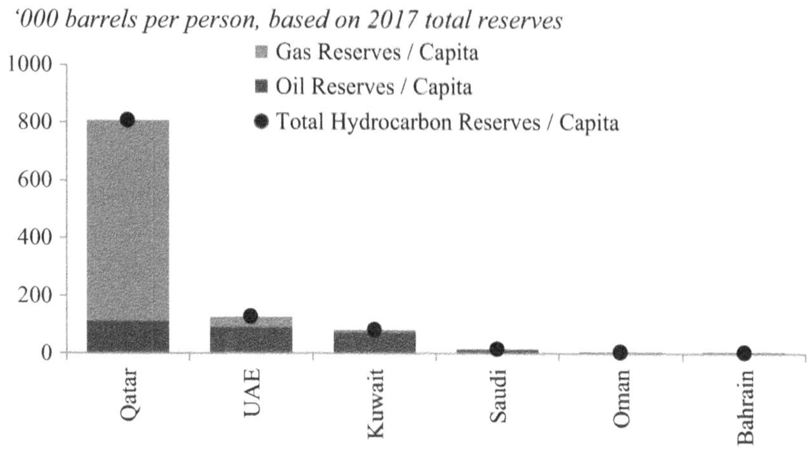


**Fig. 9.1** GCC: Hydrocarbon endowment per capita (reserves) based on national population. (Source: BP [hydrocarbon reserves data], regional statistical agencies [population data 2016–18], authors' calculations and methodology)

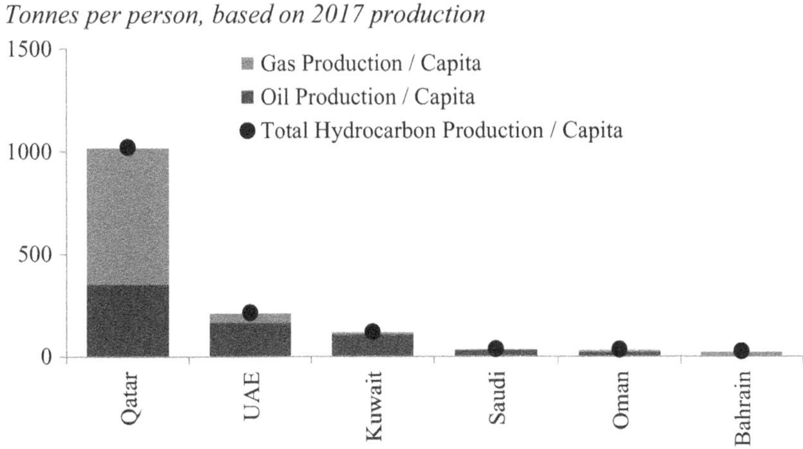


**Fig. 9.2** GCC: Hydrocarbon endowment per capita (production) based on national population. (Source: BP [hydrocarbon output data], regional statistical agencies [population data 2016–18], authors' calculations and methodology)

to build housing. There are, however, variations across the region on the level of support provided by the government to nationals, including with the differences in the pace of reforms seen since 2015 also contributing to this. Moreover, the public sector has been a key employer of nationals and there has been a limited tax base. This social contract is seeing signs of change, especially with the fall in the oil price and the rise in the domestic population. Indeed, the region has seen a pullback in subsidies and the introduction of fees and some taxes since 2015. At the same time, strong population growth means that the younger generation of GCC nationals may not fully be able to rely on the state to provide them with jobs. Nevertheless, the social contract remains largely in place despite these adjustments.

The nature of the economies and impact of the social contract has resulted in hydrocarbon richer countries generally having to spend less of their hydrocarbon income to support their populations. This has also meant that these countries have seen larger fiscal surpluses and foreign currency reserve build-ups during periods of higher oil prices. Moreover, after the collapse of the oil price at end-2014, they have largely seen smaller deficits and thus, have required lesser fiscal adjustment to balance their budgets (Fig. 9.3).

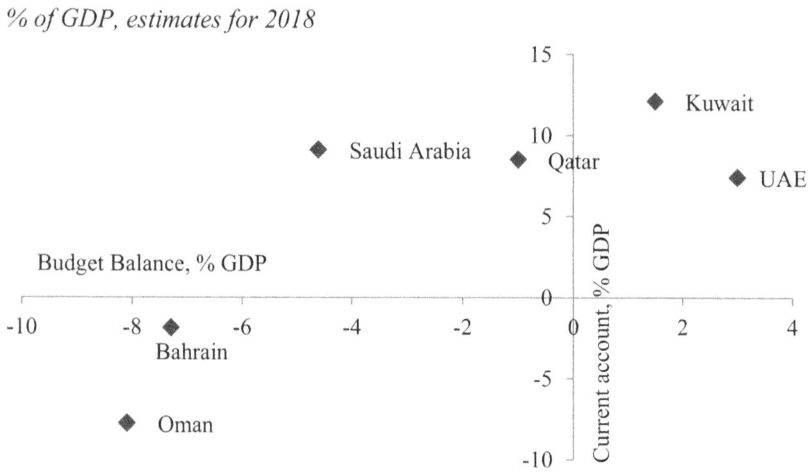


**Fig. 9.3** GCC: Hydrocarbon endowment per capita reflected in current account and fiscal balances. (Source: Regional statistical agencies, ADCB estimates [2019, 25])

## 3 Hydrocarbon Endowment Seen in Various Fiscal and Economic Indicators

The hydrocarbon endowment of the GCC countries is reflected in a number of fiscal and economic indicators, including:

- **Budget breakeven (BBE) oil price**: This is the required oil price for a fiscal budget to be balanced. GCC budget breakeven oil prices have largely fallen since the peak in 2014 until 2017 as GCC governments reduced spending and introduced fiscal reforms. The hydrocarbon richer per capita countries have lower BBE oil prices, alongside lower external breakeven oil price (the oil price needed for the current account to be balanced). This lower BBE oil price again reflects that less of the hydrocarbon resources are spent on the national population and more of the income from resources are saved. After the 2014 shock, the UAE and Kuwait remain the most comfortable fiscally and have seen an improvement in their budgetary performance with the rise in the oil price from the oil price trough; whereas Bahrain, Oman and Saudi Arabia lagged behind. It is feasible that the BBE price will again rise in the future if international oil prices

**Table 9.1** GCC: Fiscal budget breakeven oil price (USD per barrel)

| | *2008* | *2009* | *2010* | *2011* | *2012* | *2013* | *2014* | *2015* | *2016* | *2017* |
|---|---|---|---|---|---|---|---|---|---|---|
| Bahrain | 76 | 81 | 99 | 112 | 116 | 119 | 116 | 118 | 107 | 103 |
| Oman | 65 | 66 | 71 | 81 | 83 | 96 | 103 | 96 | 89 | 84 |
| Saudi | 57 | 65 | 67 | 78 | 79 | 92 | 106 | 95 | 96 | 78 |
| UAE | 44 | 66 | 70 | 74 | 77 | 76 | 83 | 65 | 60 | 65 |
| Qatar | 29 | 27 | 33 | 38 | 43 | 50 | 54 | 50 | 53 | 57 |
| Kuwait | 34 | 29 | 45 | 43 | 49 | 52 | 56 | 49 | 47 | 51 |

Source: IIF (2018, 3)

recover or stabilize, and governments look to support growth and sentiment (Table 9.1).

- **Debt to GDP:** The hydrocarbon poorer per capita countries generally have higher government debt levels relative to GDP. The GCC countries largely entered this oil price crisis with low government debt levels (with the exception of Bahrain), as they used the strong oil prices from prior years to reduce their debt levels. Again in the case of the UAE the low overall debt position relative to GDP stems from Abu Dhabi, with Dubai seeing higher levels. However, the larger fiscal deficits in Bahrain, Oman and Saudi Arabia from 2015 with the correction in the oil price have resulted in greater funding requirements and a faster acceleration of government debt. This rise in the debt stock is resulting in more government spending being allocated to interest payments and debt servicing. For most GCC countries except Bahrain, interest payments still account for a small share of overall spending (Figs. 9.5 and 9.6).
- **Foreign currency reserve position:** The hydrocarbon richer per capita countries have been able to build up significant foreign currency reserves, reflecting their larger fiscal surpluses during times of higher oil prices. Moreover, there has been less pressure to draw down these reserves to cover fiscal deficits. Income from these reserves, often invested by their sovereign wealth funds (SWF), provides an additional source of income (investment) to the government and in a way is a form of revenue diversification. Saudi Arabia was looking to sell 5% of Aramco,[1] in part to increase overseas investment and thereby boost and diversify government income. The

[1] Aramco listed its shares domestically on the Tadawul exchange on 11 December 2019, raising USD25.6 billion from a 1.5% stake sale. Subsequently, it also exercised its "greenshoe option" to sell an additional 450 million shares, taking the total stake sale to 1.725%.

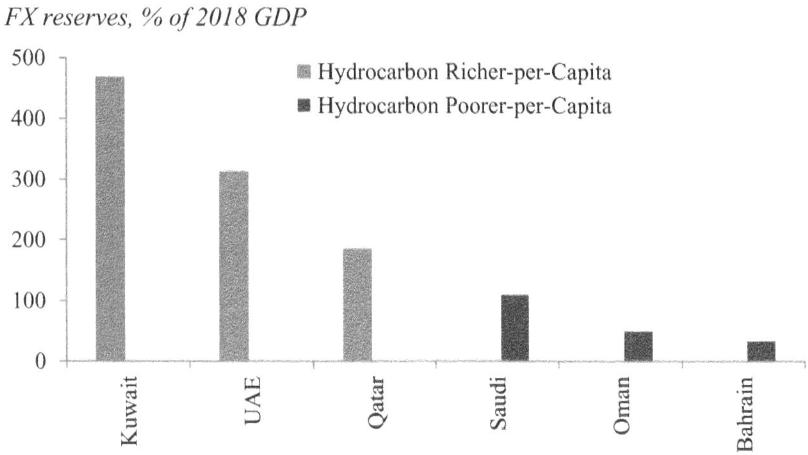


**Fig. 9.4** GCC: Hydrocarbon endowment also reflected in regional FX reserve* positions. (*Includes reserves with SWFs and held at central banks. Source: Sovereign Wealth Fund Institute, Regional Central Banks, ADCB estimates [GDP])

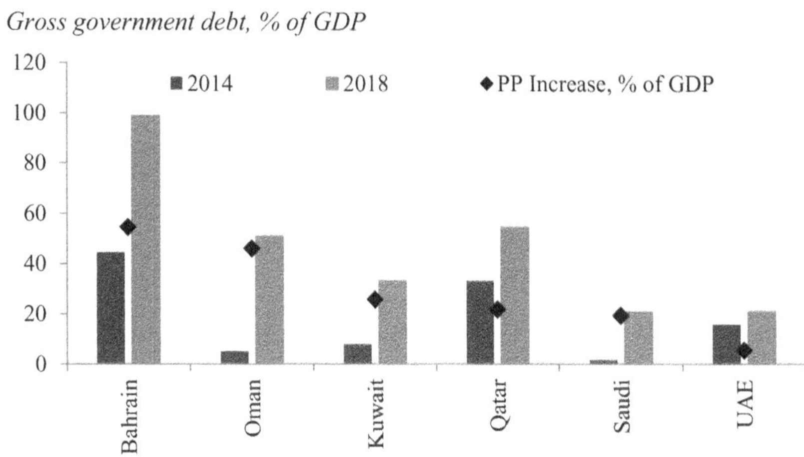


**Fig. 9.5** GCC: Hydrocarbon poorer per capita countries have largely seen a faster rise in government debt. (Source: IMF, ADCB estimates)

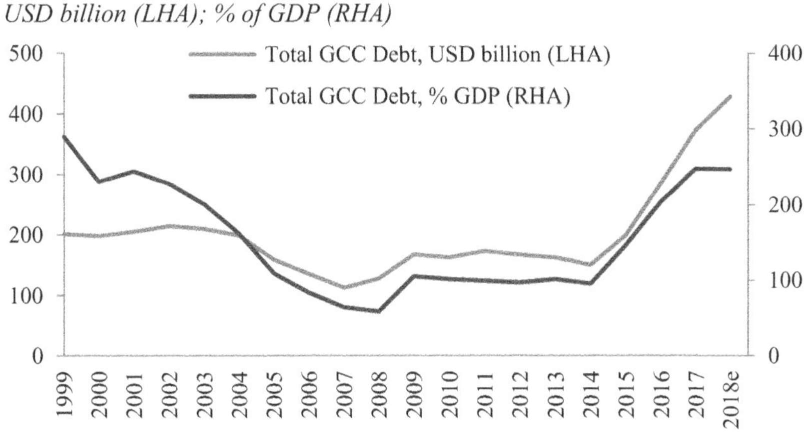


**Fig. 9.6** GCC: gross government debt has risen sharply since 2014, limited fiscal space for some countries. (Source: IMF, ADCB estimates)

reduction of foreign currency reserves since 2015 is however most clearly visible in the Saudi Arabia Monetary Authority's net foreign asset (NFA) position, which highlights the drawdown in reserves to help fund the fiscal deficit. For other GCC countries, a significant part of their foreign currency reserves are held by SWFs and thus there is limited visibility (Fig. 9.4).

- **Sovereign ratings:** foreign currency reserve positions and government debt levels are also reflected in the sovereign ratings of GCC countries. The hydrocarbon richer per capita countries have substantially higher sovereign ratings, underpinned by large foreign currency reserves and low debt levels. Notably, despite the sharp correction in the oil price, Abu Dhabi and Kuwait have not seen any ratings downgrades by the three main ratings agencies since 2014. Moreover, their ratings remain among the strongest globally. On the other hand, Bahrain, Oman and Saudi Arabia have seen multiple downgrades over this same period. The ratings are reflected in the cost of borrowing and the risk premium, with hydrocarbon poorer per capita countries more susceptible to changes in global sentiment

**Table 9.2** GCC: Sovereign rating changes by the main rating agencies since 2015–2018

| | *Fitch* | | | *Moody's* | | | *S&P* | | |
|---|---|---|---|---|---|---|---|---|---|
| | Mid-2014 | Current | Change | Mid-2014 | Current | Change | Mid-2014 | Current | Change |
| Bahrain | BBB | BB- | 4 notches down | Baa2 | B2 | 6 notches down | BBB | B+ | 5 notches down |
| Kuwait | AA | AA | No change | Aa2 | Aa2 | No change | AA | AA | No change |
| Oman | -NA- | BB+ | 2 notches down[a] | A1 | Ba1 | 6 notches down | A | BB | 6 notches down |
| Qatar | AA[b] | AA– | 1 notch down | Aa2 | Aa3 | 1 notch down | AA | AA– | 1 notch down |
| Saudi Arabia | AA | A+ | 2 notches down | Aa3 | A1 | 2 notches down | AA– | A– | 3 notches down |
| UAE | -NA- | -NA- | -NA- | Aa2 | Aa2 | No change | -NA- | -NA- | -NA- |
| Abu Dhabi | AA | AA | No change | Aa2 | Aa2 | No change | AA | AA | No change |

Source: Bloomberg

[a]2 notches down from 3 Jan 2017

[b]Ratings started from 3 June 2015

and dependent on capital inflows (to cover fiscal and current account shortfalls and to support investment and diversification programs) (Table 9.2).

- **GDP per capita and unemployment:** Reflecting the fact that fewer resources in relative terms have to be used to support their populations, the hydrocarbon richer per capita countries also tend to have higher GDP per capita and spending power. Moreover, unemployment levels for nationals tend to be lower, though regional data on this front is limited. The hydrocarbon richer countries also tend to have a larger share of expatriates in their populations, given their relatively smaller national populations relative to economy size and the ability of the public sector to employ nationals.

There are, naturally, variations within the GCC based on government policy and effectiveness. During times of high oil prices, Kuwait saw substantially larger fiscal surpluses as a percentage of GDP (2005–2008), as the government made limited progress in its investment plans, partly due to the difficult relationship between the cabinet and the National Assembly. Meanwhile, in the case of Qatar, government debt was higher at end-1990, reaching 74.4% of GDP in 1999, given borrowing to develop the gas industry. However, the rise in gas income resulted in a sharp reduction in government debt to 8.9% of GDP in 2007, and FX reserves rose. As noted earlier, GCC countries used strong oil revenues in 2002–2014 to reduce debt levels, though aggregate government debt have risen from 2014 to well above the 2002 peak in absolute terms, largely driven by the hydrocarbon per capita poorer countries. Notable, Saudi Arabia started issuing external debt from end-2016, earlier all of the government debt was SAR denominated.

## 4 Fiscal Reform Momentum

The period between 2003 and 2014 generally saw strongly expansionary government spending from the GCC countries. This pickup in government was supported by the strengthening in the oil price from around 2002, though the pace of spending increase also varied between the GCC countries. For example, Saudi Arabia initially took a more cautious approach, whilst focusing on reducing government debt levels. Notably, this debt repayment also supported liquidity in the banking sector (with all government debts being domestic at this point, including those held by banks and pension organizations), indirectly providing a boost to economic activity. Much of the initial increases in spending during this period was led by UAE (Dubai) and Qatar (including the development of the gas sector) and partly debt led, though with the broader pace rising across the GCC from around 2003.

The increase in government spending until 2011 was both on the current and the capital fronts, though overall the capital expenditure had limited impact in broadening the economic base. In many cases, a significant part of the investment was to upgrade infrastructure, compensating for weak periods of investment in the 1980s and 1990s with low and volatile oil prices, and for the rise in population. However, we also note that, in many cases, higher investment spending was implemented by government related entities (GREs), and thus not reflected in the governments' budgets. From 2012, there was a shift across the GCC toward current expenditure, likely impacted by the MENA Arab Spring developments. The overall rise in government spending was reflected in the rise in the BBE oil price of the GCC countries from 2004 to a peak in 2014, and supported by the overall rise in the oil price (ex-2009). GCC countries largely saw fiscal surpluses over this period (ex-2009), with the exception of Bahrain, which saw a sustained deficit since 2009. However, with the sharp correction in the oil price from end-2014, the focus of fiscal policy shifted towards retrenchment and reform (Figs. 9.7 and 9.8).

Since end-2014, the pace of fiscal reform in the GCC has gathered momentum. This is especially important, given (1) the large youth population and national population growth rates and (2) medium- to long-term structural challenges to the oil price, including from new technology (shale, renewable energy sources). GCC economies will have to create jobs to cater for the entry of nationals into the work force (please see our labor market and reform chapter), with educational reforms also vital for supporting the needs of the private sector and economic development. Without this job creation, greater pressure will fall on the governments to support the national populations, through jobs creation or by a support framework.

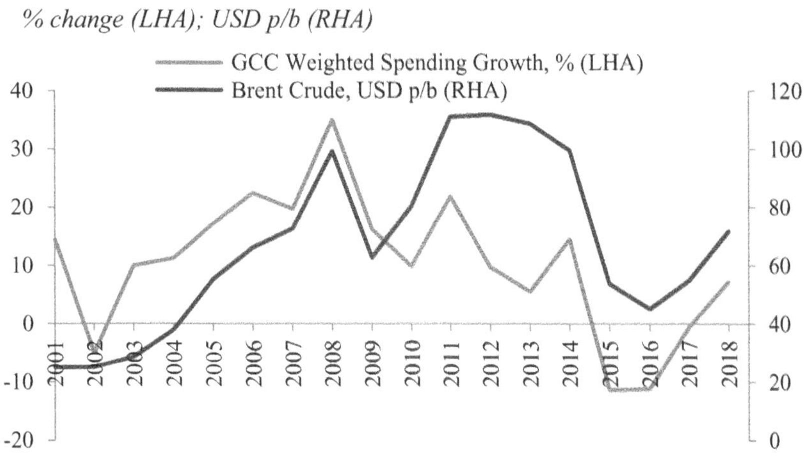


**Fig. 9.7** GCC: Government spending growth was strong between 2003 and 2014. (Source: Reuters (oil price), spending growth calculated from regional statistical agencies and IMF GDP data)

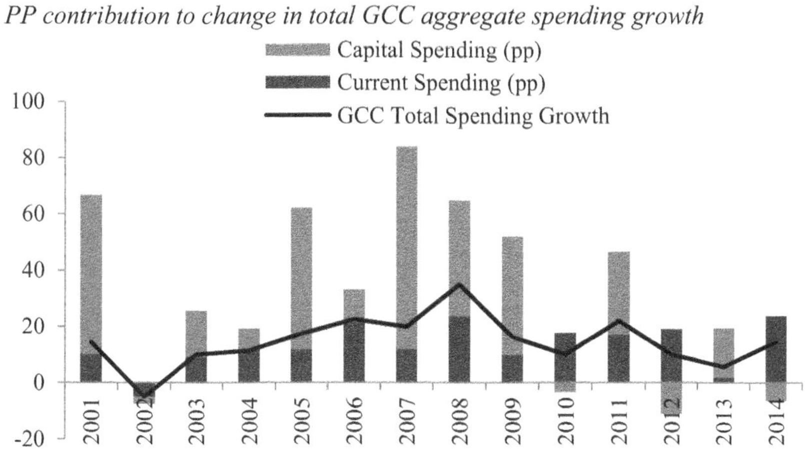


**Fig. 9.8** GCC: Drivers of total GCC spending growth, by capital and current expenditure. (Source: Calculated from regional statistical agencies and IMF GDP data)

The fiscal reforms have been much wider than seen in previous oil price downturns, indicating that areas that were previously seen as sensitive are open to change and reform. For example, subsidies have been reduced, which was not the case in the 1980s and 1990s, when the main fiscal adjustment was through the retrenchment of spending. A significant development for the GCC was the introduction of VAT in Saudi Arabia and the UAE on 1 January 2018 and in Bahrain in January 2019. This was a vital initial step in developing tax revenue in the region, which has a weak tax base.

## 5 Phases in GCC Fiscal Reform

The pace and focus of fiscal consolidation across the GCC has varied over the past few years. For the main fiscal reform measures introduced, please see Appendix A.

### 5.1 *Phase 1: Expenditure Cutbacks and Subsidy Reforms in 2015 and 2016*

The initial response by governments was to substantially cut back government expenditure, especially on the capital front. It is generally harder to retrench current expenditure in the GCC, given the dominance of public sector wages. Weighted GCC government expenditure fell by c.11.4% in 2015 and by a further 11.1% in 2016. In 2015, Kuwait, Saudi Arabia and the UAE saw a double-digit contraction in government spending. Between 2014 and 2018, Qatar and Kuwait saw a significant pullback in government spending. The UAE drop was the least among the hydrocarbon richer per capita countries, as Dubai adopted an expansionary fiscal position, whereas Abu Dhabi actually was particularly proactive in cutting government spending. The hydrocarbon richer per capita countries are generally in a stronger position to reduce government spending as a lower amount of total expenditure goes on public sector wages, reflecting the difficulties in cutting this key component of current expenditure. Nevertheless, Saudi Arabia and Oman also saw a significant retrenchment in total expenditure during this period, with only Bahrain seeing limited pullback in expenditure during this period (Figs. 9.9 and 9.10).

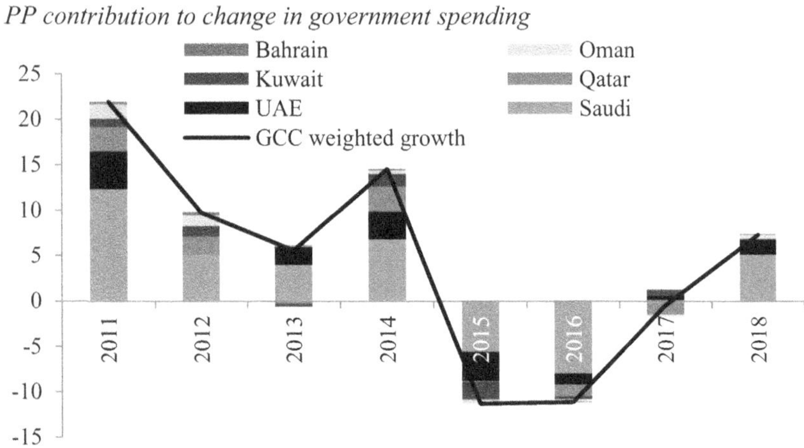


**Fig. 9.9** GCC: Government sharply pulled back spending in 2015 and 2016 in response to the lower oil price. (Source: Calculated from regional statistical agencies and IMF data)

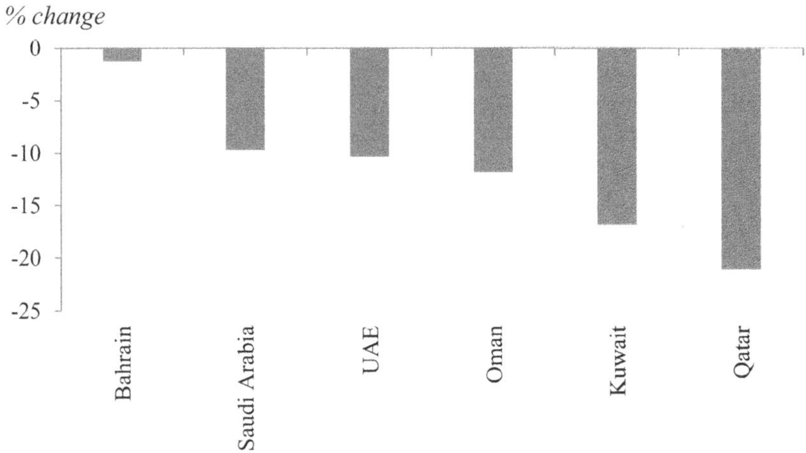


**Fig. 9.10** GCC: Change in government spending from 2014 to 2018. (Source: Calculated from regional statistical agencies and IMF data)

Alongside a direct pullback in government spending, GCC governments also adopted wider measures to strengthen their fiscal positions. First, a number of countries have also focused on streamlining government related entities (GRE), so as to reduce costs and potentially increase returns for the government. This has included job cuts alongside mergers amongst entities in similar economic areas. The objective is to boost efficiencies and extract cost synergies. This trend was most visible in Abu Dhabi, where mergers were seen in a number of areas including amongst banks and companies in the hydrocarbon sector. In Abu Dhabi, there have also been signs that GREs have had to become more reliant on raising their own funding or unlocking value of their assets to meet spending and development plans, thereby reducing the support required from the central government. A later example of this is the part privatization of ADNOC distribution in December 2017. In Qatar, the rationalization included reducing the GRE workforce, with Qatargas and RasGas also merging. The new entity, called Qatargas, started operations on 1 January 2018. There were wider adjustments, with a number of the major GCC corporates having some form or degree of government ownership, including national oil companies, national utilities companies, telecoms operators, chemicals or real estate firms.

Second, there were also increases in a number of government fees and the introduction of new ones. Importantly, progress was made on subsidy reforms—which is vital to improving fiscal sustainability, especially as the marginal ability to reduce spending moderate after the first few years. Subsidy reforms have mostly centered on rising fuel and utility prices, with the UAE kicking off the process in 2015 after some earlier price adjustments in the UAE and Qatar. Notably, these reforms have impacted nationals as well as expatriates, alongside corporates. UAE and Saudi Arabia have made some significant reductions in their subsidy framework, whilst Kuwait has seen limited reforms with ongoing opposition from the National Assembly. See Moerenhout in this volume for a detailed overview of energy subsidy reforms.

Third, there were moderate labor sector reforms. In a significant move, Saudi Arabia moved to reduce public sector wages and benefits in October 2016. This was notable, as reducing public sector wages had been/is seen as socially sensitive, especially, given the greater proportion of nationals working in the public sector versus the private. However, these were fairly short-lived.

### 5.2 *Phase 2: Reduction in Pace of Reform in 2017; Stabilization in Spending*

The pace of fiscal reform moderated markedly in 2017, which partly highlights reform fatigue and the difficulty of sustaining a multi-year reform program, in our view (Fig. 9.11). The weakening in economic momentum as a result of fiscal adjustments limited the ability of the economy to absorb new measures. This is particularly magnified in the GCC by the dominant role of the government in driving economic activity. Notably, a critical factor supporting the more gradual pace of fiscal adjustment in 2017 was a rise in the oil price, especially in 2H2017. Central to this was the change in OPEC's oil policy from one of increasing oil production to increase market share (2015 and 2016) to one of cutting output to help reduce the global oversupply and push up the price. The non-OPEC countries involved in this deal included Oman and Bahrain, alongside large global producers, such as Russia. Compliance was strong in 2017, in a large part due to greater-than-required production cuts from Saudi Arabia.

Most of the narrowing in GCC deficits in 2017 was largely due to higher oil revenue, albeit overall spending was also restrained and the pace of fiscal reform slowed. Weighted GCC fiscal spending was virtually flat in

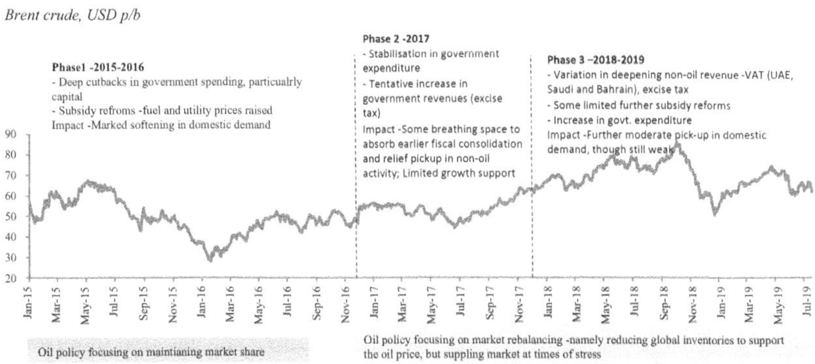


**Fig. 9.11** GCC: Phases of fiscal reforms relative to oil price developments and policy. (Source: Reuters (oil price), authors' assumptions)

2017, rising by a modest 0.9%. This spending growth calculation is based on delayed payments from previous years made by Saudi Arabia government contractors in end-2016 (around SAR80 billion) being included in the 2016 fiscal accounts, alongside for "surplus projects" (SAR25 billion) (Kingdom of Saudi Arabia 2016a, 14).

There were some limited new fiscal reforms, including the introduction of an excise tax on harmful goods (cigarettes, sugary and energy drinks) in Saudi Arabia (June) UAE (October) and Bahrain (December). Kuwait and Oman also reduced water and electricity subsidies in 2017, aimed at certain sectors (Appendix A). Saudi Arabia did not progress with further subsidy reforms that were planned for the year under its Fiscal Balance Program 2020 (Kingdom of Saudi Arabia 2016b, 38). These included linking electricity 100% to reference prices and raising gasoline and diesel prices. Both were initially planned for mid-2017. Overall, there were limited major fiscal reforms in 2017.

There were however also some reversals of earlier reforms. In Saudi Arabia, public sector wages and benefits introduced in October 2016 were reversed (April 2017). A decree reinstated all allowances, financial benefits and bonuses to public employees and military staff. Moreover, a two-month salary bonus for forces fighting on the frontline in Yemen was announced. Moreover, in December 2017, the government pushed out its balanced budget target to 2023 from the 2020 date initially set in the Fiscal Balance Program 2020 (Kingdom of Saudi Arabia 2017, 5). This allows more time to introduce reforms as well as to generate greater government expenditure. Subsidy price reforms (including fuel, gas for industry and utility prices) will be stretched over a longer period, with adjustments occurring until 2025 to reach international levels.

### 5.3 *Phase 3: Shift to Expansionary Fiscal Stance from 2018; Introduction of VAT*

After years of fiscal reforms that have placed downward pressure on economic activity, GCC governments' priorities started showing some signs of shifting to supporting growth from 2018 on. The fiscal budgets for 2018 indicated a clear shift towards expansionary spending stances to support economic activity. The stronger oil price outlook for 2018 supported the higher government spending, requiring a weaker magnitude of overall fiscal adjustment (expenditure and reforms combined). We estimate that aggregate government spending rose to 7.2% in 2018, though again there were variations is the rise in expenditure across the region. Saudi Arabia particularly saw a significant increase, with total spending rising by 10.8% in 2018. Most of the marked fiscal improvement in 2018 in the GCC fiscal positions was largely due to higher hydrocarbon revenues and GCC fiscal positions strengthened despite the pickup in government spending.

Most of the fiscal reforms across the GCC focus on deepening non-oil revenue (Figs. 9.12 and 9.13), while raising government expenditure to support economic activity and meet wider medium-term economic goals, including diversification. Nevertheless, the reform measures were more generally periodic with differences between countries. The most significant and coordinated policy was between Saudi Arabia and the UAE with the introduction of VAT. Moreover, Saudi Arabia and Bahrain also implemented another round of energy subsidy reforms. There have been some relatively small moves to boost non-oil revenue, including raising the cost of government fees and introducing new ones. As noted earlier, some GCC countries (Bahrain, Saudi Arabia and UAE) have introduced an excise tax on harmful goods, such as tobacco and sugary drinks. There have only been limited attempts at tax reform across the region, with only Oman making some progress. Oman raised the corporate tax rate from 12% to 15% for all companies in February 2017, eliminating the exemption on the first OMR 30,000 of taxable profits that was previously in place. Oman also sold 10% stake on the Khazzan gas field at the end of 2018 to Malaysia's Petronas, as a way to raise revenue.

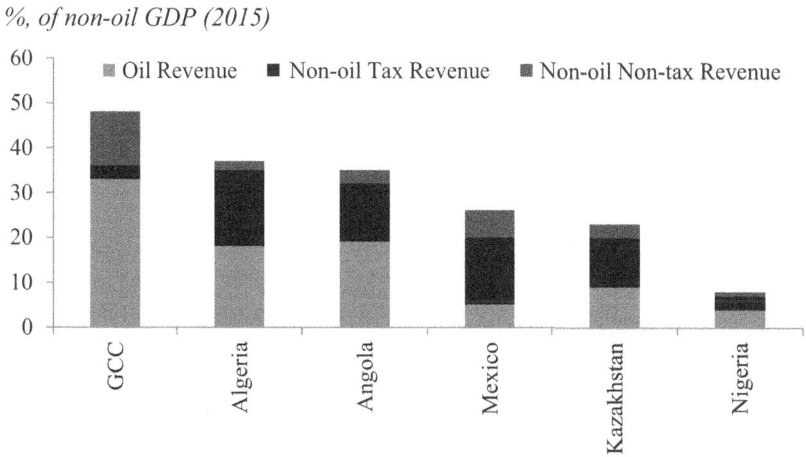


**Fig. 9.12** Global: GCC countries more reliant on hydrocarbon revenues than other commodity producers. (Source: IMF [2016, 9])

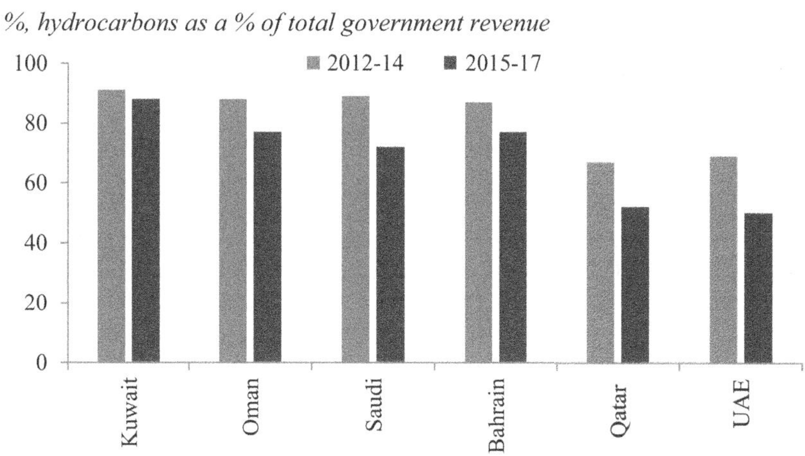


**Fig. 9.13** GCC: Hydrocarbon earnings still dominate revenues; Kuwait and Bahrain see smallest reduction. (Source: Calculated from regional statistical agencies and IMF data)

### 5.4 *VAT Introduction, the Main Reform in Phase 3*

The introduction of VAT is a highly positive development toward deepening and diversifying government revenues and deepening taxable income. The weak tax across the GCC has been a central factor for hydrocarbon revenue's dominance of total revenue despite the recent moves to reduce subsidies and broaden and increase government fees. The GCC states depend on hydrocarbon revenues more than other high commodity-exporting economies, with hydrocarbon revenues still accounting for between 50–85% of total revenues despite the fall in the oil price. We believe that the 5% standard VAT rate has been set to balance raising government revenue with limiting downside pressure on economic activity (by dampening domestic demand). The 5% rate is low on a global basis. The average VAT rate globally is c.15%, with most countries having seen a rise in their VAT rates over time.

The UAE raised AED27 billion in revenue from VAT in 2018—equivalent to around 1.8% of GDP. ADCB estimated that Saudi Arabia should have raised c.1.5–1.6% of GDP in the first year of its VAT introduction (ADCB 2017, 1). ADCB noted that the UAE and Saudi Arabia will likely see the greatest proportional revenue generation in the GCC. In the case of the UAE, this reflects the larger share of private consumption in the economy. Meanwhile, Saudi Arabia has implemented a broad VAT base aimed at maximizing revenue from the new tax, including in areas such as private education and healthcare. Revenue raised by VAT is expected to increase from the second year as private consumption normalizes.

Whilst the introduction of VAT in the GCC was meant to be coordinated to avoid competitive distortions across the region, there are indications that other GCC countries could see delays. Under the Unified VAT Agreement, which outlines a common framework, other GCC countries have 12 months to introduce the tax once the first two have done so. Indeed, by the time of publishing, only Bahrain had introduced VAT in January 2019 linked to its Fiscal Balance Program, which was announced in October 2018 and aims to balance the fiscal budget by 2022. The fiscal program is vital for the GCC five-year support package for Bahrain. However, indications suggest the VAT base is likely to have been narrower than those in UAE and Saudi Arabia, with more areas being zero-rated. For example, oil products would be exempt from VAT in Bahrain.

In Kuwait, the National Assembly's budget committee announced in May 2018 that the government will postpone the implementation of VAT until 2021. Finance minister Nayef al-Hajraf noted in an interview that the government is facing opposition from parliament to VAT (Bloomberg 2018). Qatar has as yet given no indication as to when it plans to implement the indirect tax despite having made solid progress with the framework in 2017. Meanwhile, Oman has announced that it will postpone introduction until 2021, as outlined in the July 2019 Government of Oman bond prospectus (Reuters 2019). Earlier indications suggested that Oman will likely introduce a narrow VAT base that will exclude areas such as food, included in Saudi Arabia and the UAE (ADCB 2018, 9).

### 5.5 *Measures to Support Growth*

However, Saudi Arabia and UAE both introduced measures in 2018 to bolster economic activity, which had been impacted by the fiscal policy. In the UAE, a number of wider factors were also acting as headwinds to the non-oil sectors, including the indirect impact of fiscal reforms in other countries contributing to weak external demand. We believe that the direct support in Saudi Arabia was much greater in 2019, with the measures introduced largely cancelling the fiscal reforms (VAT, subsidy reductions, rise in expatriate fees, etc.). On an individual household basis, we believe that some will not be fully insulated from the impact of fiscal reforms, particularly those working in the private sector and expatriates. In the case of the UAE, the support measures were wider based (i.e. rather than fiscal based), including changes to investment regulation. Thus, we believe that the UAE likely saw fiscal consolidation in 2018 with the introduction of VAT, while in the case of Saudi Arabia it was largely offset by the higher government spending.

In the case of Saudi Arabia, a public sector support package was announced in early January, just days after the introduction of VAT and the reduction in electricity and fuel tariffs. The one-year package was aimed at reducing the burden of the fiscal reforms implemented, with a number of corporates following the government by temporarily boosting wages for some nationals including Aramco and SABIC. The support package included a cost of living allowance and bonus for civil servants and military personnel (see later for more details). Thus the overall effect of the government's fiscal reforms measures in 2018 was muted by the measures that it introduced to shield the national population from the introduction of VAT and lower subsidies. Alongside the handout package, Saudi Arabia also introduced a Citizens Account Program—transfers to support low- and middle-income families—introduced in end December

2017. The program eventually developed into a comprehensive social security platform. Notably, the public sector handout package was extended for a further year and will also be paid in 2019. The preliminary data for 2018 indicated that the government's current expenditure (which includes wages) rose by a significant 14.3% in 2018, up SAR103.2 billion. Investment spending contracted by SAR2.6 billion in 2018, down −1.2% (Fig. 9.14).

In the case of the UAE, a number of economic support packages were announced in mid-2018—on both a country-wide and emirate levels—with additional details being announced subsequently. The support packages look to (1) reduce short-term pressure on corporates; (2) boost the competitiveness of the economy; (3) structurally strengthen the business environment and raise investment levels (domestic and international); and (5) support human capital development and accumulation, amongst others. On a country-wide basis, there have been announcements related to changes to residency and investment laws, including lengthening residency visas for up to 10 years for professionals in key sectors and for foreign investors establishing businesses in the country. Moreover, companies will be allowed to own 100% of their business outside free trade zones (the current limit is 49%) in certain sectors. The various packages also had a number of measures on the fiscal side, inclusive of reductions in fees for government services (Abu Dhabi and Dubai), alongside the outlook for higher spending in areas, such as housing, roads, infrastructure and

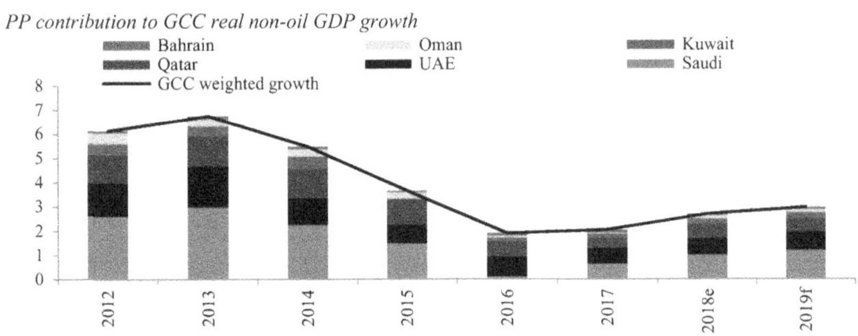


**Fig. 9.14** GCC: Gradual recovery in real non-oil GDP growth from 2017 with modest pickup in government spending and more limited fiscal reforms. (Source: Calculated from regional central banks, statistical agencies and IMF data, ADCB estimates (2019, 4))

healthcare. For more details of the measures introduced to support economic activity, please see Appendix B. There have been more announcements related to these measures in 2019, either providing more details or broadening the measures.

Fiscal developments in 2019 so far indicate an overall continuation in the 2018 fiscal stance, that is, one of expansionary spending stance, measures to support growth, with limited fiscal reforms in 2019, as governments continue to look to support their respective countries' growth outlook over further fiscal consolidation. A number of countries announced expansionary budget, with the UAE and Saudi budgets pointing to the greatest planned increase in spending. Data showed that total government spending in Saudi expanded by 8.5% in 1Q2019. Fiscal reforms in 2019 have remained patchy and relatively limited—such as the introduction of an excise tax in Qatar and Oman, rise in water prices in Oman and higher expat levies (individuals and corporates) in Saudi Arabia. However, the Saudi government announced a relief scheme for some companies to ease the impact of rising labor costs in February 2019. Saudi Arabia also announced new retail fuel prices for 2Q2019 (effective 14 April), with Aramco announcing that retail fuel prices will be set on a quarterly basis, subject to crude oil export prices. The only key exception where we expect to see rising fiscal consolidation and reforms is Bahrain, linked to its medium-term fiscal adjustment program. Meanwhile, OPEC+ countries again decided to cut output from January 2019 to support the oil price, with the production targets extended from July 2019. Nevertheless, the ability to loosen fiscal policy remains limited with the current oil price, though with the variations linked to the hydrocarbon endowment and the BBE oil price.

## 6 Greatest Adjustment by Country

Alongside the variations in the timeline of reforms, there has been varying progress in the pace of reform across the GCC as well. The UAE and to a lesser degree Saudi Arabia have been the most proactive in introducing fiscal reforms. This reflects that other economic, social and political factors are important for the ability of regional governments to progress with fiscal reforms. However, we believe that hydrocarbon endowment will likely play a strong role, especially in terms of implementing a multi-year reform program. Factors likely contributing include the variations in GDP per capita amongst the national population across the region and their ability to absorb fiscal adjustments, especially on a cumulative basis.

The UAE has frontloaded much of the fiscal adjustment and cutback in spending. The pace of subsidy reform in Abu Dhabi and the broadening and raising of government fees have occurred at a steady and staggered pace since early 2015 and early 2018 (with the introduction of VAT). The UAE, with its high hydrocarbon endowment per capita and greater pace of fiscal reform, has one of the strongest fiscal positions among GCC countries, in our view. Moreover, it benefits from the more diversified nature of Dubai's economy, which however was not immune to the fall in the oil price, given the softening in regional demand.

In Saudi Arabia, the pace of reform has been patchier despite announcing a very ambitious fiscal reform program in 2016 for a balanced budget by 2020, though this target was then extended to 2023. There have been two main phases of subsidy reforms in Saudi Arabia—(1) end-2015 and January 2016 and (2) January 2018. The impact of the second round of subsidy reforms on Saudi citizens and the introduction of VAT was dampened shortly after, with the introduction of an allowance package for public sector employees. Moreover, the reduction in public sector benefits in Saudi Arabia was short-lived and was reversed. Overall, this still reflects the sensitivity and cautiousness in implementing fiscal reforms that touch Saudi households (Table 9.3).

The other four GCC countries have seen a more moderate pace of fiscal reform, in our focus time period. There have been no major reforms in Qatar since 2017—following the regional developments, the government's focus is on stabilizing and supporting the economy. In Kuwait, there has been substantial populist opposition to fiscal reform, led by the National Assembly. This is the most independent legislature in the GCC and a central stumbling block for the government to make progress with its fiscal reforms. For Bahrain, wider GCC support was vital for the investment program, given the limited fiscal reforms. However, going forward, the pace of fiscal reforms in Bahrain is expected to strengthen for continued GCC support. GCC countries (Kuwait, Saudi Arabia and UAE) pledged USD10 billion in aid to Bahrain in October 2018, significantly reducing short-term funding pressures. The GCC support package will be spread over five years and will be linked to Bahrain's Fiscal Balance Program, which was announced straight after the aid package and aims to balance the budget by 2022. As noted earlier, VAT was already introduced

**Table 9.3** Components of allowance package, announced on 6 January 2018

| *Monthly cost of living allowance for one year* | |
|---|---|
| • Military and Civil servants | SAR1000 |
| • Retirement Pension and Social-benefit Recipients | SAR500 |
| Social Security Benefit payment | SAR500 |
| Student allowance | 10% increase |
| VAT covered by government for first home purchase for nationals | Up to SAR850,000 |
| VAT covered for private healthcare and education services for nationals | – |
| Payment to military personnel serving in Southern border | SAR5000 |
| Annual Bonus for Military and Civil servants who worked in 2017 | – |

Source: Various media sources, cited in ADCB (2018, 2)

in 2019 and the 2019 draft budget sees a 12% drop in government spending in 2019.

The hydrocarbon poorer per capita countries tend to spend more on wages and salaries as a total share of spending, whilst debt servicing costs have also been increasing with rising debt levels. Moreover, with higher GDP per capita in the hydrocarbon richer per capita countries, their populations can more easily absorb fiscal reforms, though it is also of course a function of other factors, including a change in the social contract. The larger expatriate populations relative to domestic populations in the hydrocarbon richer per capita counties also support greater fiscal reforms. On the other hand, the nationals of the relatively hydrocarbon poorer per capita countries are more impacted by the subsidy reforms, resulting in some cases in greater pushback. Moreover, the need for fiscal reform tends to be greater in these countries, though this can be a positive point in ultimately supporting fiscal reforms going forward.

## 7 Conclusion

GCC fiscal reforms have been much wider than seen in previous oil price downturns, with areas that were previously seen as too sensitive to tackle are open to reform and change. GCC countries have implemented a number of reforms since 2014, alongside a deep pullback in government spending—the main earlier method of adjusting to lower oil price. The reduction in subsidies and the introduction of taxes, including for the national population, has been significant developments in our view. Despite this, oil income continues to be the main revenue source for GCC government and the need for further reforms to deepen tax revenue remains. Thus, while GCC economies continue to be reliant on hydrocarbon revenue and the social contracts remain, we continue to see hydrocarbon richer per capita countries as being the most sustainable in fiscal terms and in the strongest position to withstand a low oil price backdrop. These governments can better support their national populations.

However, the reforms program has not been uniform, either on a country basis or in a steady implementation of the adjustment period. On a country basis, the UAE and Saudi Arabia have seen the greatest pace of fiscal reforms out of all the GCC countries since 2015—one hydrocarbon richer per capita and one hydrocarbon poorer per capita. This indicates that other economic, social and political factors are important for the ability of regional governments to progress with fiscal reforms beyond the hydrocarbon endowment per capita of countries. However, we believe that hydrocarbon endowment will likely play a strong role, especially in terms of implementing a multi-year reform program. Factors likely contributing include the variations in GDP per capita amongst the national population across the region and their ability to absorb fiscal adjustments, especially on a cumulative basis. Indeed, Saudi Arabia implemented a number of measures to limit the impact of reforms, particularly following the introduction of VAT and subsidy reforms in early 2018. In the case of the UAE, fuel prices have been liberalized and utility prices are likely close to market rates. Most of the UAE's fiscal reforms have had to be absorbed by the economy, with limited measures to dampen the impact of reforms. However, the UAE has generally had to raise prices by a lower magnitude than Saudi Arabia, as they were not so low in absolute terms to start with.

The GCC developments highlight the difficulties in implementing a multi-year fiscal reform program during times of low oil prices, whilst also looking to widen the economic base. This is especially the case when the government has largely remained the main driver of economic activity and domestic demand, either directly through government spending and policy or via government-related entities. The narrow economic (diversification) and private sector base highlights the difficulty in implementing a multi-year fiscal adjustment and austerity program. The fiscal consolidation had led to a sharp slowdown in economic activity, which in turn limits the ability of the economy to absorb new fiscal reforms. In many other regions and countries, fiscal policy is often counter-cyclical to the wider economy, whilst, in the GCC, the impact of consolidation is magnified by corporate entities are closely linked to governments and their spending plans. Moreover, with the largely ongoing social contract and the difficulties in the government to increase employment numbers, labor policies are placing additional pressures on the private sectors at times when they are expected to drive growth and support the diversification of the economy.

It remains vital for GCC countries to continue with their fiscal reform programs, to reduce the reliance on the oil sector. Two factors will be central to boosting fiscal sustainability across the region, in our view: (1) the deepening of the tax base and (2) lowering the wage composition of government spending. Hydrocarbon earnings still dominate government revenue and are the main factors driving the overall fiscal position (magnitude of surplus or deficit). GCC tax levels remain low (corporate income tax, individual income tax and VAT) on a MENA and global basis. For some countries, further subsidy reforms are also required. A number of factors will be important for developing the tax base, including reducing the role of the public sector, improving the business climate and a framework for supporting private sector growth. We focus on the issue of labor in our next chapter.

The need to progress with the fiscal reforms is vital, given the ongoing challenges to the oil sector is expected to face, including from improving technology, which will have supply and demand implications. Having a medium- to longer-term fiscal reform program, which progresses at a steady and gradual pace, could help with the implementation. At the same time, wider economic objectives could also be included to improve the sustainability of a fiscal reform framework and raise the capacity of the economies. Focus should also be on targeted measures, spending and investment to support key sectors for diversification, which will be vital in deepening the economy and absorbing further fiscal reforms. Indeed, the earlier period of high government spending growth from the early-2000s did not result in any meaningful economic diversification of the economy and the rise in government spending from 2008 has tended to be on the current side (rather than capital). Measures to attract FDI and strengthen the business environment will also be important. The gradual pace of reform should continue during times of higher oil prices and government spending should be very much targeted towards boosting investment rather than on increasing handouts, though developing an effective social protection framework to support the most vulnerable is important. Greater diversification and further developing the private sector will also be necessary for increasing the employment opportunities for nationals.

# Appendix A: Key Fiscal Reforms by GCC Countries

**Table 9.4** GCC: Summary of fiscal reform measures until mid-2019

| | |
|---|---|
| Bahrain | |
| Apr-10 | Gas price for new industrial customers increased from USD1.3 to USD2.5. |
| Jan-12 | Gas price for existing industrial customers increased 50% from USD1.5 to USD2.25 per MMBtu; price for new industrial customers remains at USD2.5 per MMBtu. |
| Oct-13 | Electricity and water tariff structure adjusted for non-domestic users, increasing tariffs for higher consumption. |
| Apr-15 | Gas price increased for industrial users to USD0.25 per MMBtu; plan for price to reach USD4 per MMBtu by April 2021. |
| Oct-15 | Bahrain removed government subsidies on meat, with the price of beef and chicken more than doubling. |
| Jan-16 | Gasoline prices increased: 60% for super, 56.3% for regular. |
| Mar-16 | Prices of water and electricity increased for expatriates, but subsidized rates continuing for nationals. Move as initial stage of four-year reform program. Will continue to increase effective from March every year till 2019 |
| Dec-17 | Bahrain introduces an excise tax on harmful products on 20 December. Tobacco and energy drinks taxed at 100% and carbonated drinks (ex-sparkling water) at 50%. |
| Jan-18 | Gasoline prices were raised: 25% for super and 12% for regular. |
| Jan-2019 | VAT introduced at 5%. |
| Kuwait | |
| Jan-15 | Kuwait doubles price of diesel; kerosene prices raised—but both these moves were subsequently reversed. |
| Apr-16 | Approves price increases to electricity and water charges for expatriates, but application will not take effect before Sept 2017. |
| Jun-16 | Raised the fees for issuing and renewing work permits. The fee for issuing an initial work permit was raised to KWD50 (from KWD2), whilst the fee for renewing a permit was increased to KWD10 (from KWD2). The cost for transferring a work permit (residence or changing employer) was raised to KWD50 (from KWD10). The decision became effective from June 1. Moreover, the Kuwaiti government agreed in October to partly compensate citizens for raising petrol in September. Nationals holding driver's licenses would be compensated with c.75 liters (20 gallons) a month of free petrol. |
| Sept-16 | Gasoline prices increased on 1 September 2016 by between 40% and 73% depending on the grade of petrol. A government committee will revise the new petrol prices every three months with regard to international oil prices. Despite the September increases, gasoline prices remain below the liberalized market price and most other GCC countries. Moreover, the government agreed from November 2016 to compensate citizens by providing 75 liters per Kuwaiti driver per month. Studies show that average monthly consumption was c. 220–240 liters per month. |

| | |
|---|---|
| May-17 | Kuwait raised electricity and water rates for the commercial sector. The new rates are 5 fils (c. USD0.016) per kWh of electricity, up from 2 fils earlier. Water charges increased to KD2 (USD6.63) per 1000 imperial gallons of water, from KWD0.8 earlier. |
| Aug-17 | Kuwait raised electricity and water rates for the "investment sector" on 22 August. The new rates are 5 fils (c.USD0.016) per kWh of electricity and KD2 (USD6.63) per 1000 imperial gallons of water. This category includes real estate properties and thus will impact the majority of expatriates, who rent properties. The increase is less than the initially proposed charges of KD0.25 per kilowatt and KD0.15 per kilowatt, respectively. |
| Oct-17 | Kuwait increased fees for using public health services for expatriates and visitors. |
| Nov-17 | Water and electricity tariffs of the government sector rose on 22 November 2017. Electricity prices increased to USD0.082 (KWD0.025) per kWh and water prices from USD2.65 (KWD0.8) to USD13.27 (KWD4) per 1000 gallons. |
| Apr-19 | Expatriates in Kuwait will now have to pay a fee of KD10 (USD32.8) per visit to access healthcare services in public hospitals. |
| Oman | |
| 2012 & 2013 | Gas price for industrial users raised by USD0.5 per MMBtu per year. |
| Jan-15 | Price of natural gas sold to industrial sector doubled to USD3 per MMBtu from USD1.5 per MMBtu, to be followed by a 3% increase per year. |
| Jan-16 | Oman raises gasoline and diesel prices—super increased 33.3%, regular 22.8%, diesel 9.6%. From February 2016, fuel prices adjusted monthly in line with international price movements. |
| Mar-16 | The Public Authority for Water and Electricity (PAEW) raised the water prices for government, commercial and industrial bodies to a flat rate of 3.5 baiza (bz) per gallon, up from the earlier 3bz. The adjusted price is still below the average cost of production and distribution of water of nearly 7bz per gallon. |
| Jul-16 | Oman's Public Authority for Civil Aviation (PACA) announces that it will raise fees charged for air traffic through the Sultanate. Outbound airline passengers will have to pay OMR2 extra as tax for every travel from Oman airports from 1 July. |
| Jan-17 | Oman introduced cost reflective electricity tariffs for large industrial, commercial and government consumers from 1 January 2017, effectively ending subsidies for such customers. A higher tariff rate will be applied to approximately 10 K electricity accounts that consume over 150 megawatts per hour each year. Official estimates suggest that this would impact around 1% of users, which together use c. 30% of the electricity output and account for 20% of government subsidies for power generation per year. The Authority for Electricity Regulation (AER) has indicated that the subsidy reform should amount to around OMR100 million, which we estimate at around 1% of 2017 GDP. |
| Feb-17 | General corporate income tax rate raised from 12% to 15%; minimum taxable income exemption of OMR 30,000 removed. |

(*continued*)

**Table 9.4** (continued)

| | |
|---|---|
| Apr-17 | Cost of short stay tourist visas was raised to OMR20 from the previous OMR5. The length of stay for the visa has been increased from 10 days to a month, with further extensions possible. |
| Jan-19 | Oman Oil Company Exploration & Production (OOCEP), a subsidiary of Oman Oil Company (OOC), sold a 10% in the giant Khazzan gas field and project (Block 61) to Malaysia's Petronas. The deal was agreed in October 2018 through the transaction likely completed either in end 2018 or early 2019 |
| Mar-19 | Revised tariffs for drinking water supplied to both residential and commercial establishments announced. Water tariff for residential estates is OMR0.002/gallon for up to 5000 gallons. Once the monthly consumption passes 5001 gallons then the cost will increase to OMR0.0025/gallon. The tariff for government institutions has increased from OMR0.002/gallon to OMR0.0035/gallon. |
| Jun-19 | Excise taxes introduced on harmful goods on 15 June. Tobacco, energy drinks, alcohol and pork products will all be taxed at 100%, with a 50% levy on carbonated soft drinks. Oman revised the imposed excise tax on alcohol to 50% for a six-month period later in June according to the Ministry of Finance. |
| Qatar | |
| Jan-11 | Pump prices of gasoline raised 25% for super 97 octane (to QAR1.1 per liter) and 30% for diesel (to QAR1 per liter). |
| May-14 | Diesel prices raised 50% to QAR1.50 (USD0.41) per liter. |
| Oct-15 | Water and electricity prices raised and tiered according to consumption. |
| Jan-16 | The price of super petrol was raised by 30%, whilst regular petrol increased by 35%. |
| May-16 | Qatar deregulates gasoline and diesel prices. |
| Aug-16 | Passenger leaving Qatar from Doha's Hamad International Airport, including transit passengers, will be charged QAR35 (USD9.61) for using airport facilities. The charge will apply to tickets issued after Aug. 30 and for any travel starting on Dec. 1 onwards and would be used to "further increase the airport's capacity and invest in new infrastructure". |
| Jan-2019 | The prices of alcohol, energy drinks and tobacco products have since risen by 100% and sugary drinks by 50%. |
| Saudi Arabia | |
| Jul-10 | Average price of electricity sold to non-individual users raised more than 20%. Some 43% of total consumption impacted. |
| Dec-15 | Price of water raised to SAR9 (from SAR4) per cubic meter for corporate, industrial and government entities. |
| | Fuel prices raised, including 95 octane gasoline by 50% to USD0.24 per liter and lower-grade 91 octane by 67%. |
| Jan-16 | Prices of natural gas, kerosene, crude oil, heavy fuel oil, ethane and butane increased. |
| | Electricity tariff rates for residential, commercial and government users raised by an average of roughly 43%. Low-consumption users exempt. |

| | |
|---|---|
| Jun-16 | Saudi Arabia's cabinet approved a tax on undeveloped urban land, which had been under discussion for over a year. A 2.5% tax will be levied on plots exceeding 10 K sqm. The Ministry of Housing will specify which land falls under the new tax law, and a fair value for the land will be determined by a committee. Land owners are required to register their plots within six months of the release of the final regulations of the tax. |
| Oct-16 | Fees increased for a number of government services. This included raising the costs of ports, passports, car driving licenses, car transfers to new buyers, traffic violations, renewal of residence permits for domestic workers, and customs tariff protection for 193 commodities. Moreover, the price of a three-month multi-entry visa was increased to SAR500 (which was previously the cost for a six-month one). The new fee for a two-year multi-entry visa was raised to SAR8000 (USD2133). |
| Oct-16 | Changes to public sector wages and benefits:<br>• Salaries of ministers (or those of ministerial rank) reduced by 20%;<br>• A 15% reduction in the annual subsidy granted to Shoura Council members for housing and furnishing;<br>• A 15% reduction in the lump sum paid to Shoura Council members for their car maintenance and fuelcosts for four years;<br>• Overtime bonuses were curbed at between 25–50% of basic salaries;<br>• No annual increment for next year;<br>• Hiring and renewal of contracts for expatriate workers in non-essential sectors suspended;<br>• All appointments in vacant posts halted;<br>• Provisions of vehicles for senior officials suspended until next year;<br>• Annual leave may no longer exceed 30 days;<br>• Monthly transportation allowance for employees during vacation days stopped. |
| Apr-17 | Cuts to public sector wages and benefits introduced in September 2016 were reversed via a royal decree by King Salman bin Abdulaziz Al Saud on 22 April. The decree reinstated all allowances, financial benefits and bonuses to public employees and military staff. Moreover, a two-month salary bonus for forces fighting on the frontline in Yemen was announced. |
| Jun-17 | Saudi Arabia introduces an excise tax on harmful products on 11 June. Tobacco and energy drinks taxed at 100% and carbonated drinks (ex-sparkling water) at 50%. |
| Jun-17 | All allowances and bonuses to civil servants and military personnel that were cancelled in September 2016 are retroactively reinstated in June 2017. This was announced at the same time of Prince Mohammed bin Salman's appointment as crown prince in June 2017. |

(*continued*)

Table 9.4 (continued)

| | |
|---|---|
| Jul-17 | Saudi Arabia also started raising fees on the dependents of expatriate workers from 1 July 2017, at SAR100 per month. The fees will have to be paid in advance at the time of renewal of the work visas. |
| Jan-18 | VAT introduced at 5%. |
| Jan-18 | The Saudi government raised electricity (residential and commercial) and fuel prices on 1 January 2018, with the magnitude of the rise considerably greater than in the first round of subsidy reforms in 2016. The price of 95 octane petrol jumped by 126%, whilst lower grade 91 octane increased by 82% (both including VAT). New residential electricity tariffs will see costs double or triple for lower-use customers compared with the previous rates. |
| Jan-18 | Fee for companies on expatriate labor introduced. For 2018, the fee is SAR300 per month on the number of expatriates equal to nationals and SAR400 per month on expatriates more than nationals. The fee is due to increase in 2019 and 2020. |
| Feb-19 | Ministry of Municipal and Rural Affairs (MMRA) has approved executive regulations for levying fees on various municipal services starting 3 February. Annual fees will be imposed for garbage collection, including waste from fuel stations, apartment buildings, hotels and resorts. Service fees will also be levied for issuing licenses for the construction of new buildings or for the expansion of existing buildings. The regulations will streamline the licensing of cinema halls and theaters to bring them in line with other commercial activities. |
| Apr-19 | Retail fuel prices were raised—by 4% for 95 octane petrol to SAR2.1 per litre (from SAR2.02) and 5.1% for 91 octane to SAR1.44 per litre (from SAR1.37). New prices effective 14 April 2019. Retail fuel prices will subsequently be set on a quarterly basis, subject to crude oil export prices. |
| May-19 | Saudi Arabia has imposed a special tax on electronic cigarettes and sugary drinks, extending similar excise taxes introduced in 2017. A 100% tax would be levied on electronic cigarettes and products used in them, and a 50% tax on sugared drinks. Saudi already had a 100% tax on cigarettes and tobacco products, a 100% tax on energy drinks and a 50% one on fizzy drinks. |
| UAE | |
| 2010 | UAE raises fuel prices at petrol stations twice in 2010 in initial gradual moves to liberalize the market. Gasoline prices rise by some AED0.35 per liter. |
| Jan-11 | Electricity and water consumption tariffs in Dubai increased 15%. |
| Jan-15 | Water and electricity prices raised in Abu Dhabi for nationals and expatriates. Water price for expatriates raised by 170%, electricity by 40%. |
| Aug-15 | UAE deregulates gasoline and diesel prices. Gasoline prices initially increase some 25% (depending on grade); diesel prices cut by some 29%. Fuel prices adjusted monthly in line with international price movements. |
| Jan-16 | Water and electricity prices raised for higher-consumption expatriates in Abu Dhabi. |

| | |
|---|---|
| Apr-16 | Abu Dhabi's ADNOC Distribution increased the retail price of unsubsidized liquefied petroleum gas (LPG) cylinders for April 2016. The cost of an 11.3 kg cylinder was increased by 9.5%, whilst a price of a 22.7 kg cylinder was raised by 10.8%. The new price for LPG will be announced on the 10th of every month based on global prices. |
| Apr-16 | Abu Dhabi introduced a 3% municipal fee—will be charged on the annual value of expatriates' rental contracts. The minimum charge will be AED450 per year, with the fee backdated to the beginning of March. |
| May-16 | Dubai increased parking fees, with the price doubling in certain areas (effective 28 May). Police to start charging for some services. |
| Jun-16 | Abu Dhabi introduces a 4% municipality fee on hotel bills and an AED15 charge per night per room. The fees will be collected by the Abu Dhabi Tourism and Culture Authority (TCA) and added to the government budget of the Department of Municipal Affairs. Dubai introduced a fee of between AED7 for budget hotels and guesthouses and AED20 for five-star hotels and luxury hotel apartments in 2014 per room per night. |
| Jun-16 | Passengers traveling from an airport in Abu Dhabi and Dubai pay an AED35 departure fee from 30 June. Sharjah had announced a similar move in April 2016. |
| Jul-16 | Sharjah Municipality increases tenancy attestation fee to 4%, from 2%. |
| Jan-17 | Water and electricity prices raised for consumers, corporate and government entities. |
| Jun-17 | Abu Dhabi Municipality increases the fares for taxis. The starting fare has been increased to AED5, from AED3.5 earlier. The charge for every kilometer has risen to AED1.82 from AED1.6. |
| Oct-17 | Excise tax introduced on harmful products on unhealthy products—100% tax on tobacco products and energy drinks, with carbonated drinks (ex-sparkling water) at 50%. The tax is expected to raise c.AED 7 billion per annum. |
| Jan-18 | VAT introduced at 5%. |
| Jun-18 | Abu Dhabi residential municipality fees raised to 5% of the value of an annual rental contract for villas and apartments, up from 3%. |

Source: IMF, various media sources, ADCB estimates

## Appendix B: UAE Key Policies Announced to Support Economic Activity

**Table 9.5** UAE: Key policies announced in 2018

| | |
|---|---|
| UAE wide | |
| May 18 | Legislation introduced for a VAT reverse-charge mechanism for the wholesale trade in gold and jewelry to support the diamond and gold sectors. |
| May-18 | A one-month salary bonus for all government employees announced to celebrate 100 years since the birth of the UAE's Founding Father, Sheikh Zayed. The bonus was for all serving and retired government employees (civilians and military) and beneficiaries of social welfare. The bonus was based on the basic monthly salary with a maximum of AED50K and a minimum of AED5K with disbursement made in early-June. The bonus payments totaled AED1.6 billion (USD436 million), which we estimate will equate to c.0.1% of GDP. |
| May-18 | UAE cabinet approved changes to rules governing expatriate residency and foreign ownership of companies. The changes will allow 100% ownership of businesses by global investors and permit residency visas for up to 10 years for professionals in certain sectors and foreign investors. Five-year student visas were also announced with exceptional university graduates eligible for 10-year visas. Reforms are to be introduced by end-2018. |
| Jun-18 | A number of visa reforms were adopted. The previous mandatory deposit of AED3000 per worker in the private sector was replaced by a new insurance scheme that will cost only AED60 annually per worker (effective October 2018). This will reduce the burden on employers by around AED14 billion, according to official estimates. In a further bid to support tourism, transit passengers will be exempt from all entry fees for the first 48 hours of their visit to the UAE. The visa may be extended for up to 96 hours for a fee of AED50. |
| Jun-18 | UAE Central Bank issues new fee caps (43 categories) for retail banking fees to be reviewed annually. |
| Sep-18 | VAT refund scheme for tourists announced, due to be introduced in end-2018. |
| Sep-18 | Retiree visa announced, which will allow expatriates aged 55 and over to obtain a five-year residence visa. The law will come into effect in 2019. Applicants will be required to have (1) an investment in a property worth at least AED2 million; (2) financial savings of no less than AED1 million; or (3) an active income of no less than AED20K per month. |
| Nov-18 | Federal Law issued regarding foreign direct investment (new FDI law) announced in September, which went into force in November upon its publication in the Official Gazette. The new FDI law is the latest in a number of policies implemented to liberalize domestic markets and increase the UAE's competitiveness globally. The new FDI law looks to attract foreign investment in certain specified business sectors to be tentatively identified in early 2019 (positive list), which will be ineligible for majority foreign ownership. |

| | |
|---|---|
| Jan-19 | The UAE Cabinet approved long-term visa system for investors, entrepreneurs, specialized talents and researchers in the fields of science, knowledge and outstanding students. The new system defines two categories for investors: (1) investors in a property of a value of AED5 million or more will be granted a residence for five years; and (2) a renewable 10-year visa will be provided to foreigners with investments in the UAE of at least AED10 million, as long as non-real estate assets account for at least 60% percent of the total. The amount invested shall be wholly owned by the investor and not loaned. Investors can bring spouses and children into the country. Other rules offer five-year visas to entrepreneurs and 10-year visas for scientists and researchers with top qualifications. Outstanding students can stay for five years. The new visas were announced in November 2018 with implementation from January 2019. |
| May-19 | Announces permanent residency framework (Gold Card scheme) for investors and exceptionally skilled professionals. |
| May-19 | Fess for 1500 government services performed by the interior, economy and human resource ministries were cut or amended to help lower the cost of doing business. (effective 1 July) The move was aimed to improve the business environment and boost growth. |
| Jun-19 | Broadening of the 10-year visa program (Gold Card scheme) announced for skilled professionals, from the standard of three years. The criteria for eligibility: (1) monthly salary of at least AED30,000 (USD8174); (2) a bachelor's degree; and (3) at least five years of work experience in the UAE. |
| Jul-19 | The UAE Cabinet has announced 13 broad sectors eligible for up to 100% foreign ownership under a law ratified last November. The list includes 122 economic activities across 13 sectors, including renewable energy, space, agriculture, manufacturing, transport, logistics, hospitality, food services, information and communications, amongst others. Previously, foreign investors could hold up to 49% of a company registered in the UAE, unless it was in a designated free trade zone, and would have to partner with an Emirati investor who would hold the remaining 51%. Each emirate will decide the maximum foreign ownership allowed within its jurisdiction. |
| Jul-19 | The Ministry of Human Resources and Emiratisation (MOHRE) reduced fees for 145 services and transactions, including 50% to 94% cut in fees for 17 services. |
| Abu Dhabi | |
| Jun-18 | Abu Dhabi has announced an AED 50 billion (USD 13.6 billion) package to boost economic growth over the next three years. Abu Dhabi's Executive Council is tasked with preparing detailed execution plans over the following 90 days. |
| Jun-18 | Abu Dhabi cuts the tourism fee applied to hotel rooms and outlets from 6% to 3.5%, whilst the municipality fee was halved to 2%. |

(*continued*)

Table 9.5 (continued)

| | |
|---|---|
| Sep-18 | Abu Dhabi Department of Economic Development started rolling out dual licenses allowing companies in free zones to establish an onshore branch. The first phase is open to companies headquartered in Abu Dhabi and based in one of its free zones. The second phase will allow more companies to qualify for a dual license. The dual licenses are one of 10 strategic initiatives linked to a three-year AED50 billion stimulus package for the emirate. Physical locations will no longer be required for dual licenses, thereby reducing the cost of a dual license by c.80% from a standard license, according to the department. |
| Sep-18 | Greater details about Abu Dhabi's three-year AED50 billion package, called Ghadan 2021 (Tomorrow 2021), were approved. Of this package, AED20 billion will be allocated for 2019, whilst all undisputed private sector dues and receivables will be settled before mid-November 2018. In 1Q2019, three to five projects with a value of over AED3 billion will be announced in areas such as housing, roads, infrastructure and healthcare. A credit guarantee program is expected to be launched later in 2018 to facilitate Dh10bn in financing for SMEs from local banks over the next three years and a PPP law is expected soon. |
| Dec-18 | Service fees for the registration of properties, rental contracts and other transactions at the Abu Dhabi Municipality have been reduced by up to 50%, according to a new government resolution. Abu Dhabi has also exempted all businesses issued with new licenses from local fees for two years as the Capital looks to attract investors. In the new resolution, 75 municipality services fees were cancelled and 23 municipality fees were reduced by 10% to 50%. |
| Jan-19 | Industrial products and materials shipped into Abu Dhabi will be exempt from customs tax from 15 January 2019. The exempted imports include items such as raw materials, machinery, equipment and spare parts. The move forms part of the AED50 billion stimulus program announced in 2018 and aimed at boosting the industrial sector. The new waivers will apply largely to goods headed for the emirate's industrial free zones, such as the Khalifa Industrial Zone and the Khalifa Port Free Trade Zone. |
| Mar-19 | Approved a series of incentive packages worth AED1 billion (USD272 million) to support agricultural technology projects. The incentives aim to encourage the companies to build and grow a presence in Abu Dhabi, establishing the emirate as a global center for desert environment agriculture innovation. |
| Mar-19 | Abu Dhabi Airports Free Zone (ADAFZ) has reduced business setup costs by more than 65%, with the aim of enhancing its regional competitiveness and attracting new sources of Foreign Direct Investment. |
| Apr-19 | All foreigners will be entitled to own the freehold of land and properties purchased in investment zones. Previously, this was only permitted for UAE and GCC nationals, with foreign investors generally limited to leasehold arrangements with 99-year leases. |
| May-19 | State-run Abu Dhabi Investment Office launched an AED535 million fund (USD145.7 million) to support venture capital activities and start-ups. The 'Ghadan Ventures Fund' is to increase the availability of capital to start-ups based in Abu Dhabi and to attract fund managers. This is a part of Abu Dhabi's AED50 billion Ghadan 21 program. |

| | |
|---|---|
| May-19 | Housing loans worth AED3.4 billion (USD0.92 billion) to be distributed to nationals in the emirate as part of the first installment of the 2019 housing program, as well as government houses and residential plots. |
| Jun-19 | Abu Dhabi announced further details of measures linked to its Ghadan 2021 growth support program, initially announced in 2018. The measures announced included: (1) credit guarantees for SMEs; (2) instant business licenses; (3) variable electricity tariffs for the industrial sector; and (4) support for ecotourism in the emirate. |
| Dubai | |
| Mar-18 | Dubai announces no increase in government fees for next three years. |
| Apr-18 | Dubai unveils a new economic plan intended to boost growth, attract new investment and reduce the cost of doing business. New economic plans and measures to boost growth discussed. |
| Jun-18 | Dubai's Executive Council approved a number of measures to boost growth and attract investment, including:<br>• Reducing municipal taxes on businesses from 5% to 2.5%<br>• Waiving 19 fees related to the aviation sector and airplane landing<br>• Waiving 4% fines on late property registration for 60 days and<br>• Freezing private school fees for the coming academic year (2018–19). |
| Jun-18 | Municipality fees on sales at restaurants and hotels lowered from 10% to 7%. |
| Jun-18 | Dubai's Department of Economic Development (DED) launched a package to make it much easier for businesses to clear fines and renew their licenses. The measures will allow businesses to pay their fees and fines in easy installments, freeze their trade license for a year and seek an amicable settlement with DED on commercial violations. |
| Dec-18 | DEWA announced that it will waive charges for new connections of up to 150 kW for commercial and industrial customers for the next two years. |
| Jan-19 | Dubai announced that it would maintain the facility of a 50% reduction in fines for businesses in 2019 |
| Jan-19 | Dubai announced that companies in the tourism industry will soon be able to get back the bank guarantees they had earlier set aside when they started doing their business in the emirate, releasing c.AED250 million. New businesses will no longer need to provide a guarantee before securing a trade license |
| Mar-19 | Dubai Department of Finance launched a second package of economic growth initiatives aimed at boosting economic activity, following the earlier package by DED. The package includes initiatives to support SMEs and public-private partnership. |
| Mar-19 | Dubai Healthcare City Authority (DHCA) announced revised fees for several of its healthcare, commercial, and research and education offerings. The initiative aims to reduce the cost of doing business and enhancing economic growth. |

Source: Various media sources, ADCB estimates

## References

ADCB Economic Research. 2017. GCC Economic Update – Assessing the economic impact of VAT. (Published on 14 June 2017, authors Monica Malik and Shailesh Jha).

ADCB Economic Research. 2018. Oman Economic Update – Previous investment paying dividends and boosting export capacity. (Published on 3 September 2018, authors Monica Malik and Thirumalai Nagesh).

ADCB Economic Research. 2019. GCC Economic Update – Fragile non-oil growth recovery with limited pick-up in momentum. (Published on 13 February 2019, authors Monica Malik and Thirumalai Nagesh).

Bloomberg. 2018. Odd One Out? Kuwait Is Struggling to Pass VAT Agreed by Gulf (published 10 May 2018, authors Fiona MacDonald and Zainab Fattah).

IMF. 2016. Diversifying Government Revenue in the GCC: Next Steps. (Date 26 October 2016) https://www.imf.org/external/np/pp/eng/2016/102616.pdf.

Institute of International Finance. 2018. Hydrocarbon Exporters Breakeven Oil Prices have declined. (Published on 1 February 2018, authors Garbis Iradian and Boban Markovic).

Kingdom of Saudi Arabia, Ministry of Finance. 2016a. 2017 Budget. https://www.mof.gov.sa/en/financialreport/budget2017/Documents/The_National_Budget.pdf.

Kingdom of Saudi Arabia. 2016b. Fiscal Balance Program, Balanced Budget 2020. https://vision2030.gov.sa/en/bb2020.

Kingdom of Saudi Arabia. 2017. Fiscal Balance Program, 2018 Update. https://vision2030.gov.sa/en/bb2020.

Reuters. 2019. Oman to delay VAT to 2021 amid sluggish growth (published 30 July 2019, author Davide Barbuscia) https://www.reuters.com/article/us-oman-economy-vat/oman-to-delay-vat-to-2021-amid-sluggish-growth-idUSKCN1UP1TM.

CHAPTER 10

# GCC Fiscal Reforms and Labor Market Policies

*Monica Malik and Thirumalai Nagesh*

## 1 Introduction

The social contract between the government and the population in GCC countries is central to the issues of fiscal reform and adjustment to the lower oil price, as we highlighted in our earlier chapter. The social contract is based on a framework of distribution of hydrocarbon earning and wealth to the national population, including areas, such as employment, healthcare, education, housing and subsidized fuel and utilities. A significant fiscal reform program, both on the revenue and expenditure sides, will result in a re-defining of this contract. The ability to continue to provide the same degree of support by GCC governments to the national population is both impacted by the fall in hydrocarbon prices from end-2014 and the strong population growth over the last few decades. However, as noted earlier, there are marked variations on the ability to support the national population based on hydrocarbon endowment per capita.

M. Malik (✉) • T. Nagesh
The Hills Compound, Abu Dhabi, United Arab Emirates
e-mail: monica.malik@adcb.com; Thirumalainagesh.Venkatesh@adcb.com



G. Luciani, T. Moerenhout (eds.), *When Can Oil Economies Be Deemed Sustainable?*, The Political Economy of the Middle East,
https://doi.org/10.1007/978-981-15-5728-6_10

One of the main pillars of the social contract is the role of the public sector in the employment of nationals, whilst conversely, the private sector has benefited by the ability to employ cheaper expatriate labor. Historically, with the rise in oil revenue, particularly from the 1970s, though also earlier, the influx of expatriate labor has been central for the quick transformation of the economy and the development of infrastructure. The IMF noted that productivity growth in the region has been unsatisfactory, and the strong economic expansion has been fuelled by low-cost foreign labor. This expatriate labor force has also been highly flexible and elastic, adjusting to economic growth cycles. Meanwhile, the public sector employment of nationals was central to the quick rise in the living standard, alongside the deployment of wider social distribution. The IMF noted that for the Middle East Region "countries in the region have used public employment and compensation policies to achieve multiple socioeconomic objectives, including employment and redistribution of wealth" (IMF 2018, 9). Public sector employment has generally been better paid than the private sector, with the tendency for jobs for life, generous pension schemes and perks, such as greater holiday benefits (Figs. 10.1 and 10.2).

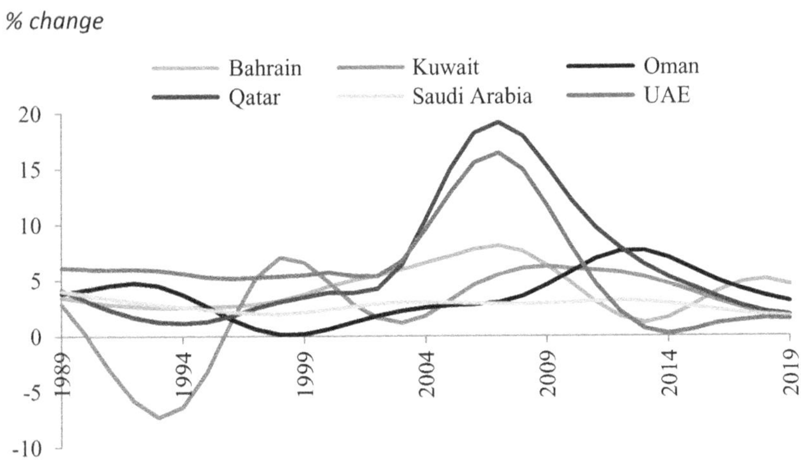


**Fig. 10.1** GCC: Total population growth reflects economic cycles and the flexibility of foreign labor. (Source: United Nations)

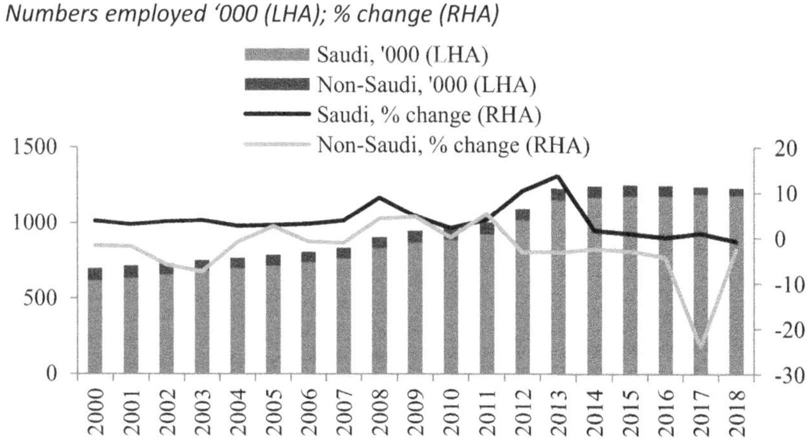


**Fig. 10.2** Saudi Arabia: Employment dynamics in the public sector. (Source: SAMA)

With the lower oil price in the 1990s, GCC governments increasingly turned policy discussion towards diversification of the economic base and private sector development, in part to create job opportunities for nationals outside the public sector. Moreover, there were a number of initiatives to raise private sector employment directly, such as labor nationalization programs, and indirectly, educational reforms. However, the preference to work in the public sector remained, and with the sharp increase in the oil price in the 2000s, government hiring and spending on wages rose. Again, the region saw strong expatriate population growth as the higher oil revenue supported an upgrading of infrastructure, alongside wider spending. The IMF noted that the rise in the public sector wage bill was attributable to both increase in government employment and high public sector wages (IMF 2014, 13). Furthermore, the IMF highlighted that GCC governments' public sector wage bills are considerably larger than in other oil-exporting countries. Since 2014, the emphasis on private sector developments and diversification has reemerged and is being stressed, though it had remained a central tenet of medium-term development objectives (Figs. 10.3 and 10.4).

The need to create jobs for the youth population and, in some cases, to reduce the current levels of national unemployment remains one of the key focuses of domestic policy and arguably the greatest economic

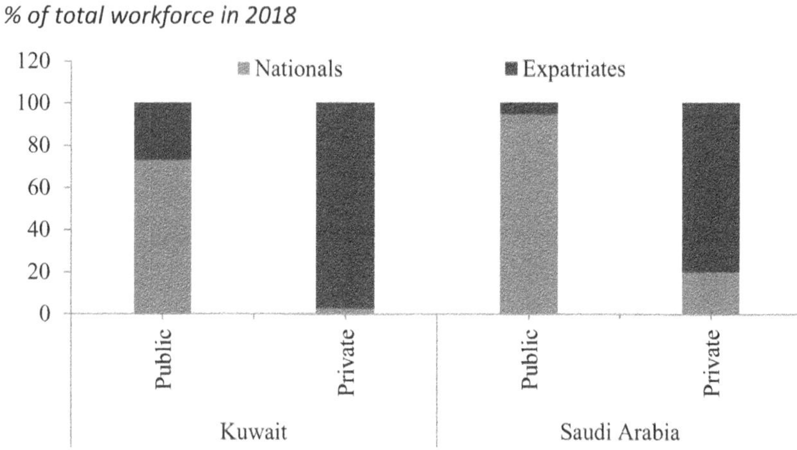


**Fig. 10.3** Saudi Arabia and Kuwait: Significant variations in composition of public and private sector labor force. (Source: PACI, Saudi Arabia General Authority of Statistics)

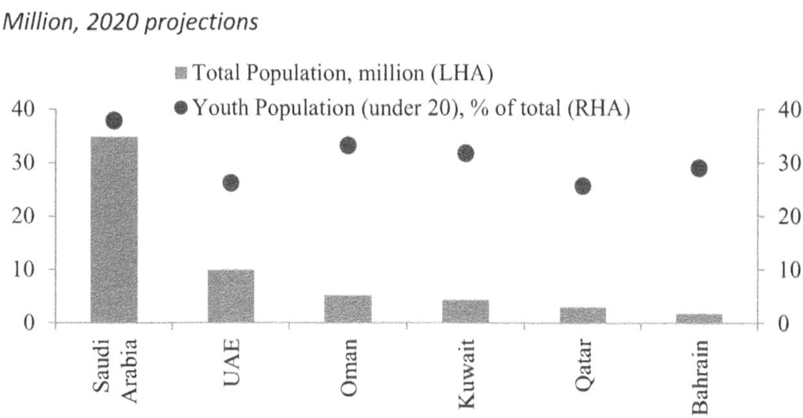


**Fig. 10.4** GCC: Large youth population requires strong job creation. (Source: United Nations)

challenge. We had argued in one of our earlier research reports on the Vision 2030 plan that "Ultimately, creating jobs for younger people and diversifying the economy will be the key litmus test to determine whether or not the transformation process has been successful" (ADCB 2017b, 3). Youth unemployment is already high in some countries, notably in Saudi Arabia and Oman, which are both relatively hydrocarbon poorer per capita countries with less diversified economies. The World Bank noted that "The main social concern for Oman is the lack of jobs and the adverse effects of subsidy reform on vulnerable households (World Bank 2018). Indeed, the greater social pressures in these two countries are reflected in their recent labor and fiscal policies. The national unemployment rate in the hydrocarbon richer per capita countries is lower, with a relatively small population which can more easily be absorbed by the public sector. Nevertheless, the young demographic profile across the region points to the need for strong job creation over the next five to ten years.

In this chapter, we develop the themes from our previous one—on hydrocarbon endowment per capita and fiscal reform—focusing on the context of population dynamics and the labor market. We note that the fiscal reform program this time has impacted the national population, making it significant and broader than the previous rounds. Nevertheless, any changes in the social contract were always going to be gradual and take time to redefine. There have been instances of measures introduced to mitigate the impact of the reforms on the domestic population and reduce short-term social pressures. Alongside the fiscal reforms, a number of labor reforms have also been introduced to support job creation in the private sector for nationals. With the rise in oil price from 2017, there have also been some tentative signs of greater hiring by the government, alongside raising support in other areas, such as housing (Figs. 10.5 and 10.6).

## 2 National Population Impacted by Reforms; Greatest Burden on Expatriates

The broader nature of the fiscal adjustment has meant that the national population has been directly affected, making this reform cycle notable, especially compared to the previous one. The national population has been impacted by measures, such as the higher prices linked to the reductions in subsidies (fuel and utilities), the introduction of VAT and the excise tax. The fact that fiscal reforms have included areas that were

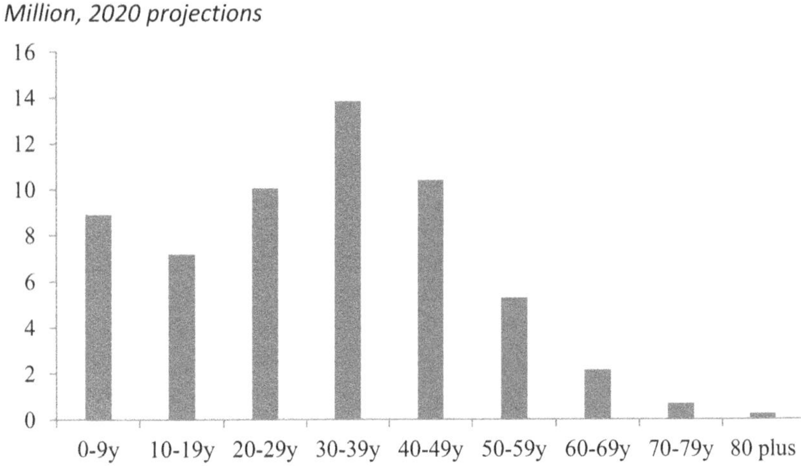


**Fig. 10.5** GCC: Demographic profile of GCC population dominated by young and working age population. (Source: United Nations Population Division)

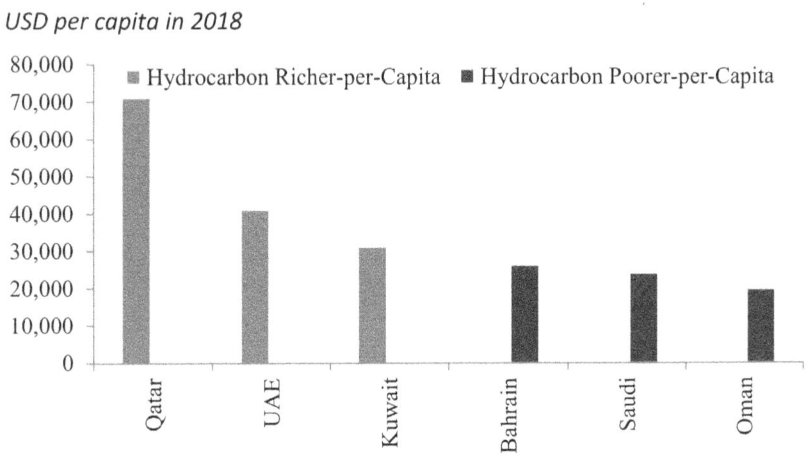


**Fig. 10.6** GCC: UAE and Qatar have higher GDP per capita among the GCC countries. (Source: IMF, WEO April 2019)

previously seen as being too sensitive to change or tackle is a significant step, even though the social contract remains in place. This potentially opens the door for deeper and wider fiscal reforms going forward in the medium- to longer-term. Nevertheless, overall a greater burden of the fiscal reforms has fallen on the expatriate population and in some cases corporates, with the national population relatively more shielded, especially from 2017. With the corporate sector impacted by fiscal reform, it raises questions over the short-term ability of the private sector to drive real GDP growth and support the diversification of the economy. In some cases, the private sector has faced additional pressure due to labor market reforms. The development of a framework to boost private sector activity, including through PPP schemes, remains vital.

The trend to target more of the reform burden at expatriates was seen in a number of instances in the region, including in the last two rounds of utility subsidy reductions in Abu Dhabi. In January 2016, utility prices were increased just for higher-consumption expatriates. In January 2017, prices were raised for both expatriates and nationals, alongside corporate and government entities, though with the cost markedly higher for expatriates than nationals in absolute terms. According to the official statement announcing the January 2017 utility price changes, the amended tariffs reflect the actual cost of supplying water and electricity to all categories of customers. This implies that prices have been largely liberalized (Table 10.1). In Kuwait, utility reforms of August 2017 targeted corporates, government entities and the investment sector. The investment sector includes real estate, and thus the higher prices affect the majority of expatriates who rent properties. When presenting the draft budget for FY2018–2019, the Kuwaiti minister of finance noted that the main reform "starts with curbing spending, while maintaining a healthy rate of capital expenditure on infrastructure and minimizing the impact of our fiscal reforms on citizens".

As mentioned in our previous chapter, the Saudi government moved to reduce the impact of price increases on nationals following the rise in utility and fuel prices and the introduction of VAT on 1 January 2018. A package of hand-outs for public sector employees was announced within one week of the 1 January price increases and came after signs of disappointment with the latest austerity measures and after the cash handouts from the Citizens Account introduced at end-December 2017 fell short of covering rising living expenses. The Citizens Account is a means-tested cash-transfer program to support low- to middle-income households from

**Table 10.1** Abu Dhabi: Water and electricity tariffs for households

| | *Pre Jan-2015* | *Jan-15* | *Jan-16* | *Jan-17*[a] |
|---|---|---|---|---|
| *Nationals* | | | | |
| Power usage below 30 KwH/day for apartments and 400KwH/day for villas | 5 fils/ kWh | | | 6.7 fils/ kWh |
| Power usage above 30 KwH/day for apartments and 400 KwH/day for villas | 5 fils/ kWh | 5.5 fils/ kWh | | 7.5 fils/ kWh |
| Water usage below 700 liters/day in apartments and 7000 liters in villas | AED0 | AED1.7 | | AED2.09 |
| Water usage above 700 liters/day in apartments and 7000 liters in villas | AED0 | AED1.89 | | AED2.6 |
| *Expatriates* | | | | |
| Power consumption below 20 KwH/day in apartments and up to 200 KwH/day in villas | 15 fils/ kWh | 21 fils/kWh | | 26.8 fils/ kWh |
| Power consumption above 20 KwH/day in apartments and up to 200 KwH/day in villas | 15 fils/ kWh | Not announced | 31.8 fils/ kWh | 30.5 fils/ kWh |
| Water usage below 700 liters/day in apartments and 5000 liters in villas | AED2.2 | AED5.95 | | AED7.84 |
| Water usage above 700 liters/day in apartments and 5000 liters in villas | AED2.3 | AED9.9 | AED10.55 | AED10.41 |

Source: Reuters, various media sources, cited from ADCB (2017a, 11)

[a]Price increases effective 1 January 2017

the negative consequences of fiscal reform. The first payment was on 21 December 2017 ahead of the introduction of VAT and the second round of subsidy reforms. The swift response by the government in announcing the handout package is particularly notable given the wider reforms (including social), alongside developments, such as the anti-corruption drive in November 2017. We argued in our earlier research that economic reforms "are creating and redefining the economic, social, cultural and political structures in the Kingdom, as well as their relationships to each other" (ADCB 2017b, 2). As such, the government was quick to announce a package of support measures to reduce the impact of fiscal reforms for some nationals (particularly in the public sector) at a time of significant social, economic and political change. ADCB estimated that c.70% of Saudi households likely benefited from the government allowance packages in 2018, with c.45% likely seeing higher net income from the Citizen Account program (low-income Saudi households) (ADCB 2018, 1). Thus, expatriate households will be the most impacted by the reforms, especially when considering the levy on dependents from mid-2017 (see below). Notably, the public sector handout package for citizens (introduced in January 2018) was extended for 2019, alongside annual public sector bonuses for all state employees from the start of 2019.

In Oman, the government announced mitigation measures to support the vulnerable population in 2018. These included i) allocating OMR100 million (USD260 million) to support vulnerable households, ii) a new fuel subsidy scheme where households with an income below OMR600 will receive 200 liters of petrol per month at a subsidized rate (World Bank 2018). As highlighted in our earlier chapter, Oman has seen relatively limited fiscal reforms in this cycle, which perhaps reflects the greater social pressures linked to high unemployment rate, especially amongst the youth. Oman has focused on progressing with projects, in part to create jobs, alongside maximize revenue from each barrel of oil and raising gas output. Outside the hydrocarbon and petrochemical side, key areas of investment include logistics (ports, airports, etc.), tourism, utilities for industrial needs and other related infrastructure (road and rail). Foreign investment has been vital for supporting the investment drive, whilst also relying on raising debt and drawing down FX reserves to help cover the fiscal deficit. Out of the GCC countries, Oman has made the most progress with broadening the economic base since 2014. Bahrain also has made limited progress until 2019. These additional fiscal support packages, announced for public sector employees, reinforce the preference for nationals to work for government entities. Support measures have also been seen in other areas of the GCC. However, the UAE unified holiday between the private and public sectors in March 2019, with the aim of enhancing the attractiveness of the private sector for Emiratis.

## 3 Hydrocarbon Endowment Per Capita and the Labor Market

Alongside the fiscal reforms, a number of policy measures have been introduced targeting the labor market to increase job opportunities for nationals in the private sector. Private sector job creation will be vital for (1) reducing social pressure, (2) the number of jobs required to be created by the public sector and (3) the ability to introduce further subsidy and tax reforms going forward. Some countries have introduced or widened policies that target reducing the size of the foreign workforce, with the aim that these jobs will be replaced by nationals. The success of the programs has varied, though they have not addressed underlying factors, such as the cost of labor and skills mismatches. Within the GCC, the labor market pressures differ greatly, again reflecting the differences in hydrocarbon endowment per capita. The lower share of government spending on wage growth in the UAE and Qatar relative to other GCC countries (Figs. 10.7 and 10.8) reflects the ability for the government to absorb the domestic labor force and support spending in wider areas, bolstering non-hydrocarbon economic activity. Reflecting this, expatriates make up a greater proportion of total population in hydrocarbon richer per capita countries versus the relatively poorer per capita ones. In the case of the UAE, the more diversified economic base also lowers reliance on government jobs. In the case of Bahrain, the relatively more diversified economic backdrop supports the greater share of foreign employment in the labor force, versus Saudi and Oman. Kuwait stands out amongst the countries with higher hydrocarbon endowment per capita. We believe that the relatively high share of government spending to the total likely mirrors (1) the preference and ability of the government to employ nationals, (2)

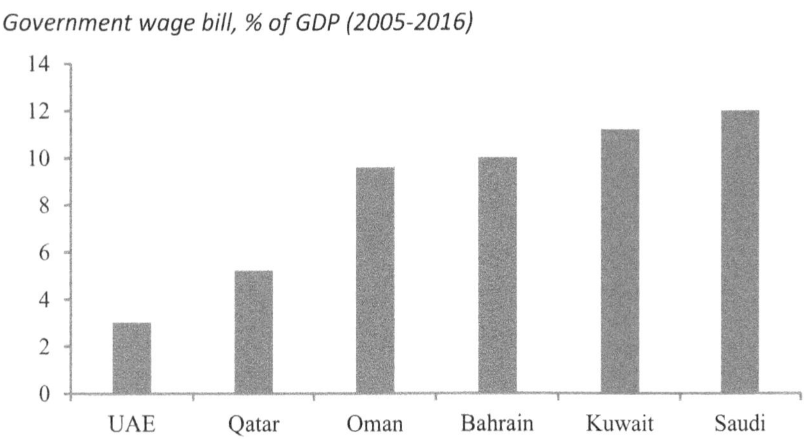


**Fig. 10.7** GCC: Government spending on wages high in the region, though lower in the UAE and Qatar. (Source: IMF (2018, 9))

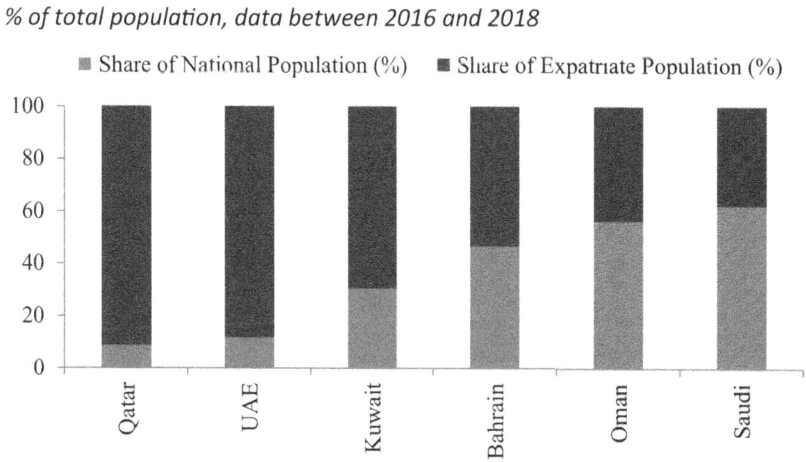


**Fig. 10.8** GCC: Hydrocarbon richer per capita countries tend to have larger expatriate population in relative share. (Source: Calculated from regional statistics agencies and World Bank for UAE data)

relatively limited progress with the government investment program and (3) a less diversified economy. Government subsidies, handouts and public sector wages in Kuwait are some of the most generous in the region.

All GCC countries already had employment nationalization program to some degrees. The hydrocarbon richer per capital countries generally have lower employment targets and quotas for nationals in the private sector. The need for job creation among nationals remains high in the hydrocarbon poorer per capita countries, with the public sector unable to absorb all job seekers. Saudi Arabia and Oman have particularly focused on policies to replacing foreign labor with national in this round of reforms, in line with their challenging employment positions, and particularity from 2017 (Table 10.2).

The latest data for Saudi Arabia show that the unemployment rate for nationals stood at 12.5% in 1Q2019 (male at 6.6% and female at 31.7%), up from 11.5% in 4Q2015. The youth unemployment rate is significantly higher, at 36.3% in 1Q2019 for the 20–24 age range (ADCB 2019, 1). Going forward, Saudi Arabia could face added demand for jobs from greater female participation, which stood at around 20.5% in 1Q2019. The Saudi government has introduced a number of measures to support

**Table 10.2** GCC: Summary of key labor market nationalization policies

| | *Saudi Arabia* | *UAE* | *Kuwait* | *Oman* | *Bahrain* | *Qatar* |
|---|---|---|---|---|---|---|
| Minimum quota of nationals | Typically 30% in private sector, rising to 100% for certain sectors depending on company size | Only outside of free zones. 2% for commercial entities, 4% for banks, 5% for insurance companies. Target to increase Emirati employment to 6% of total workforce by 2021 | 100% in public sector by 2028. Typically 30–60% in the private sector, although as low as 3% in some industries (manufacturing, agriculture) and as high as 66% (banking) | Typically 35% for private sector, 100% for public sector roles | Typically 20–35% for most businesses | Target of 20% by 2030, limited enforcement |
| Levies on expat workers | Companies employing more foreign nationals pay a monthly fee of SAR600 per foreign national employee, falling to SAR500 if they employ a majority of Saudi nationals | No | No. remittance tax under discussion | 7% levy on foreign workers' wages for training programs for nationals | Yes. (USD26.50, or USD13.25 for small businesses) | No |
| Fees on expat dependents | Expat levy (SAR300) rising to SAR400 by 2020 per dependent | No | No | No | No | No |
| Visa restrictions | Bans in sectors such as retail implemented in 2018, but partially relaxed since | No | All expats under 30 (initially from July 2018, currently postponed) | Ban on expat visas in 87 professions | No | No |

Source: Moody's Investors Service (2019, 9)

female employment, including setting quotas for certain sectors, as well as introducing childcare programs. In Oman, the World Bank (citing ILO data) indicated that unemployment "was 17% in 2017, while youth unemployment is approximately 49% — a pressing challenge in Oman where over 40% of the population is under the age of 25" (Figs. 10.9 and 10.10).

## 4 Labor Market Reforms and Policies to Reduce Foreign Labor

Below we highlight some of the main labor market reforms introduced in this economic cycle. There are also some signs of greater public sector hiring in Saudi Arabia, potentially reflecting the fact that the private sector is not generating sufficient jobs to reduce current unemployment levels meaningfully. It could also reflect the hiring by institution driving the transformation program such as the PIF, alongside the development of new institutions around key diversification projects. The number of Saudis employed in the civil services was, however, lower than in 1Q2016 by some 4.2 K—though insufficient to reduce the government wage bill, especially when considering the pay increases. Meanwhile, regarding the

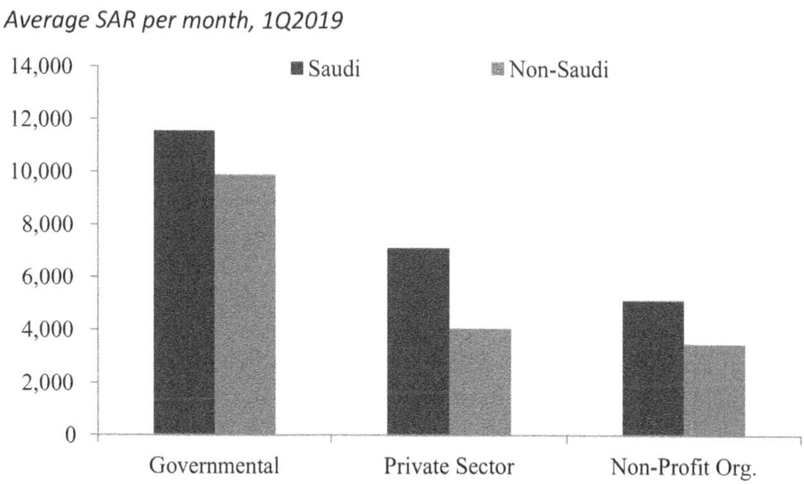


**Fig. 10.9** Saudi Arabia; Higher wages and benefits in the public sector support demand for government jobs. (Source: Saudi Arabia General Authority for Statistics)

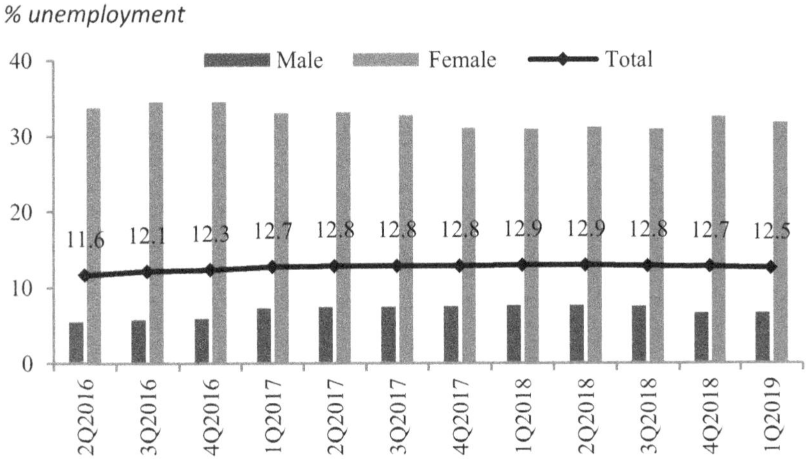


**Fig. 10.10** Saudi Arabia: Unemployment rate amongst Saudi nationals remains high, especially females. (Source: Saudi Arabia General Authority for Statistics)

hydrocarbon richer per capita countries, Kuwait has seen the most populist pressure to reduce foreign employment levels, despite the low unemployment level and generous public sector packages, even on a GCC basis.

**Saudi Arabia**: The government has increased the number of sectors that require 100% Saudiization, particularly in the retail sector. This started with 100% Saudiization in mobile phones retail shops in September 2016, followed by car rental companies in March 2018. Also, in early 2018, Saudi Arabia's Ministry of Labor issued a decree to implement 100% Saudiization in 12 new, mostly retail areas, including those concerned with watches, eyewear, medical equipment, electrical and electronic appliances, auto parts, building materials, carpets, cars and motorcycles, home and office furniture, children's clothing and men's accessories, home kitchenware and confectionery. Implementation started at end-September 2018 in three phases (to end by January 2019) though the Saudiization target was scaled down to 70%, from the 100% initially outlined (Albawaba 2018). Notably, the move to nationalize jobs in areas such as retail indicates that Saudis are accepting non-managerial positions, which perhaps would not have been the case in previous decades.

Saudi Arabia also announced its decision to Saudiize management and specialist professions in the tourism hospitality sector in July 2019. The implementation will come into effect starting 28 December 2019 for operating and specialized jobs, 22 June 2020 for supervisors and assistant managers, and 16 December 2020 for managers. The decision includes hotels classified as 3-star and above, resorts, hotel suites, and villas classified as 4-star and above. Jobs in booking, marketing and front offices will be 100% Saudiized, whilst 70% will be required for sales managers and events and conferences sales managers. Tourism and hospitality are expected to be key sectors of jobs creation, going forward with a number of developments planned (NEOM, Red Sea, Al-Ula, Amaala, etc.). These Saudiization measures are likely aimed at ensuring that a significant part of new jobs being created as this sector develops go to nationals.

Moreover, the Saudi government has also raised fees and levies on foreign labor aimed at achieving two policy goals: (1) raising more non-oil revenue and (2) creating job opportunities for Saudis in the private sector. Saudi Arabia introduced a tax in July 2017, whereby expatriates have to pay SAR100 per month per dependent, including children and domestic staff. The fee was expected to raise SAR1 billion in 2017. These were some of the main fiscal reforms introduced by Saudi Arabia in 2017, alongside an excise tax on harmful products. This fee will rise by SAR100 per year from 2018 to 2020 and will mostly affect professional expatriates. Saudi Arabia also announced a levy on expatriate employees at companies from January 2018, where foreign labor is equal to or exceeds national employees. Again the fee will rise on an annual basis, though by SAR200 per year. The wage (and skills) gap between nationals and expatriates will likely not be bridged by this levy and thus will be an additional cost for companies to absorb (Table 10.3) (Figs. 10.11 and 10.12).

**Oman**: To support job creation for Omanis, a six-month ban on hiring expatriates began in January 2018 in 10 different industries, including media, information technology, marketing, insurance and aviation. The expatriate visa ban was extended for an additional six months at end-May, and extra sectors were added, including carpentry, metal, aluminium workshops and brick factories. The Omani government has also been looking to increase the number of nationals in the public sector. The visa ban was extended for a fourth time in mid-2019, again with four more

**Table 10.3** Schedule of fees and levies on expatriates

| | *Dependents for Expats*[a] | *No. of Expats equal to no. of Saudis in a company*[b] | *Expats more than Saudis in a company*[b] |
|---|---|---|---|
| 2017 | SAR 100 per month from July onwards | | |
| 2018 | SAR 200 per month from July onwards | 300 per month from January onwards | 400 per month from January onwards |
| 2019 | SAR 300 per month from July onwards | 500 per month from January onwards | 600 per month from January onwards |
| 2020 | SAR 400 per month from July onwards | 700 per month from January onwards | 800 per month from January onwards |

Source: Saudi Arabia Fiscal Balance Program 2020 (2016b, 51)

[a]To be paid by foreign nationals sponsoring dependents

[b]fess on corporates

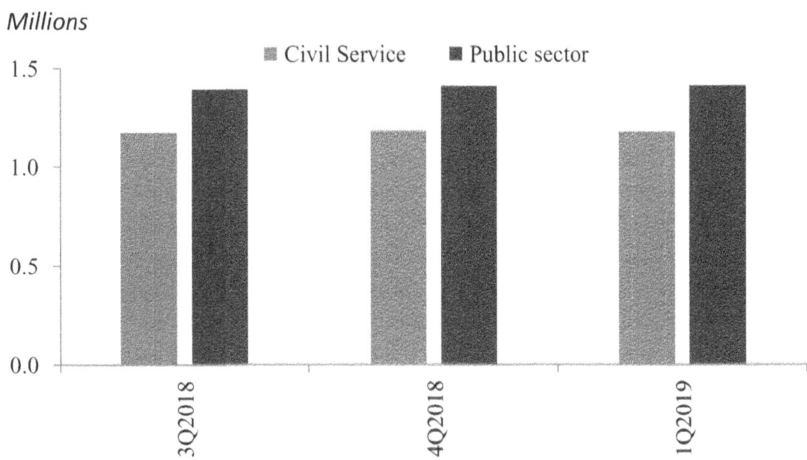


**Fig. 10.11** Saudi: Public sector employment seen rising though civil services hiring remains broadly steady. (Source: Saudi Arabia General Authority for Statistics)

categories added. Omanization quotas have also been increased in certain sectors, including industry and tourism. On 3 October 2017, Oman announced that 25 K jobs would be created for nationals in both the private and public sectors starting from December.

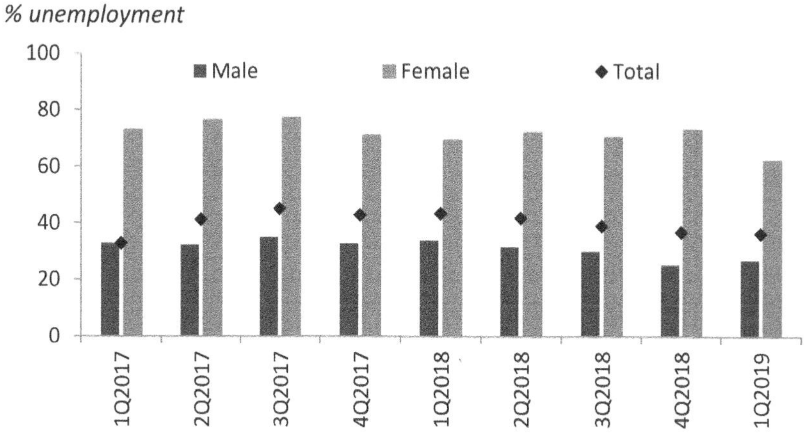


**Fig. 10.12** Saudi: Unemployment rate amongst Saudi youth (20–24 years) high at 36.3% in 1Q2019. (Source: Saudi Arabia General Authority for Statistics)

**Kuwait**: Some resentment against the expatriate population has been visible in Kuwait, including among some MPs in the National Assembly looking to capitalize on growing resentment against non-nationals. MP Safa Al-Hashem was quoted as saying that "Before asking citizens to pay, the government should reform the population mix by levying taxes on foreigners" (Bloomberg 2017). The populist belief is that lowering expatriate numbers will help to reduce the subsidy burden. The government introduced a framework in 2017 to reduce the number of foreign workers in ministries and government institutions to create jobs for nationals. According to Kuwait Times, some 2799 expatriate jobs were terminated from the government sector in the first year of implementation of the Kuwaitization push to replace expat employees with nationals (Kuwait Times 2018). The government issued a decree to reduce the number of expatriates in all government departments by fiscal year 2019–2020 to reach a 90% Kuwaitization rate (Arab Times). However, the president of the Civil Service Commission indicated that Kuwait is not in a position to replace all expatriates working in the public sector due to a shortage of potential Kuwaiti employees (Gulf News 2018). Meanwhile, a ban on recruiting expats under the age of 30 holding diplomas and higher degrees came into effect in July 2018 in an attempt to increase the hiring of younger Kuwaitis, though this has been currently postponed (Moody's 2019, 9).

Elsewhere, Abu Dhabi also announced plans to create at least 10 K jobs for Emiratis in the private and public sectors over the next five years in June 2018, as part of a multi-year stimulus plan (Ghadan 21). Dubai's 2019 budget looks to create 2498 additional positions. In Bahrain, the cabinet approved hikes in fees imposed on foreign labor licenses under a job nationalization program to BHD500 (from BHD200) in February 2019 within the draft budget (Banker Middle East 2019). It also proposed higher unemployment benefits for Bahrainis. On the other side, a key component of Bahrain's Fiscal Balance Program is to introduce a voluntary retirement scheme to reduce government spending on wages. However, the progress with this is likely to be gradual.

GCC governments have also looked to progress with housing and education projects for nationals, alongside schemes to promote housing loans (via the government entities or in conjunction in the private sector). A number of housing projects have been announced in the UAE for nationals. In Saudi Arabia, an objective of Vision 2030 is to raise home ownership by nationals and a number of affordable housing programs have been introduced. In January 2018, SAMA raised the maximum loan-to-value rate for mortgages for first-time homebuyers to 90% from 85% in an effort to stimulate mortgage lending. Banks are also partnering with Real Estate Development Fund (REDF) to help develop the affordable housing segment, with the REDF providing support to eligible customers in the form of down payments, guarantees or full or partial payments on the interest on the mortgage.

## 5 Pressure on Expatriate Population and Economic Impact

Oman and Saudi Arabia have particularly seen large declines in expatriate populations in the past few years. In the case of Saudi Arabia, the data points so far to a relatively limited number of jobs being created for nationals, especially given the magnitude in the contraction of expatriate employment. In the case of Oman, official statement indicates some positive trends in private-sector job creation, suggesting that there has been some success in Omani nationals replacing expatriates. However, a question remains over labor policy and the private sector's ability to expand and drive economic activity and create new jobs in the medium to longer term. Looking ahead, the hydrocarbon weaker per capita countries will continue to face the twin challenges of narrowing the fiscal deficit and ensuring that the economy creates sufficient job opportunities for the youth.

The expatriate labor force in Saudi Arabia declined in 2017 and 2018 with the dependent's fee, corporate levy on expatriate labor and rising Saudiization requirement. Nevertheless, as noted earlier, the unemployment rate for Saudi nationals has remained high. Around 1.53 million expatriates left the labor market in 2017 and 2018 combined, whilst the number of Saudis employed in the private sector rose by just 52.9 K over the same period. The rise in expatriates jobs in 1Q2019 by some 223.8 K to 9.65 million was solely driven by domestic workers, with foreign labor working in the private and government sectors continuing to decline (ADCB 2019, 1). The fact that expatriate workers are not always being replaced by nationals partly reflects the skills mismatch among domestic workers and the higher costs of national labor (ADCB 2018, 7). However, there are signs of a rise in public sector employment of Saudis from 3Q2018 to 1Q2019 (period data available) (Figs. 10.13 and 10.14).

In Oman, official statement indicated that 64,386 jobs were created for Omanis in the private sector in 2018 amid the expatriate visa ban. Meanwhile, official data show that a total of 67.4 K expatriates left the

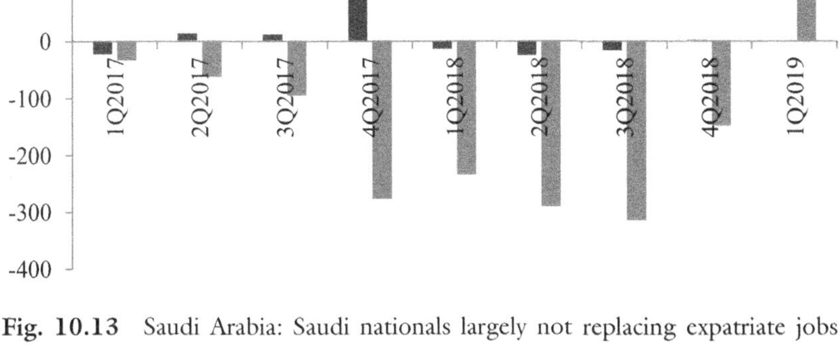


**Fig. 10.13** Saudi Arabia: Saudi nationals largely not replacing expatriate jobs lost. (Source: Calculated from Saudi Arabia General Authority for Statistics data)

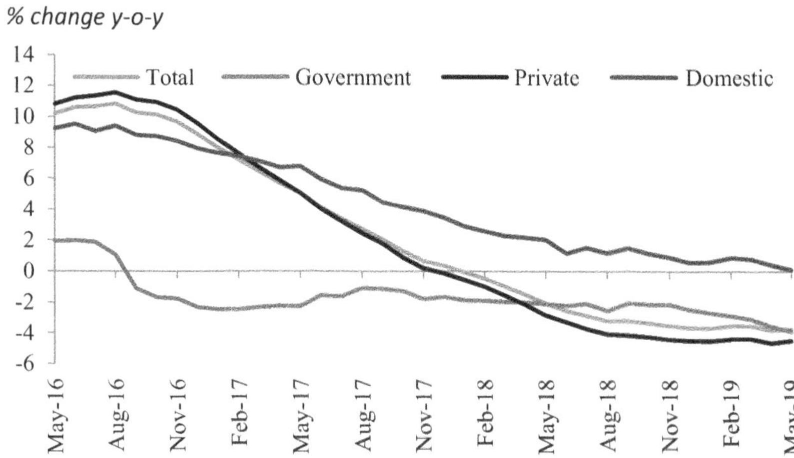


**Fig. 10.14** Oman: Expatriate labor force contracted in 2018 with visa ban on a number of sectors. (Source: Calculated from Oman's Ministry of National Economy data)

private labor force in 2018, indicating that Omani nationals have largely been replacing expatriates in this year. The latest data continues to point to a further decline in the expatriate labor force in 2019, contracting by 3.7% y-o-y in May (public sector: −3.9%; private sector: −4.5%). The sector seeing the greatest fall in 5M2019 was construction, alongside mining, utilities, transport and storage and financial.

Any moderation in the unemployment rate among nationals will be positive for the consumption outlook, especially given the propensity to save and the high level of remittances associated with expatriate labor. However, in the more immediate term, the fall in the expatriate population is contributing to weakness in private consumption. Moreover, the fall in expatriate workers in Saudi Arabia, Oman and Kuwait is resulting in a fall in rental prices, quite marked in some areas. The IIF noted that the "Weakness in the real estate market has both reflected and contributed to weak economic growth in much of the GCC" (2019, 1). Investment in the real estate sector is popular with nationals; the fall in population and rental prices is impacting income of individuals and, in some cases, corporates. The Times of Oman indicated that some areas of Muscat were seeing rental drops as much as 40–60%, with indications of vacant properties in both Kuwait City and Muscat (Kuwait Times 2019, Times of Oman

2019). In the UAE and Qatar, the lack of private sector job growth has also contributed to the fall in rental prices, though the sharp rise in supply has also been a key factor. GRE restructuring has largely been aimed at extracting cost efficiencies, though it has also resulted in labor market weakness and job losses largely of expatriates.

More detrimental to regional economies would be a skills shortage and/or a moderation in productivity. Indeed, in Oman and Saudi Arabia, who have particularly seen a significant loss in expatriate labor, there have been indications that the private sector has faced rising costs from this—including having to find replacement labor, skills mismatches and the generally higher cost of national labor versus expatriate labor. In Saudi Arabia, in sectors seeing 100% Saudiization, a number of companies have gone out of business, especially SMEs, given the higher cost of hiring nationals. In Saudi Arabia, the average wage per month of an expatriate working in the private sector was around SAR4060 in 1Q2019 and SAR7099 for a national. Around 3000 mobile phone retailers closed down following the 100% nationalization of labor in this sector in September 2016 (EFG Hermes 2018, 10). The introduction of a fee on expatriate labor in January 2018 in Saudi Arabia is an extra burden (Saudi Gazette, 14 February 2018). The pressure on corporates has been compounded by a shift in the payment system, from an annual work permit renewal to a one-time lump-sum payment at the beginning of the year for all foreign workers employed—that is, collective invoicing. The loss of jobs in the private sector is reducing its productive capacity and further suppressing household demand, at a time of either fiscal austerity or weak government spending growth. Private sector companies have had to discount prices to support sales, thus adding additional pressure on margins, alongside the rise in costs of labor.

Recognizing that the increase in expatriate labor costs has been one of the many challenges facing the private sector, there have been some adjustments to the plan to reduce the impact on corporates. Businesses with fewer than nine employees are exempt from the expatriate levy for their first five foreign workers. Moreover, in early 2019, the government announced a scheme to reimburse some firms for the rising cost of expatriate labor fees, with SAR11.5 billion (USD3.1 billion) being allocated. However, the payment mechanism and eligibility criteria will be central in our view. The decree indicates that the scheme will only be available to companies with a higher or equal number of Saudi employees versus expatriates. Thus, a number of labor-intensive sectors that are struggling to attract Saudi employees or are impacted by the higher cost of Saudi labor might see limited benefits.

## 6 Conclusion

Reducing the wage composition of government spending will be critical to boosting fiscal sustainability across the GCC, especially for the hydrocarbon poorer per capita countries, and reducing their budget breakeven oil process. Raising the skills set of the population and human capital to meet the need of the private sector will be important for reducing dependence on government employment. This process will take time and is required at all educational levels, though it will still have to adapt to changing landscapes, including technological advancements and digital transformation. Factors such as labor market dynamics, productivity levels and mismatch in worker skills have also impeded FDI into the region, alongside private sector development. Without upgrading the human capital base of the economies, there is a risk that investment will mostly be on infrastructure without developing or attracting businesses and resulting in job creation. Better aligning the benefits (salaries, holidays, etc.) of the public in line with the private sector should also reduce the demand of working for the government. The youth population in GCC countries tends to be high, thus the employment demand is expected to rise, especially in Oman and Saudi Arabia. Creating private sector jobs will be central for reducing the reliance on the state for employment, which in turn will allow a greater scope for reducing wage spending.

Providing institutional support to the private sector during the transition period will also be important. Non-oil activity has remained weak across the GCC with the fiscal austerity of the last few years. Even though the pace of fiscal reforms has slowed and there has been a shift to expansionary government spending from 2018, oil prices have not recovered to a point for governments to provide a meaningful stimulus. The private sector is expected to provide jobs, new jobs at a time of weak economic activity and growth and, in some cases, with or after rising cost of doing business including labor. As noted, in the case of Saudi Arabia, the marked drop in the expatriate labor force, which is not being replaced by nationals, is resulting in an erosion of the production base. The private sector in the regions has traditionally relied on cheap foreign labor, and has arguably also been a part of the social contract, and had to also adjust. As we noted in our earlier chapter, the UAE has introduced a number of measures to support the private sector, including reduction in fees and regulatory reforms. This is supported by its stronger fiscal position in the region. However, other countries are not in a position to do this and still have significant fiscal reforms to introduce. Developing further training and apprentice schemes between the government and the private sector could be a way to complement educational reforms, support job creation and help reduce the burden on the private sector until workers reach the skill requirement.

Thus, the development of a framework to boost private sector activity, including through PPP schemes, remains vital. Moreover, in Saudi Arabia, the Public Investment Fund (PIF) is playing a central role in the transformation plans and the development of new industries (tourism, finance, industrial, etc.). This includes leading the development of major projects (such as NEOM, Al-Ula and Amaala) and the redevelopment of cities such as Jeddah and Medina. The development of these new sectors and the role of the PIF could potentially create or define a new private sector with a different relationship with the state, albeit still connected. The inclusion of the current private sector in these developments and projects will also be critical on how it develops and their role in creating new jobs. Looking forward, with increased globalization and for expanding the non-hydrocarbon and petrochemical export base, the domestic labor force will have to compete internationally on cost and productivity. This will especially be the case for the relatively hydrocarbon weaker per capita countries, where their economies will depend more on domestic labor. Any higher cost of labor will have to be supported by productivity gains. This level of productivity, alongside the sectors that will drive diversification away from hydrocarbons, has further significance for the level of wages and spending capacity of household, as the private sector will increasingly have to play a greater role in job creation. If the private sector does not provide adequate employment opportunities, then benefit payments and/or public sector wage would have to increase further.

## References

ADCB Economic Research. 2017a. GCC Economic Update: Higher oil prices to provide some relief, though growth recovery to be weak (Published on 16 February 2017, authors Monica Malik and Shailesh Jha)

ADCB Economic Research. 2017b. Saudi Arabia Economic Update—Winds of change. (Published on 16 November 2017, authors Monica Malik and Thirumalai Nagesh).

ADCB Economic Research. 2018. Saudi Arabia Economic Update—Weather watching as winds change (Published 9 December 2018), authors Monica Malik and Thirumalai Nagesh).

ADCB Economic Research. 2019. Saudi Arabia Economic Update—Gradual fall in 1Q national unemployment rate; challenges remain (Published 11 June 2019), authors Monica Malik and Thirumalai Nagesh).

Albawaba. 2018. 70 Percent Saudization in These 12 Sectors (published on 29 August 2018) https://www.albawaba.com/business/70-percent-saudization-these-12-sectors-1178800.

Banker Middle East. 2019. Bahrain projects $1.6 billion budget deficit in 2020. (Published on 26 February 2019) https://www.bankerme.net/en/home/articles/bahrain-projects-1.6-billion-budget-deficit-in-2020.html?%3Fcountry-tag=.

Bloomberg. 2017. In One of the World's Richest Nations, Anti-Immigrant Tide is Rising (published 13 April 2017, author Zainab Fattah).

EFG Hermes Equity Research. 2018. Saudi Consumer and Retail—Eyes on retail consolidation plays, some food names offer good value but lack catalysts. (Published on 31 May 2018, authors Hatem Alaa, et al).

Gulf News. 2018. Replacing public sector expats 'too difficult'. (Published 20 May 2018) https://gulfnews.com/world/gulf/kuwait/replacing-public-sector-expats-too-difficult-1.2224119.

IMF. 2018. Regional Economic Outlook Middle East, North Africa, Afghanistan, and Pakistan. (Published May 2018) https://www.imf.org/en/Publications/~/media/E7F9D70809DF41A8A55D8046A13480F0.ashx.

Kingdom of Saudi Arabia. 2016b. Fiscal Balance Program, Balanced Budget 2020. https://vision2030.gov.sa/en/bb2020.

Kuwait Times. 2018. 2,799 expatriates terminated in first year of replacement plan. (Published on 22 December 2018) https://news.kuwaittimes.net/website/2799-expatriates-terminated-in-first-year-of-replacement-plan/.

Kuwait Times. 2019. Apartments vacant as expatriates leave Kuwait amid. (Published on 4 July 2019) https://news.kuwaittimes.net/website/apartments-vacant-as-expatriates-leave-kuwait-amid-hardship/.

Moody's Investors Service. 2019. 2019 outlook is stable, but fiscal reform, geopolitics and unemployment pose challenges. (Published on 16 January 2019, authors Thaddeus Best, Alexander Perjessy, et al.).

Times of Oman. 2019. 65,000 expats left Oman in a year (published on 16 July 2019). https://timesofoman.com/article/1617683.

World Bank: Oman's Economic Outlook. 16 April 2018; https://www.worldbank.org/en/country/gcc/publication/economic-outlook-april-2018-oman.

CHAPTER 11

# Economic Diversification and Job Creation in the Arab Gulf Countries: Applying a Value Chain Perspective

*Martin Hvidt*

## 1 Introduction

This chapter aims to discuss economic diversification in the Arab Gulf countries with special emphasis on job creation. The article will use the concept of value chains to provide insight into the type of jobs and especially the knowledge content of these jobs that is likely to be most beneficial to pursue in order for the Gulf countries to further develop their societies. As such, value chains are used solely to illustrate the job categories in the production process. As such, it is not the intention to discuss the Arab Gulf economies within the framework of Global Value Chains (GVCs) by, for example, discussing the path of upgrading their GVC participation or how the Gulf states could reap further benefits from the global value chains.

M. Hvidt (✉)
Odense M, Denmark
e-mail: hvidt@sdu.dk



G. Luciani, T. Moerenhout (eds.), *When Can Oil Economies Be Deemed Sustainable?*, The Political Economy of the Middle East,
https://doi.org/10.1007/978-981-15-5728-6_11

The issue of diversification has once again reached the top of the political agenda in the region. The '2014 oil price collapse,' which halved the income from oil in year 2015 to 2016 compared to the previous years, demonstrated how volatile the income from oil is and, not least, just how closely tied the performance of the Gulf economics are to the oil income. The crisis, thus, has served as a stark reminder how profoundly dependent the Gulf countries continue to be on oil and gas incomes, and thus how little diversified their economics are.

In addition to discussing the diversification of their economies away from a near total dependency on the revenues from oil and gas, this chapter discusses a second key challenge to the Gulf economies, namely creation of jobs to accommodate their fast-growing population. Contrary to earlier, job creation for the national population has recently become a political aim of the diversification effort.

Normally diversification is viewed within a framework of risk reduction. In economic terms, risk is reduced when an economy changes from relying on one or a few sources of income to a broader variety of sources. Diversifying the Gulf economies thus, usually translates into policies directed at reorienting production from oil and oil-based activities into non-oil activities. This emphasis is in line with the way economists measure diversification, namely as the size of the output from—or the contribution of—the non-oil sector (Jörg 2012, 41).

This chapter, however, argues that, in order to address the issues of employment for a fast-growing number of young entrants to the working age, diversification within the GCC states should rather focus on the content, that is, knowledge and technological content of the jobs created, than on whether or not the jobs are located in the oil sector. In this respect, it does not matter whether or not the jobs are created within or outside the oil sector.

When it comes to job creation, the Gulf countries are facing two significant challenges: first, the demographic challenge visible in the significant growth of the national labor forces. Approximately 500,000 nationals will enter working age (25 years) each year during the next decade and a half in the Gulf countries, and even though not each and every one of these individuals will seek employment (salaried work), this still represents a formidable challenge.[1] Saudi Arabia, with a national population of approximately 20 million, is expected to have 400,000 new entrants a year and estimates that 4.5 million new jobs will need to be created before 2030 to meet this challenge (McKinsey Global Institute 2015, 5), while Kuwait, with a native population of a mere 1.3 million, is expected to have 28,000 entrants a year and estimates that 384,000 jobs are required before year 2030 (Tony Blair Associates 2009, 199). Note that these numbers only refer to the national population, thus excluding the migrant workforce.

The second and related challenge is that the public sector, which, until now, has been the main employer of the national labor force, cannot any longer act as the sole provider of jobs for the national population. In other words, the many new entrants among the working age youth must find jobs in the private sector. Thus, the challenge is of a double nature. First, it is to invigorate the private sector so its demands for labor increases and, secondly, to ensure that the national labor forces are willing and able to take the jobs offered in the private sector.

## 2 Diversification in the Gulf Countries

Diversification entails a broad societal process, which transforms a country from a single source of income, in this case oil or gas, to a society where multiple sources of income are generated across the primary, secondary, and tertiary sectors, and where large sections of the population participate. Private sector involvement is generally seen as a necessity for attaining diversification. United Nations Economic and Social Commission for Western Asia (ESCWA) defines diversification as:

> Within the context of the GCC countries, economic diversification means reducing overly dependence of the oil sector by developing a non-oil economy, non-oil exports and non-oil revenue sources. By implication, it also means reducing the leading role of the public sector in the GCC economies by promoting the growth of the private sector. (ESCWA 2001, vii)

And the benefits of diversification are, as stated by the General Secretariat for Development Planning in Qatar,

[1] Annual number of GCC nationals reaching working age (25 years) over the coming 15 years: Bahrain 13,642; Kuwait 28,434; Oman 54,360; Qatar 7236; Saudi Arabia 402,186; UAE n.a. Calculated by the author on behalf of GLMM data in Hvidt (2018a, 17).

> A more diversified economy is inherently more stable, more capable of creating jobs and opportunities for the next generation and less vulnerable to the boom and bust cycles of oil and natural gas prices. (GSDP 2011, 10)

Diversification is not a new strategy among the GCC states. It has been on the political agenda since oil and gas became the main and almost sole source of income in these countries some half a century ago (Henry and Springborg 2001, 26ff; Hvidt 2007; Luciani 2012b; McBrierty and Al Zubair 2004, 116ff; Niblock and Malik 2007; Rivlin 2009, 218ff; Seznec and Kirk 2011). Notable projects like aluminium smelting in Bahrain, the industrial cities of Yanbu and Jubail in Saudi Arabia, and the ports in Dubai were established in the 1970s with the specific aim of diversifying the economies by means of investing oil money in productive assets.[2] In addition, extension of education, health services, infrastructure development, and so on has been seen as key elements of the diversification strategy from the earlier days (ESCWA 2001, 8).

Historically, there have been different reasons for diversification. In the 1970s, the finite nature of the oil resources led to a quest to build viable economies which could secure income in the post-oil era. During the 1980s and 1990s, oil prices showed significant volatility and the focus of diversification shifted to counteract fluctuations in state incomes (ESCWA 2001, 3; Hvidt 2013, 11ff).

As I interpret it, today's quest for diversification includes both elements. The finite nature of oil has resurfaced in the form of climate considerations, which aim to 'green' the energy mix. While there is no immediate threat to oil producers from renewable energy systems (e.g. solar panels combined with electric cars) and $CO_2$ neutral nuclear energy, the global climate considerations have and do inflict uncertainties upon the analysts and decision makers in the Gulf as to the long-term call for oil and gas.[3]

The second element is clearly present in today's diversification efforts. For the Gulf countries, their dependence on oil incomes has in fact increased, not decreased over the past two decades, which is reflected in the rising fiscal break-even price of oil. Fifteen years (1999–2014) of rising

[2] The Emirate of Dubai possesses meager oil reserves and, as such, the ports were built in order to secure a stable income in the future.

[3] Luciani (2012a, 184). He argues, that even though demand for fuel might decrease, the global refining industry will continue to demand sizable inputs of oil and gas in the future.

oil prices have led to significantly higher levels of government spending in the region, and so did attempts to weather the effects of the Arab spring in 2011, which manifested itself as knee-jerk reactions to increase public spending (foremost on public sector salaries), in order to preempt actual or potential public uprisings. Both factors, in combination with increasing population sizes, have resulted in the simple fact that it has become considerably more expensive to run the government in each of the countries. As such, government spending has increased by an annual average of 11 per cent in real terms in 2003–2014.[4] According to IMF statistics, the break-even oil prices peaked in 2014, with Bahrain at $103 per barrel, Kuwait $54, Oman $94, Qatar $56, UAE $79, Saudi Arabia $105,[5] but have, since then, generally fallen as a consequence of the attempt to weather the 2014 oil price collapse through substantial ad hoc measures to cut public spending, that is, within public services (among them hospitals) and development projects and some more permanent features related to reduction in subsidies, for example, on fuel, water, and electricity.[6]

A third element in the historical reasons for diversification seems recently to be present, namely job creation. This is most visible for three countries: Saudi Arabia, Bahrain, and Oman. See especially the *Saudi vision 2030*.

On a more aggregated level, multiple problems exist in the Gulf economies, which arise from the developmental issues particular to the oil- and gas-driven economies. The oil-based or distribution state model applied in all Gulf states relies on the sale of hydrocarbons, is state-led and state-driven, emphasizes wealth distribution, makes extensive use of migrant labor, and is characterized by a significant underdevelopment of productive assets. However, with rapidly increasing population sizes, and rising popular demands on the governments, this model can no longer adequately support these societies.

Furthermore, this model suffers from problems such as low growth rates, lack of public and private incentives to accumulate human capital, lack of competitiveness in manufacturing, and so on, the likelihood of shocks and spill-over effects in the economies, and various rentier effects.

[4] https://www.e-marmore.com/Blog/Economy/January-2018/Why-are-the-breakeven-oil-prices-coming-down-for-G.

[5] IMF (2018, Table 6)

[6] Saudi Arabia and UAE have, furthermore, implemented 5 per cent VAT in 2018, and the other Gulf countries are to follow.

All these factors make it imperative for the countries to pursue economic diversification strategies (El-Kharouf et al. 2010, 135ff).[7]

As such, this model generally fails to support further development of the GCC states in two important respects. First, it has not shielded the economies from fluctuations in incomes, and second, and of most interest to this chapter, it has failed to create sufficient job opportunities to accommodate the many new entrants to the job market (Hvidt 2011, 88ff, 2015, 30ff).

The current development plans published in the region all emphasize diversification and underline the significant pressure to reform their economies. The *Vision 2030* launched in Saudi Arabia in 2016 is an example of the urgency of this reform process and is a statement of how job creation for nationals have accented on the political agenda (Govt. of Saudi Arabia 2016).

Generally, the countries which have the lowest income per capita are in most need of diversification. Saudi Arabia, Bahrain, and Oman are most challenged, while the so-called super rentiers Kuwait, UAE, and foremost, Qatar, experience less urgency to diversify.

## 3 Value Chains

Value chains have generally been studied with the purpose of identifying increased profitability in a given company, either by making each part of the chain more efficient or by adding or deleting parts of the chain. Adding more parts of the value chains hold the potential of reaping more of the value added. In the oil value chain, for example, moving from only undertaking oil production to also, for example, include refining diversifies the product (from crude oil, to a variety of usable products, e.g. gasoline, naphtha, jet fuel, kerosene, etc.). In this perspective, adding steps of the value chain will imply a diversification of the product, and thus of the markets.

And of primary importance for the argument in this chapter, this process of product diversification furthermore holds significant implications for a diversification of job content, and thus skill levels of the workers undertaking the jobs: The more diversification takes place within the value

[7]The authors (pp. 135–136) provide a detailed literature review of various economic reasons that make a
diversification strategy imperative for the resource-based GCC states.

chain (adding more stages in the production chain), the greater varieties of jobs will be demanded. Generally, the more primary productive activities are pursued, the higher the demand for supporting activities, which implies jobs related to services. And service-related jobs generally have a higher knowledge content and provide more value added than jobs in 'pure' manufacturing or oil extraction.

The concept of value chains was first introduced by Michael E. Porter in 1985 in his book "Competitive Advantage: Creating and Sustaining Superior Performance." In his book published in 1998, he further develops the concept of a value chain as the tool to analyze the sources of competitive advantages of a firm and define a value chain as the collection of activities that are performed within a firm to design, produce, market, deliver, and support its products (Porter 1998, 36) (Fig. 11.1).[8]

The value chain displays total value and consists of value activities and profit. Value activities are the physically and technologically distinct activities a firm performs (Porter 1998, 38). For companies that produce

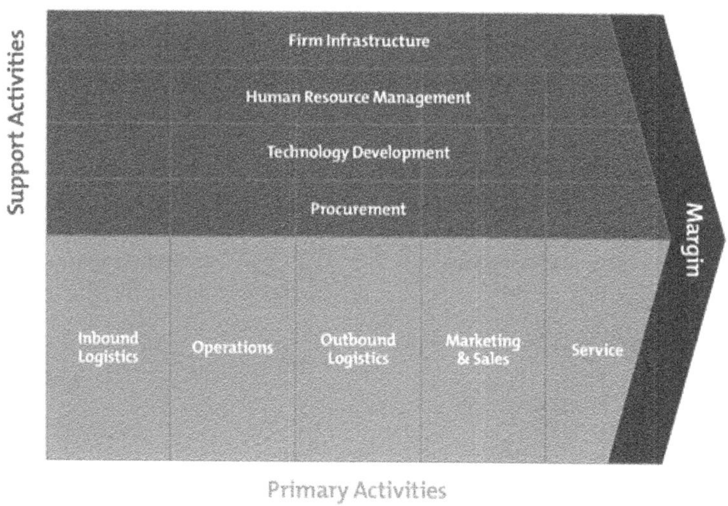


**Fig. 11.1** Porter's value chain. (Source: Porter 1998, 7)

[8] Porter (1998, 39) writes that the "value chain is a theory of the firm that views the firm as being a collection of discrete but related production functions, if production functions are defined as activities."

goods, the value chain starts with the raw materials used to make their products and consists of everything added before the product is sold to consumers.

As seen in the illustration above, Porter divides a business's activities into two main categories: primary and secondary activities, where the secondary activities are seen to support or to service the primary activities.

Building on this insight, but with a less rigorous division between primary and secondary activities, the so-called 'Smiley-model' was proposed by Stan Shih, founder of Acer computer company in 1992 (Ye et al. 2015, 2) (Fig. 11.2).

The 'Smiley—model' depicts information of the magnitude of value-added as it relates to various stages in the value chain. The insight derived from this model is that the activities related to manufacturing and assembly, that is to production in itself, yields the *least* value added (most likely due to the standardization of those processes, and thus exposure to high levels of competition), while the secondary activities related to the product, namely to design, R&D, innovation, logistics, marketing, and so on, yields high levels of value added. A classic illustration of this relationship is that the manufacturing costs of a jacket, that is, the cost of all inputs and the cost of processing (e.g. cost of labor and use of production facilities)

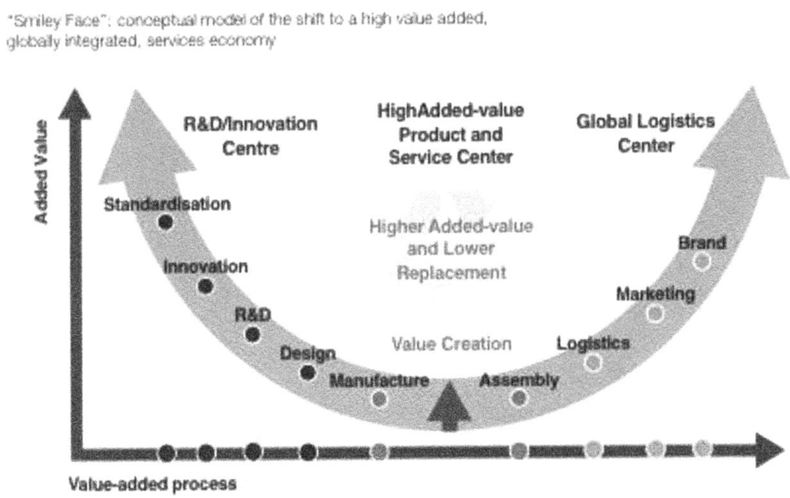


**Fig. 11.2** The smiley model. (Source: WTO Director of Trade Abdelhamid Mamdouh Presentation, Doha 27 November 2016)

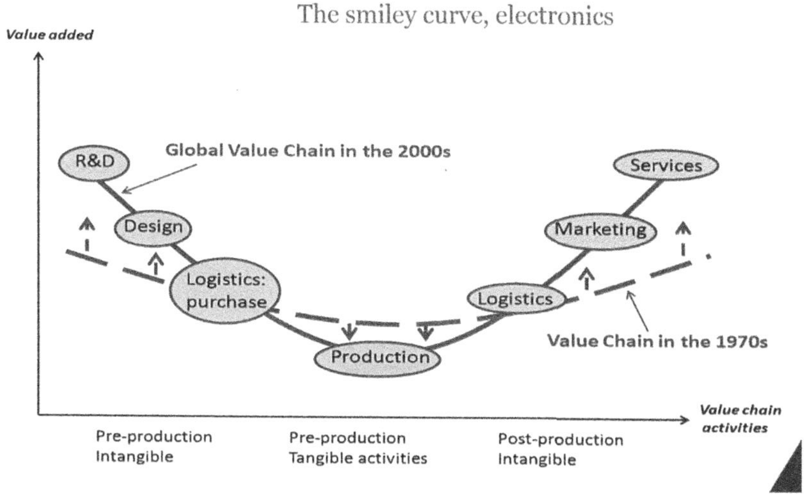


**Fig. 11.3** Transition to the smiley curve. (Source: (De Backer 2013, Slide 16))

only make up 9 per cent of the price of the final product. Thus, the secondary services and profit make up the remaining 91 per cent (WTO Director of Trade, Presentation, Doha 27 November 2016).

Furthermore, as shown in Fig. 11.3, the Smiley in this particular value chain has attained a more pronounced U-shape over the last four decades. In other words, the value-added in the manufacturing activities compared to the innovation and marketing activities of the chain is getting wider. This development is believed to take place in most value chains under the impression of improved manufacturing processes and increased global competition, thus providing even further incentives for the Gulf states to empathize job creation in the services part of the value chains.

Dubai has spearheaded development toward these goals, for example, through its aviation industry, airports, ports, the financial institutions, for example, Dubai International Financial Center, and so on.

## 4 The Challenge of Job Creation in the Gulf

Diversification is an activity that is pursued both within the oil and gas sectors and outside it, that is, within the oil value chain (upstream, midstream, or downstream) or by economic activities outside the oil value chain, which could be either import substitution activities, which are activities that aim to produce the goods or services, which the country otherwise would have to import (building materials, aluminium frames, cladding, cement, steel rebars) (El Beblawi 2011, 185ff), or it could be international services, or advance business services (i.e. logistics, banking, finance, aviation, etc.). For these three categories of economic activities, the aim related to diversification is to attain more jobs within the job categories that provides high value added to the various economic sectors.

### 4.1 *Diversification Within the Oil and Gas Sector*

As mentioned in the Gulf countries, the oil and gas industry continues to be the dominating source of export earnings and income for the public sector (Hvidt 2013, 13).

The general structure of the Gulf oil sector is that the Gulf countries export the bulk of the oil production basically untreated, namely as crude oil.[9] This implies an industry where relatively few highly skilled engineers and technicians plus larger contingents of blue-collar workers lift the oil from the ground and transport it to the shipping terminals for export. In a value chain perspective, thus the national oil companies only reap a relatively modest part of the potential profit, but furthermore, lots of potential jobs with high knowledge content (service jobs) associated with the downstream activities (refining, delivery, sales to consumers) are left to other firms and countries. To my knowledge, Kuwait is the only Gulf state which operates the full value chain in the international market. Under the brand of Q8, it operates more than 4400 filling stations in Europe, owns and operates a refinery in Rotterdam, and owns a tanker fleet, and so on.[10]

However, reality is more complex. While it is true that national oil companies in the Gulf countries primarily export their oil as crude, they, however, depart from this picture in two ways. First, the Gulf countries have built structures and capacities to supply their own markets with refined oil

[9] In average the Gulf countries exported 78 per cent of their oil production as crude in 2017. This number does not include Bahrain. The number is derived from a comparison between crude oil exports and crude oil production for each country, see OPEC (2018, Table 3.6 and 5.3).

[10] See www.q8oils.com/EN/Mainnavigation/Aboutus/Q8.aspx, accessed 30-05-2018.

products and, as such, operate refineries, filling stations, and handles the logistics related to marketing and transport, but only within each country (Hvidt 2018b).

Second, the Gulf countries, especially Saudi Arabia, Abu Dhabi, and Kuwait have built a substantial refining and petrochemical business aimed first and foremost at the global market (Luciani 2012a). While the share of total crude and gas that is used as feedstock in the petrochemical industry remains relatively low (Luciani 2012a, 184), the Gulf countries have risen to become the world's leading suppliers in petrochemicals (basis products for rubber, plastics, gels, fibers, etc.) (Seznec 2011, 30ff). Since much of the expansion in the Gulf-owned petrochemical industry, both within the Gulf countries and especially on a global scale, are done with foreign partners and through mergers and acquisitions (208ff), it is most likely that the number of nationals employed in this sector, especially in the high-level technological jobs, is relatively small. However, as pointed out by Luciani (2012a, 184), the Gulf countries could and should use their oil and gas resources to develop industrial and progressively technological capabilities for the benefit of the countries, by investments in such high tech industries.

This leads to the conclusion that diversification has in fact been pursued actively *within* the oil sector in the Gulf countries.

### 4.2 *Moving Downstream Via Technological 'Upskilling' in the Gulf Countries*

The stated political aim in the Gulf countries is to pursue a diversified knowledge economy according to their development plans. This aim seems to fit well with the fundamentals of the Gulf region. Besides Saudi Arabia, the Gulf countries are made up of small states with tiny native populations. As such, a development strategy which aims to create sizeable numbers of industrial jobs with low or medium knowledge content makes little sense, because the workforce would largely have to be imported.[11] But to aim for fewer jobs with higher technological or knowledge content,

[11] A similar argument is made in GSDP, Qatar National Development Strategy 2011–2016, March 2011.

(Doha: Qatar General Secretariat for Development Planning, 2011), 55, and by M. A. J. Althani, The Arab.

Spring & The Gulf States: Time to Embrace Change (London: Profile Books, 2012), 52.

which yields a better pay targeted at the regions' own population, seems logical.

The labor markets of the GCC countries are of a dual nature wherein nationals, as a part of the social contract, primarily seek employment in the public sector, which is better paid, and offers both a higher level of job security and shorter working hours than jobs in the private sector (McKinsey Global Institute 2015, Exhibit E3). Private sector employment not only entails lower pay, longer working hours, and less job security but is also more competitive, meritocratic, and takes place in a gender-mixed environment, which might present a cultural challenge to segments of the national population (Ulrichsen 2018, 12ff; World Bank 2017, 4).

The key reason for this is that the private sector is overwhelmingly characterized by jobs that are low-skilled and pays low wages, which is facilitated by state policies permitting the sourcing of cheap migrant labor. As such, private sector employment is less attractive to nationals than public sector employment.

In a report titled *The Jobs Agenda for the Gulf Cooperation Council Countries*, the World Bank (2017, vii) argues that in order to make the jobs more attractive in the longer run salaries must be increased in the private sector and suggests that this should be achieved through reforms that increase the productivity of the private sector by shifting economic activity to "higher value-added sectors, more technology-intensive production, diversified and more sophisticated exports, and technology-driven foreign direct investment (FDI)." As documented by IMF, relative to other countries, productivity gains in the GCC countries have contributed little to growth since 1970. Rather, growth has been attributable to "hiring more hands" and thus neither to capital nor total factor productivity (IMF 2017, 26).

The inclusion of more nationals in the private sector remains however a contentious subject, which relates not only to the salaries paid but also to the motivation, willingness, and ability of the nationals to take a job in this sector (World Bank 2017, vii).

According to ability, that is, the educational qualifications of the nationals, the issue is whether or not the nationals will be able to compete on qualifications with imported skilled workers?[12] A review of the available indicators of educational achievement in the GCC countries highlights structural problems within and around the educational system that lower

[12] Assuming the same salary.

the quality of teaching, minimize research outputs, and lessen the usefulness to society of the education provided. In this respect, it is questionable how well the current educational system prepares its graduates to play an active role in a future knowledge economy (Hvidt 2015, 31ff). At the international level, a link between high incomes from natural resources and lower political emphasis on education is proven (Gylfason 2001).

A second issue is related to the likely success of the political initiatives to make private sector employment attractive to nationals. In all Gulf states, there are policies of localization often termed Emiratization, Omanization, Bahrainzation, or Saudization, which aim to pressure and/or encourage nationals to take jobs in the private sector on the one hand and, on the other, to open the private sector to employment of nationals, which, as mentioned above, not always is the first choice of employees. One such initiative is the drive to localize production. For example, Saudi Arabia aims to localize 50 per cent of defense spending and 75 per cent of the oil and gas industry, meaning that, all inputs should come from local sources and that a very high percentage of the higher skilled jobs within these sectors should be undertaken by Saudi nationals (*Vision 2030*). Other measures include implementation of quotas for employing nationals within various categories of private sector firms (e.g. the Saudi *Nitaqat* system) or even more ridged, restricting certain job functions to nationals only. For example, in January 2018, the Saudi government announced it would add to its growing list of Saudi-only jobs by including the sale of watches, eyewear, medical equipment and devices, electrical and electronic appliances, auto parts, building materials, carpets, cars and motorcycles, home and office furniture, children's clothing and men's accessories, home kitchenware, and confectionery (Young 2018, 17), and Oman which has placed a temporary ban on visas for 87 job categories within information and technology, accounting and finance, marketing and sales, administration and human resources, insurance, information and media profession, and medical professions (Khaleej Times 2018).

Such policies are implemented across the six countries, however, primarily in the three countries with the highest challenge related to unemployment for nationals, namely Saudi Arabia, Oman, and Bahrain.

This chapter focusses solely on the national population. However, it is recognized that presently the migrant labor force outnumber the nationals in the Arab Gulf countries. In Saudi Arabia and Oman, approximately half the jobs are held by migrants, while in the smaller Gulf states, such as Qatar and UAE, 80–90 per cent of all jobs are held by migrants.[13] And these migrants do not only hold jobs as construction workers, drivers, and nannies. An estimated 1/3 of them classify as highly skilled migrants and as such undertake jobs with a high knowledge content both in the private and the public sectors of the Gulf economies (Hvidt 2019, 76). It is generally these jobs which would be of interest to the Gulf economies to replace with their nationals.

Besides the obvious point that it is positive for the national economy if jobs can be undertaken by a national compared to a migrant worker, who usually transfers a substantial part of his/her salary and savings out of the country, the localization efforts of the jobs should be seen as a long-term strategy of 'upskilling' the national labor forces to undertake the kind of jobs to be offered in the high-end of the private sector.

High value-added jobs have, as seen above, a much larger potential for income generation and thus play an active part in attracting the national labor force, but, furthermore, high value-added jobs or high technology jobs come with a much larger development potential, in that jobs with high skill levels or knowledge content not only increases technological change, the key factor behind development, but also increases the ability to quickly adjust to new and changing demands in a rapidly changing world (Hvidt 2019, 76). As pointed out by Ian Bremmer in his latest book *Us* vs. *Them: The Failure of Globalism* (2018, 45), not only are the profits larger in the technology driven sectors, but, in the next decade, the world will experience an even higher rate of technological change than seen in the previous decades. The technologies of artificial intelligence (AI), robotics, and machine learning have now reached a stage where they profitably can be applied across sectors. These technologies are likely to transform the job market as we know it, placing an even higher premium on the jobs which develop, interact with, and control the digital processes, including the increasingly cheaper and more sophisticated robots. The implication is that the Smiley curve is likely to attain an even more pronounced U-shape, thus leaving countries and sectors which do pursue technological upskilling behind (Bremmer 2018, 97). Stated differently by UNIDO and World Economic Forum (2014, 8) today, competitiveness of nations or

[13] GLMM Demographic and Economic Data base, table titled: "GCC: Total population and percentage of nationals and non-nationals in GCC countries (national statistics, 2010–2017)," available at http://gulfmigration.org; accessed October 1, 2018.

city states is less about decreasing costs related to production or transactions, and more about harnessing added value through improving capabilities.

### 4.3 *Upskilling of the National Labor Force*

Mueller's (2012) analysis of GCC Industrial development is useful to illustrate the process of upskilling the labor force. He points out that non-oil manufacturing industry is likely to make a substantial contribution to further development in the Gulf countries because (1) it is an essential ingredient of socio-economic development responsible for 1/3 of variation in per capita income across developing countries; (2) a broader manufacturing base is better able to take advantage of new opportunities as compared to an economy which is specialized in a narrow range or resource-based products; and (3) manufacturing industries are drivers of innovation and main source, user, and diffuser of technological progress (Mueller 2012, 140). He notes that the manufacturing sectors in GCC are comparatively weak when it comes to technological status and capabilities, but elaborates on the crucial role that research-intensive industries play for growth in both the manufacturing and the service sectors. In other words, the two sectors are interlinked:

> research-intensive industries generate higher growth rates in terms of productivity, value added and high-skilled employment. Besides, research-intensive industries are the source of new cross-cutting technologies, which contribute to enhancing the growth potential of the economy as a whole. The expansion of industrial R&D activities also stimulates the development of knowledge-intensive services, and it is the interplay of industrial and services sectors that determines the technological capabilities of an economy. (Mueller 2012, 149) (my emphasis)

As such, the process of 'upskilling' the national workforce should be seen as a process, which generally raises the knowledge level in all economic activities throughout a society. This is generally what we understand as a knowledge economy (Hvidt 2015, 27–28).

As mentioned, all Gulf states have statements in their development plans of transforming their economies into such knowledge economies and have taken steps to fulfill those aims (Hvidt 2015, 24ff). However, to transform into a knowledge-based economy is neither an easy nor a fast process. It is a comprehensive process, in which appropriate incentives for the population to upgrade their skills and become active participants in this economy have to be established. Among such efforts, the economy must adjust simultaneously both on the supply and on the demand side. As such, private business owners should be able to see their benefit in raising the technological level, while, at the same time, potential employers should approach the labor market with the skills and motivation to adequately fulfill such jobs. And, in all cases, there should be a possibility to make a profit.

A good number of initiatives have been taken over the past two decades to foster this development. For example, the internationalized service sectors, like the aviation industry, the financial sector, shipping services. In addition, the transformation process toward higher-knowledge level jobs is manifest in the stated-owned industries like the aluminum smelters, the oil industry, the shipyards, the property developers, and so on. However, as pointed out above, in discussion of the World Bank report related to the labor market, generally the jobs, especially in the private sector, remain characterized by low efficiency and thus low knowledge content.

## 5 Conclusion

This chapter has discussed diversification and job creation in a value chain perspective. Instead of viewing diversification in a narrow sense, as often done by economists, merely as the percentage of exports deriving from non-oil sectors, the value chain perspective as applied in this article directs the focus for both scholars and decision-makers toward analyzing the *type of job* and the *knowledge content* of the jobs, which the policies are attempting to create.

The incorporation of the value chain perspective leads to the understanding that, for diversification to be successful, it is not of particular interest whether or not a newly created economic activity falls within or outside the oil sector, but whether the new activity fosters jobs in the low value-added segment or in the high value-added one. Job creation in the high value-added segments potentially stimulates a virtuous process, which, on the one hand, necessitates upskilling of the labor force but, at the same time, fosters further upskilling. As such, this process includes a larger potential for growth and economic development in the long-term perspective. For the virtuous process to gain momentum, the governments in the region need to adjust their policies, for example, related to access to cheap foreign labor, in order to create the right incentives.

## References


Backer, Koen De. 2013. *Presentation OECD: Interconnected Economies: Benefitting from Global Value Chains—The Czech Republic*. OECD,: Presentation Prague, 25 June 2013.

Beblawi, Hazem El. 2011. "Gulf industrialization in perspective." In *Industrialization in the Gulf: A Socioeconomic Revolution*, edited by Jean-Francois Seznec and Mimi Kirk, 185–197. London: The Center for Contemporary Arab Studies, Georgetown University and Routledge

Bremmer, Ian. 2018. *Us vs. Them: The Failure of Globalism*. London: Portfolio/Penguin.

El-Kharouf, Farouk, Sulayman Al-Qudsi, and Shifa Obeid. 2010. "The Gulf Corporation Council Sovereign Wealth Funds: Are They Instruments for Economic Diversification or Political Tools?" *Asian Economic Papers published by The Earth Institute at Columbia University and the Massachusetts Institute of Technology* 9 (1): 124–151.

ESCWA. 2001. *Economic Diversification in the Oil-producing Countries: The Case of the Gulf Cooperation Council Economies*. New York: United Nations, Economic and Social Commission for Western Asia (ESCWA).

Govt of Saudi Arabia. 2016. *Saudi Vision 2030*. Riyadh: Government of Saudi Arabia.

GSDP. 2011. *Qatar National Development Strategy 2011–2016*. Vol. March 2011. Doha: Qatar General Secretariat for Development Planning.

Gylfason, Thorvaldur. 2001. *Natural Resources and Economic Growth: From Dependence to Diversification*. Paper prepared for an Expert Group Meeting on Economic Diversification in the Arab World organized by The United Nations Economic and Social Commission for Western Asia (UN-ESCWA) in cooperation with the Arab Planning Institute (API) of Kuwait, held in Beirut, Lebanon, 25–27 September, 2001

Henry, Clement M., and Robert Springborg. 2001. *Globalization and the Politics of Economic Development in the Middle East*. Cambridge: Cambridge University Press.

Hvidt, Martin. 2007. "Public—private ties and their contribution to development: The case of Dubai." *Middle Eastern Studies* 43 (4): 557–577. https://doi.org/10.1080/00263200701348862.

———. 2011. "Economic and Institutional Reforms in the Arab Gulf Countries." *The Middle East Journal* 65 (1): 85–102.
———. 2013. *Economic Diversification in the GCC Countries—Past Record and Future Trends, Research Paper No. 27: Kuwait Programme on Development, Governance and Globalisation in the Gulf States*. London: London School of Economics and Political Science (LSE)
———. 2015. "The State and the Knowledge Economy in the Gulf: Structural and Motivational Challenges." *The Muslim World* 105 (1): 24–45.
———. 2018a. *Socio-Econonomic Outlook for the Arab Gulf Countries: Figures & Forecasts*. Abu Dhabi: Emirates Policy Center and Arab Strategy Forum.
———. 2018b. "The United Arab Emirates: Modernity and Traditionalism in Petroleum Sector Management." In *Public Brainpower: Civil Society and Natural Resource Management*, edited by Indra Overland, 311–328. Cham: Palgrave Macmillan.
———. 2019. "Exploring the Nexus between Highly-Skilled Migrants, the *Kafala* System, and Development in the UAE." *Journal of Arabian Studies* 9 (1): 75–91. https://doi.org/10.1080/21534764.2019.1649827.
IMF. 2017. *Regional Economic Outlook: Middle East and Central Asia. October 2017, World Economic and Financial Surveys*. Washington D. C.: International Monetary Fund.
———. 2018. *Regional Economic Outlook: Middle East and Central Asia. Statistical Appendix 0518*. Washington D. C.: International Monetary Fund.
Jörg, Beutel. 2012. "Conceptual problems of Measuring Economic Diversificaiton, as Applied to the GCC Economies." In *Resources Blessed: Diversification and the Gulf Development Model*, edited by Giacomo Luciani. Germany: Gerlach Press.
Khaleej Times. 2018. "Oman temporarily halts expat visas for 87 job roles." In *Khaleej Times*. Dubai.
Luciani, Giacomo. 2012a. "GCC Refining and Petrochemical Sectors in Global Perspective." In *Resources Blessed: Diversification and the Gulf Development Model*, edited by Giacomo Luciani, 183–212. Germany: Gerlach Press.
———. 2012b. "Introduction: Resources Blessed: Diversification and the Gulf Development Model." In *Resources Blessed: Diversification and the Gulf Development Model*, edited by Giacomo Luciani. Germany: Gerlach Press.
McBrierty, Vincent, and Mohammad Al Zubair. 2004. *Oman: Ancient Civilisation: Modern Nation. Towards a Knowledge and Service Economy*. Dublin and Muscat: Trinity College Dublin Press and The Bait Al Zubari Foundation.
McKinsey Global Institute. 2015. *Saudi Arabia Beyond Oil: The investment and Productivity Transformation*. Riyadh: McKinsey Global Institute / McKinsey&Company.

Mueller, Hans-Georg. 2012. “GCC Industrial Development.” In *Resources Blessed: Diversification and the Gulf Development Model*, edited by Giacomo Luciani, 139–181. Germany: Gerlach Press.

Niblock, Tim, and Monica Malik. 2007. *The Political Economy of Saudi Arabia*. London: Routledge.

OPEC. 2018. *Annual Statistical Bulletin*. Geneva: OPEC.

Porter, Michael E. 1998. *Competitive Advantage: Creating and sustaining superior performance: With a new introduction*. New York: The Free Press.

Rivlin, Paul. 2009. *Arab Economies in the Twenty-First Century*. Cambridge: Cambridge University Press.

Seznec, Jean-Francois. 2011. “Financing industrialization in the Arab-Persian Gulf.” In *Industrialization in the Gulf: A Socioeconomic Revolution*, edited by Jean-Francois Seznec and Mimi Kirk, 30–43. London: The Center for Contemporary Arab Studies, Georgetown University and Routledge

Seznec, Jean-Francois, and Mimi Kirk (Eds.). 2011. *Industrialization in the Gulf: A socioeconomic revolution*. London: Routledge in Association with the Center for Contemporary Arab Studies, Georgetown University.

Tony Blair Associates. 2009. *Vision Kuwait 2030. Final Report*. Kuwait City: Tony Blair Associates.

Ulrichsen, Kristian Coates. 2018. *Economic Inclusion in Gulf Cooperation Council (GCC) states: Findings from an expert survey*. Washington D.C. : James A. Baker III Institute for Public Policy of Rice University.

UNIDO, and World Economic Forum. 2014. *The Future of Manufacturing: Driving Capabilities, Enabling Investments*. Geneva, Switzerland: World Economic Forum.

World Bank. 2017. *The Jobs Agenda for the Gulf Cooperation Council Countries*. Washington D. C.: World Bank Group.

Ye, Ming, Bo Meng, and Shang-jin Wei. 2015. “Measuring Smile Curves in Global Value Chains.” In *IDE Discussion Paper No. 530*. Japan: Institute of Developing Economies.

Young, Karen E. 2018. *The Difficult Promise of Economic Reform in the Gulf*. Washington D.C.: James A. Baker III Institute for Public Policy of Rice University.

CHAPTER 12

# Climate Strategy for Producer Countries: The Case of Saudi Arabia

*Jim Krane*

## 1 Introduction

The simultaneous collapse of oil production and social stability in Venezuela has provided a vivid preview of a potential pathway for oil-export states deprived of hydrocarbon rents to fund their national budgets.

While Venezuela's predicament in 2019 was based on domestic policy-making, some observers suspect that climate action could provoke similar results, leaving oil exporting states with insufficient revenues to maintain public order (Van de Graaf and Verbruggen 2014, 2015, 456–62; Van de Graaf 2018, 97–121).

The notion of an encroaching threat to hydrocarbon demand and rents—as opposed to the debunked "peak oil supply" conundrum of the prior decade—is leading policymakers to reassess long term assumptions about the oil business in two ways: by promoting diversification into alternate businesses, and by protecting and enhancing the competitiveness of their oil industries.

J. Krane (✉)
Rice University's Baker Institute for Public Policy, Houston, TX, USA
e-mail: jkrane@rice.edu



G. Luciani, T. Moerenhout (eds.), *When Can Oil Economies Be Deemed Sustainable?*, The Political Economy of the Middle East,
https://doi.org/10.1007/978-981-15-5728-6_12

The two strategies appear compatible. The first prepares the economic landscape for a day when oil rents no longer dominate the state's fiscal revenue, whether from a plateauing of global oil demand or any other reason. The second seeks to insulate flows of oil and gas rents against the more direct challenges emanating from climate policy.

Saudi Arabia, the global oil market's largest supplier, has taken steps in both directions. Its initiatives toward diversification are well known and covered elsewhere.[1] This chapter looks at the other, less explored climate strategy: near-term actions that the kingdom and other producer states have taken or may take in the next few years to maintain oil exports amid the emergence of restrictions on fossil fuels.

Some of the strategies that Saudi Arabia has developed are altering the nature of its participation in the oil business. From simply supplying energy commodities, the kingdom is increasing its involvement in importing markets and in bolstering oil-consuming technology.

One strategy involves investment into refining and other infrastructure ties with developing states where expectations for growth in oil demand are high. A related effort emphasizes low-emission and non-combustion uses for crude oil, which are consistent with a transitioning energy system. Another has the kingdom increasing cooperation with the global climate regime, pursuing a commitment to energy efficiency which does double duty in domestic oil demand management. Yet another strategy seeks to lobby the international community to moderate its targets for greenhouse gas emissions, and accept a higher level of human climate damage.

While this chapter applies these strategies to Saudi Arabia, other producer countries should be expected to take similar steps. The realization that the world is over-endowed with hydrocarbon reserves that may never be produced is forcing a strategic alteration in oil market behavior. Oil and

[1] Sources on the subject are too numerous to cite, but a few bear mentioning: Martin Hvidt, "Economic Diversification in GCC Countries: Past Record and Future Trends," *Research Paper*, Kuwait Programme series, 2013; Ali Aissaoui, "Saudi Arabia's Economic Diversification: Progress in the Context of the GCC and Challenges," in *Resources Blessed: Diversification and the Gulf Development Model*, ed. Giacomo Luciani (Berlin: Gerlach Press, 2012), 1–24; Bassam A. Albassam, "Economic Diversification in Saudi Arabia: Myth or Reality?," *Resources Policy* 44 (2015): 112–17; Tim Callen et al., *Economic Diversification in the GCC: Past, Present, and Future* (International Monetary Fund, 2014); Giacomo Luciani, "Resources Blessed: Diversification and the Gulf Development Model," in *The Gulf Region: Economic Development and Diversification*, ed. Giacomo Luciani et al., 4 vols. (Berlin: Gerlach Press, 2012).

gas markets appear likely to grow more competitive, with producer states vying for market share and differentiating products based on environmental criteria.

What this portends for markets and demand over the longer term is unclear. One envisions the eventual emergence of multiple potential pathways. A cooperative path would see producers collude to manage reductions in production so that prices remain above their cost of extracting the marginal barrel. One can also imagine the emergence of a more hostile dynamic, pitting fossil fuel producers against the efforts of climate-focused states and international organizations. The second pathway might be accompanied by a "green paradox" conundrum, where low-cost producers maintain or even ramp up production, driving down oil prices and helping fossil fuels compete against alternatives.[2] Given that global carbon dioxide emissions that had fallen from 2014 to 2016 returned to the path of growth in 2017—rising by 1.6% that year and again by 2.7% in 2018 (Jackson et al. 2018)—an intensification of international efforts to discourage fossil fuel use appeared likely.

## 2 Contrasting Strategies

This chapter focuses on three types of *nearer-term* producer country climate strategies outlined above. Although they sometimes overlap, I have titled them "Dig In," "Join In," and "Throw In."

- **DIG IN**: Producer governments overtly defend their oil and gas export industries. Their primary concern is not with the direct effects of climate change but indirect effects of climate action on global demand for exportable hydrocarbons, and the socioeconomic damage posed by loss of economic rents. By "digging in," states assume GHG accords such as the 2015 Paris Agreement remain aspirational rather than binding, and act to insulate or harden the hydrocarbon sector against the aims of such accords. Such actions include steps to enhance competition with rival producers and to protect technologies

[2] For more detailed discussion of possible oil market reactions to peak demand, see Anupama Sen and Bassam Fattouh, "Economic Diversification in the MENA in the Context of Peak Oil and the Energy Transition," in *Workshop on Sustainability in the GCC* (Gulf Research Meeting, University of Cambridge, UK, 2018).

linked to continued demand. This includes developing uses for crude oil that remain viable in a climate-constrained market.

  - **LOCK IN**: An important subset of the "Dig In" strategy is the bilateral "Lock In" of markets in key import states through direct investment.

- **JOIN IN**: Producer governments "join in" to pursue economically rational domestic energy policies that provide secondary benefits in reducing greenhouse gas emissions. The Paris Agreement's Nationally Determined Contributions (NDCs) provide useful political cover for unpopular—albeit environmentally beneficial—actions like Saudi Arabia's reform of energy subsidies. Domestic reforms have the added benefit of freeing up oil and gas that can be exported at market prices.
- **THROW IN**: The "Throw In" strategy suggests the adage "throw in the towel." Here, elites sympathetic to the interests of producer states and fossil fuel companies concede that climate change is inevitable and argue that damage caused by human greenhouse gas emissions is preferable to costly GHG mitigation in line with the Paris goals.

## 3 Saudi Arabia's Climate Profile

How does Saudi Arabia fit within the taxonomy above? Like other large producers of carbon-intensive fossil fuels, Saudi Arabia finds itself at the center of the global climate conundrum, increasingly recognized as major fossil fuel *producer*, *exporter*, *subsidizer*, and *consumer*. Saudi Arabia also stands to become an early and significant *victim* of climate change, since its arid geography and harsh summer climate is highly vulnerable to damage (Pal and Eltahir 2016, 197–200).

Saudi Arabia also carries an outsized legacy of GHG emissions, which began in the pre-nationalization era, due to large-scale methane venting that far outweighed its relatively minor emissions of carbon dioxide of the period. While Saudi $CO_2$ emissions have grown nearly 6% per year, roughly the same rate as its primary energy demand (EDGAR 2017), the kingdom's methane emissions peaked during the 1970s, a period when the

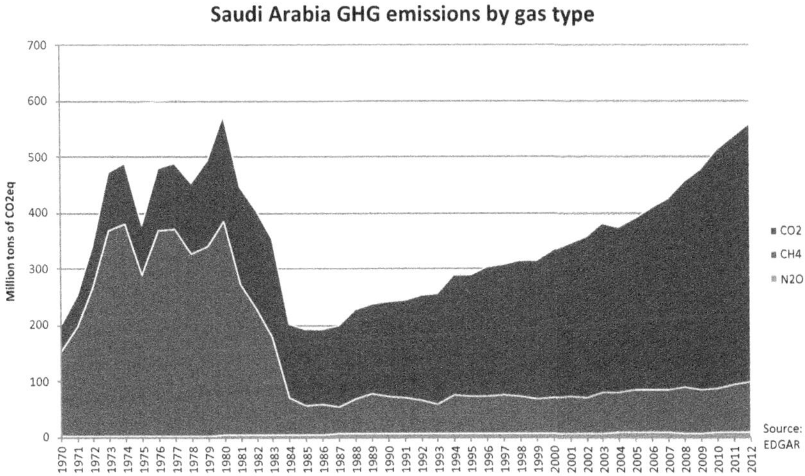


**Fig. 12.1** Saudi Arabia GHG emissions by gas type

kingdom's overall GHG emissions reached levels that were only surpassed within the last few years (Fig. 12.1).[3]

At the time, the national oil company, then known as Aramco, was owned and operated by four US companies. After the Saudi government took full control in 1980, GHG emissions dropped as Saudi Aramco built up its Master Gas System based on capturing waste methane and redirecting it to the power sector (Krane 2014). Fugitive (non-combustion) $CO_2$ from the oil industry was another major source of emissions which was curtailed after nationalization, dropping below 50% of total $CO_2$ emissions by 1979 and reaching 2.2% in 2008.

Once upstream fugitive emissions were reduced, the power sector grew to become the kingdom's largest source of carbon dioxide emissions. By 2008, electricity generation was responsible for 43% of GHG emitted, followed by transport at 27% and industry at 17%. Fugitive emissions from the Saudi energy industry totaled 10.5m metric tons of $CO_2$ equivalent in

[3] Atmospheric methane has far more potent heat-trapping properties than $CO_2$, but, after a decade or two, its harm is reduced as it decays into $CO_2$. Over a decade, methane warms the planet by 86 times as much as $CO_2$. See: Gayathri Vaidyanathan, "How Bad of a Greenhouse Gas Is Methane?" *Scientific American*, Dec. 22, 2015. https://www.scientificamerican.com/article/how-bad-of-a-greenhouse-gas-is-methane.

2008, amounting to just 3% of the kingdom's entire GHG emissions of 382m metric ton (EDGAR 2011).

Today's Saudi emissions flow mainly from Saudi energy consumption. By 2017, Saudi Arabia—the world No. 1 oil exporter—had become the world No. 5 oil consumer, after the United States, China, India, and Japan. It consumed more oil than much larger countries, including Russia, Brazil, or Germany (BP 2018). As such, Saudi Arabia is a leading global emitter of GHGs. The kingdom was the world No. 9 carbon emitter in 2016, according to BP data, ahead of populous oil producers like Brazil, Mexico, and Indonesia, as well as developed states like Canada and Australia, but behind Iran, South Korea, Germany, and Japan. However, BP data only reflect $CO_2$ emitted from fossil fuel combustion, and ignore other sources. When all GHGs are factored in, Saudi Arabia ranked as the No. 15 emitter in 2012, according to the most recent data collected by the EU's Emissions Database for Global Atmospheric Research (EDGAR). EDGAR GHG data rank the kingdom *behind* Brazil, Mexico, and Indonesia, as well as Canada and Australia (Table 12.1 and Fig. 12.2).

When compared with its peers in the Gulf Cooperation Council (GCC), Saudi Arabia was the least prolific per-capita emitter, far outpaced by

**Table 12.1** Saudi Arabia benchmarked against selected countries in terms of GHGs, population, and economy

| *Country* | *Crude & NGL export (KBD)* | *2012 population (million)* | *GDP (US$ bn 2011)* | *2012 GHG emissions per capita (ton of $CO_2eq$)* | *Average annual growth rate of GHG emissions since 1970* |
|---|---|---|---|---|---|
| Australia | 281 | 23 | 967 | 26.4 | 1.70% |
| UAE | 2625 | 9 | 532 | 25.3 | 2.20% |
| Canada | 2516 | 35 | 1452 | 20.6 | 1% |
| USA | 424 | 314 | 15,863 | 19.5 | 0.20% |
| Saudi Arabia | 7442 | 29 | 1444 | 19.2 | 2.50% |
| Russia | 4858 | 143 | 3602 | 15.9 | 0.60% |
| Iran | 1371 | 76 | 1271 | 10.3 | 2.50% |
| Mexico | 1333 | 121 | 1972 | 6 | 2.60% |
| Brazil | 532 | 200.5 | 3032 | 5.6 | 2.80% |
| Indonesia | 315 | 249 | 2302 | 3.3 | 3.40% |

Source: IEA, World Bank and EDGAR

2012 data (most recent GHG emissions data available)

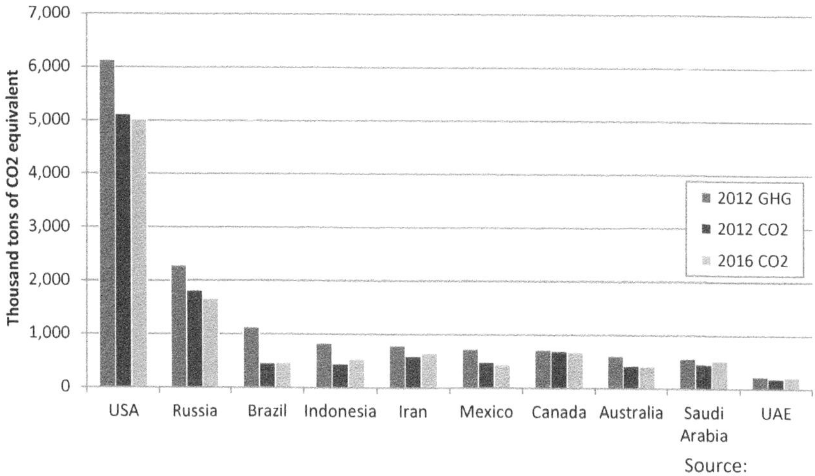


Fig. 12.2 GHG and $CO_2$ emissions in selected countries

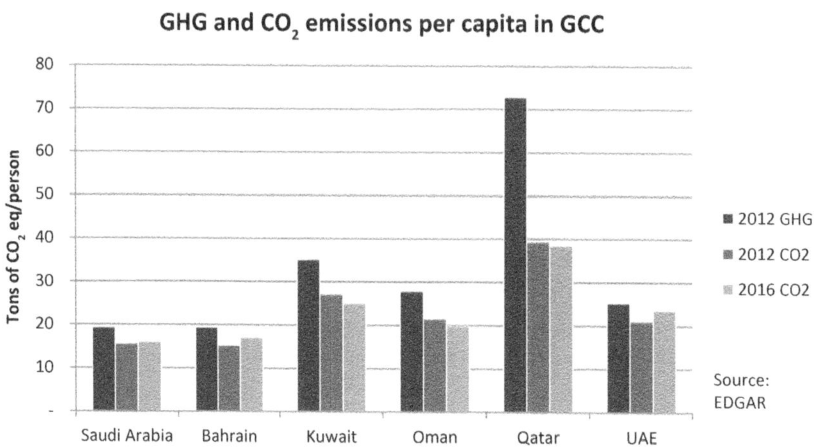


Fig. 12.3 GHG and $CO_2$ emissions per capita in GCC. (Source: EDGAR)

wealthy Qatar and Kuwait, and even ranking behind the United Arab Emirates, Oman, and Bahrain (Fig. 12.3). The high levels of per-capita GHG emissions in the smaller monarchies are partly due to small populations relative to the size of energy-intensive industries. Qatar's natural gas sector, including its LNG and gas-to-liquids industries, is a major emitter of both carbon dioxide and fugitive methane.[4] Other factors include high per-capita incomes, large homes, low electricity prices, and a hot climate that generates demand for indoor cooling (Krane 2015b). Electric power in the residential sector is nearly free in Kuwait, and is free in unlimited quantities for Qatari citizens.

Of course, most of the emissions from Saudi Arabian energy products are tallied in the countries where final combustion takes place. When aggregated, the carbon emitted from Saudi Aramco's oil production has been identified as the world's No. 1 current and No. 2 historical source of GHGs among all commercial fossil fuel companies. Hydrocarbons produced by Saudi Aramco are responsible for about 4.3% of current global GHG emissions and about 3.2% of aggregate atmospheric accumulations.[5] Chevron, the former Standard Oil of California, which discovered oil in Saudi Arabia in 1938 and held the original Saudi concession, is responsible for about 1.2% of current emissions and 3.5% of historic emissions, the highest share among historical emitters.[6]

## 4 The Importance of Oil in Saudi Political Economy

The GHG emissions attributed to Saudi Aramco are not only a byproduct of the company's business, but of the commercial activity that underpins the kingdom's economy, its governance structures, and its stature in international relations.

Saudi concerns around climate change center on the reduction or loss of potentially enormous oil rents. The oil sector in Saudi Arabia, like those of other low-cost producers, is extraordinarily profitable. Crude oil that costs less than $10 to produce and export sold for $60 per barrel at the

[4] Fugitive methane constituted nearly half of Qatar's 2012 GHG emissions, or about 70 of 150 megatonnes of $CO_2$ equivalent, according to EDGAR's dataset. EDGAR, "Fossil $CO_2$ & GHG Emissions of All World Countries, 2017."

[5] These figures include flaring and venting of GHGs during production. See: Benoit Mayer and Mikko Rajavuori, "National Fossil Fuel Companies and Climate Change Mitigation under International Law," *Syracuse J. Int'l L. & Com.* 44 (2016): 55.

[6] Emissions are from upstream production and final consumption of oil and gas products produced by Chevron. Mayer and Rajavuori. See also: Richard Heede, "Tracing Anthropogenic Carbon Dioxide and Methane Emissions to Fossil Fuel and Cement Producers, 1854–2010," *Climatic Change* 122, no. 1–2 (2014): 229–41.

time of writing. Subtracting investment and other costs along with a reasonable rate of return, a substantial portion of the sales revenue—perhaps 80%—could be described as economic rent. A threat to oil rents thus constitutes a threat to the Saudi government's operating budget and to the nature of tribal-autocratic governance that controls the state.

The profitability of oil also undercuts producers' appetite for diversification. The long term persistence of rents is a nearly unique aspect of the oil business and is partly due to production constraints that maintain high prices. In other sectors, rents are typically driven out over time by new entrants and competition. The durability of rents undermines the already lackluster determination for economic diversification, since it entails acceptance of lower profits and a loss of rents. A thorough diversification might even force the state to augment its income through taxation, undermining the rentier polity.

Finally, carbon extraction and marketing are even of strategic importance for Saudi Arabia, given that oil exports form the basis for the kingdom's geopolitical stature and its important partnership with the United States, which provides hard security for the region. A reduced role in oil markets would weaken Saudi Arabia's external security and downgrade its international standing.

Therefore, Riyadh, like producer governments elsewhere, should be expected to protect an industry so central to the survival of its regime, even to the detriment of the climate affecting its national territory. The following sections describe the shape such actions might take.

## 5 Strategy No. 1 "Dig In": Reduce Vulnerability of Oil Sector to Climate Action

To executives of a national oil company (NOC) such as Saudi Aramco, the default concern about climate change represents qualms around "security of demand" for crude oil within the ongoing clean energy transition (Smith 2010). Securing demand and markets takes several forms.

### *5.1 Climate Obstructionism*

In the past, Saudi concern was manifested in an obstructionist stance in international climate fora. Depledge argued that Saudi Arabia joined UN-led climate treaty negotiations because its leadership feared a successful agreement and wished to thwart, delay, or weaken it. The kingdom's confrontational stance was based on fears that mitigation policies would harm the kingdom more than the damage of climate change itself (Depledge 2008, 9–35).

Saudi tactics documented by Depledge involved the following:

- Allowing its position to be determined and led by the national petroleum ministry
- Highlighting skeptical views on climate science and downplaying the harmful effects of atmospheric carbon accumulations
- Employing negotiating techniques aimed at postponing or blocking progress
- Aligning with coal lobbies and other interests which oppose climate action
- Using financial incentives to persuade delegations from poorer countries to back the Saudi position.

As the global consensus on climate became more urgent, obstructionist Saudi tactics triggered negative publicity and damaged the kingdom's public image.[7] During the 2018 UN climate talks in Poland, the Saudi team reprised its obstruction tactics, backed by negotiators from the United States, Russia, and Kuwait. The quartet blocked a measure that would have supported the findings of an Intergovernmental Panel on Climate Change (IPCC) report (Plumer and Friedman 2018). But the kingdom has also turned to a more sophisticated set of strategies to protect its oil business.

### 5.2 *Petrochemicals and Non-Combustion Uses for Crudes*

First among these is to focus on applications for oil and gas that do not involve burning it. The most important of these "non-combustion uses" is in producing petrochemicals. Demand for petrochemicals has grown more robustly than demand for oil-based fuels[8] and with higher shareholder

[7] See, for example: Jad Mouawad and Andrew C. Revkin, "Saudis Seek Payments for Any Drop in Oil Revenues." *New York Times*, Oct. 13, 2009; https://www.nytimes.com/2009/10/14/business/energy-environment/14oil.html; and Gerard Wynn, "U.N. climate talks threaten our survival: Saudi Arabia." *Reuters*, April 8, 2009.

[8] The IEA in 2018 estimated that petrochemicals formed the fastest-growing demand source for crude oil, with an estimated 25% of overall growth focused on the sector by 2023.

returns than other sectors, including, at times, the upstream oil and gas business that provides petrochemical feedstocks (Ezekoye et al. 2018). Attractiveness for investors is augmented by the perceived longevity of the petrochemical business, since petrochemical-based plastics have few substitutes and remain vital to post-transition energy systems. For instance, plastic is crucial for the manufacture of solar panels, wind turbine blades, thermal insulation, batteries, and other components of electric vehicles (IEA 2018).

Thus the conversion of crude oil and natural gas into chemical products comprises perhaps Saudi Aramco's most promising climate hedge. Chemicals represent a growing "climate proof" use for hydrocarbons, through which oil and gas feedstocks are converted into precursor resins and polymers that form the basis for finished products ranging from auto parts to cushions, paint, and toothpaste. The carbon is sequestered in the finished product, rather than released upon combustion, as is the case with gasoline and other fuels.

Saudi-based companies have made major investments in petrochemical plants, both inside the kingdom and outside. Domestic investment includes the $20 billion Sadara joint venture with Dow Chemical, the largest single-phase chemical plant ever built. Internationally, Saudi Aramco and SABIC, the state-held petrochemical giant, have proposed joint venture projects in the United States, China, Malaysia, and India.[9] Demand for plastic goods is closely correlated with GDP growth, with large markets emerging in developing countries, where populations are moving into the middle class. To increase its competitiveness in the sector, Saudi Aramco is investing in technology for direct conversion of crude oil into chemicals, bypassing the refining phase and reducing energy inputs (Brelsford 2018).

Lubricant production is another climate-compliant use for crude oil, since lubricants are (ideally) not combusted. Improvements in lubricant

See: "Oil 2018," International Energy Agency, March 5, 2018. https://www.iea.org/oil2018/.

[9] Examples include an ethylene plant proposed in Texas in 2017 that would combine investments from SABIC and Exxon Mobil; a Saudi Aramco-Sinopec-ExxonMobil integrated refining and petrochemical plant operating in Fujian, China since 2009. Further, in 2018, Saudi Aramco announced a combined refining-petrochemical venture in India that would link it with three Indian firms; and in Malaysia, the Saudi company announced joint ventures with Petronas, Malaysia's state-owned oil company, for an integrated refinery and petrochemical project in southern Malaysia.

performance have also decreased fuel consumption by reducing friction and improving the efficiency of machine operations (Boonen and Vanderreydt 2017). As with plastics, lubricating oil can be recycled, which improves environmental efficiency, but reduces growth prospects of the market.

### 5.3 *Differentiating Among Crude Oil Grades by Carbon Intensity*

Saudi Arabia and other big Middle East oil producers have long enjoyed low average costs of producing their oil reserves, in comparison with other oil producing states. Now it turns out that low-cost producers like Saudi Arabia enjoy a related advantage: crude oil with low carbon intensity.

That is because cost of production is partly based on the energy expended in lifting crude oil from the reservoir to the surface, and in processing and transporting it. Lifting costs are lowest in reservoirs with low water content and high levels of natural drive pressure, requiring minimal use of energy-intensive recovery techniques such as steam flooding (Masnadi et al. 2018, 220).

A further factor affecting carbon intensity of oil production relates to the flaring of natural gas produced in association with crude oil. Flaring—or burning off gas at the wellhead—releases even more $CO_2$ into the atmosphere. Saudi Arabia and some of its neighbors, such as the UAE, Qatar, and Kuwait, flare very little natural gas, preferring to capture the gas for domestic use. Other producers like Russia, Iran, Iraq, and the United States flare enormous amounts of gas, exacerbating the carbon footprint of their crude oil (Fig. 12.4).[10]

Altogether, producing, transporting, and refining crude oil accounts for 15% to 40% of the entire lifecycle GHG emissions—including final combustion—of oil-based transport fuels. Among major producers, crude oil from Saudi Arabia exhibits the lowest carbon intensity. Other relatively low-carbon oil is produced by Ghana, Norway, Azerbaijan, and Kuwait. At the high end of the scale is ultra-heavy crude from Canada's oil sands and

[10] Russia was No. 1 in flaring in 2016, followed by Iraq, Iran, Venezuela, Algeria, and the United States. Big oil producers in the Persian Gulf flared much smaller amounts. Saudi Arabia was No. 13, Kuwait was No. 25, Qatar was No. 28, and the UAE was in 31st place, with less gas flared than even minor oil producing countries such as the United Kingdom and Vietnam. See: World Bank, "Global Gas Flaring Reduction Partnership (GGFR): Upstream Gas Flaring;" http://www.worldbank.org/en/programs/gasflaringreduction#7.

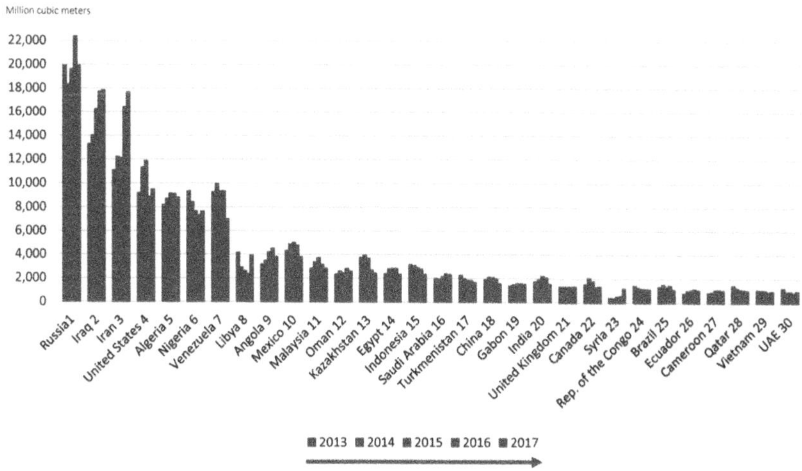


**Fig. 12.4** Top 30 flaring countries. (Source: NOAA/GGFR)

Venezuela's Orinoco Belt, where energy-intensive steam flooding is used to increase viscosity of crude oil that, in its natural state, appears closer to a solid (Masnadi et al. 2018).

Given such a timely environmental advantage, Saudi Aramco will begin highlighting the low carbon intensity of its crude oil as a marketing strategy (Interview with Saudi oil official, Feb. 13, 2018). Low carbon intensity could even translate into a price advantage in countries that levy carbon taxes, if carbon taxes were designed to differentiate among crude grades by carbon intensity. More typically, carbon taxes apply an average value to oil products irrespective of origin.

As Table 12.2 shows, Saudi medium crude priced at \$70/barrel with a \$25/ton carbon tax would cost \$81.64 per barrel. A barrel of Venezuelan Orinoco crude would be priced at \$85.98, a \$4.34 premium. At a \$50 carbon tax, the effect would be magnified. The Saudi barrel would be nearly \$9 cheaper, at \$93.29, versus \$101.97 for the Orinoco.

### 5.4 *Backing Internal Combustion Engines over EVs*

The kingdom also plans strategic investments to prolong the use of gasoline in transportation, by maximizing the efficiency of the internal combustion engine, so that oil-fueled transportation remains cost-competitive

Table 12.2 Carbon taxes on Saudi and Venezuelan crudes at $70 per barrel

| *Crude oil source* | *Upstream GHG intensity (g $CO_2eq$/ MJ)* | *Upstream GHG tax per barrel @ $25 tonne* | *Upstream GHG tax per barrel @ $50 tonne* | *Total GHG tax @$25 tonne** | *Total GHG tax @$50 tonne* | *Oil price @ $70 bbl + $25 tax* | *Oil price @ $70 bbl + $50 tax* |
|---|---|---|---|---|---|---|---|
| Saudi average | 3.5 | $0.54 | $1.07 | $11.64 | $23.28 | $81.64 | $93.28 |
| Venezuela Orinoco | 31.9 | $4.88 | $9.76 | $15.98 | $31.97 | $85.98 | $101.97 |

Source: Baker Institute using $CO_2$ intensities from Masnadi et al. (2018)

*Includes upstream $CO_2$ emissions as well as those from transport, refining and final combustion; Venezuelan heavy crude oil typically sells at a discount to more valuable lighter grades, a distinction that is not captured in these figures

with electric vehicles. From the Saudi perspective, an efficient hybrid vehicle is preferable to an all-electric vehicle that uses no petroleum. Saudi Aramco touted its vehicle research—and the "unparalleled energy density of hydrocarbons"—at the North American International Auto Show in Detroit in 2018. On display were high-efficiency gasoline compression ignition engines as well as a car with on-board carbon capture and storage capability (Interviews with Saudi Aramco, Feb. 13, 2018).[11] In August 2018, Saudi Aramco announced it would cooperate with Japanese auto manufacturer Mazda to develop more efficient engines and gasoline that would reduce GHG emissions from the transport sector (Saudi Aramco 2018). These developments would improve petroleum's competitiveness versus alternative fuels and technologies.

### 5.5 *The 'Lock In' Strategy*

A subset of the "Dig In" strategy is the "Lock In" approach, where NOCs purchase oil and gas infrastructure inside importing countries. Refineries have been the main target, and Saudi Aramco the largest proponent.

Aramco has created joint ventures that provide influence or even control over the refinery configuration and, by extension, the supply of crude

[11] See also: "Aramco showcases flagship technologies at Detroit auto show," Saudi Aramco press release, Jan. 31, 2018. http://www.saudiaramco.com/en/home/news-media/news/flagship-technologies-detroit-auto-show.html.

oil to the refinery process. Aramco has bought stakes in refineries in China,[12] South Korea (S-Oil n.d.), Japan (Showa Shell Sekiyu K.K. 2016), and the United States (Gramer 2017) and was in the process of negotiating further purchases in India (El Gamal 2018) and Malaysia (Lee and Cheong 2018). Similarly, the Kuwait Petroleum Co. has purchased a 35% ownership and operational stake in a refinery in Vietnam configured around Kuwaiti crude (Vu 2018).

At times, when oil markets are in oversupply—perhaps due to weakening demand amid a transition to cleaner energy—Saudi Aramco can ensure that Saudi crudes have preferential access to Saudi-owned refineries (Krane 2015a, 99–104). The tactic is reminiscent of that adopted in the early twentieth century by the Seven Sisters cartel, which controlled market entry through integrated ownership of upstream, midstream, and downstream infrastructure. Other states, which lack the "integration" pursued by Aramco could find ready markets lacking.

## 6 Strategy No. 2 "Join In" Climate Action

For now, climate change in producer states is primarily seen as an indirect threat to demand for oil and gas. But producer countries are also vulnerable to the direct threat of climate change, and to the multilateral call to "join in" the campaign against global warming. As international resolve has coalesced around the desirability of GHG mitigation, the Saudi climate negotiating stance has grown more accepting of the scientific consensus and the need for action. The 2015 Paris Agreement saw Saudi Arabia and its neighbors declaring nationally determined contributions (NDCs) to reduce emissions of GHGs.

Notwithstanding the backtracking described above, Saudi Arabia has, at least in public, shifted its stance to one of support for climate action.

[12] Aramco owns a share of the Fujian Refining & Petrochemical Co. (company website, http://www.frep.cn/en/view-category?categoryId=ff808081 29f00d8e0129f00d94e31100) and is in the process of purchasing stakes in two more, as well as entering long-term supply contract with a private Chinese refiner. See: Rania El Gamal, "Saudi Aramco eyes partnerships as it expands refining, petrochems." *Reuters*, June 12, 2018; https://www.reuters.com/article/us-saudi-aramco-downstream-exclusive/exclusive-saudi-aramco-eyes-partnerships-as-it-expands-refining-petrochems-idUSKBN1J81M9; See also: "New Private Chinese Refinery In Talks For Long-Term Saudi Oil Supplies," *Oilprice.com*, May 31, 2018; https://oilprice.com/Latest-Energy-News/World-News/New-Private-Chinese-Refinery-In-Talks-For-Long-Term-Saudi-Oil-Supplies.html.

Energy minister Khalid Al-Falih issued a statement in support of the 2016 Congress of Parties (COP) 22 in Marrakesh. "We view the Paris Agreement as balanced and fair, and this will pave the way to effective implementation in addressing our climate goals and sustainable development goals holistically," al-Falih said. "It is encouraging to note that the Paris Agreement has achieved the threshold for entry into force, and Saudi Arabia is determined to see it implemented." (Ministry of Energy, Industry & Mineral Resources of Saudi Arabia 2017)

The revised Saudi posture accepts the necessity of reducing GHG emissions, but in a selective way. Domestically, the kingdom has launched two reforms of fossil fuel subsidies, raising prices and reducing demand for transport fuels, electricity, and desalinated water. These reforms have economically rational goals of reducing government spending on energy provision and decreasing the "cannibalism" of exportable energy commodities. But they serve double-duty as environmental policy, since they also reduce growth in the kingdom's GHG emissions.

Internationally, Saudi Arabia promotes an altogether different strategy, featuring efforts that protect the interests of oil-exporting states in ways that do not harm demand for fossil fuels. Supported strategies include:

- **Carbon capture and storage (CCS)**: CCS involves capturing carbon emissions after combustion and storing them permanently underground. CCS technology has progressed slowly, and adoption has been undermined by expense and inefficiency, and the slow uptake in carbon pricing required to spur adoption. CCS actually *increases* fossil fuel input for the same energy output because capturing and compressing $CO_2$ requires combusting additional fuel. Saudi Arabia and the UAE are members of the Carbon Sequestration Leadership Forum, and the kingdom has pledged to capture and use $CO_2$ in various applications, including in petrochemical production and EOR.
- **Flaring reductions**: As mentioned, Saudi Arabia has succeeded in curtailing wasteful flaring of natural gas, but other countries—including the United States—have been less successful. Altogether, 3.5% of global gas consumption, or 149 billion cubic meters—more than Japan's entire 2017 consumption—is burned as waste (World Bank n.d.). Many fossil fuel proponents acknowledge that "cleaning up" the sector itself would reduce pressure to curtail final consumption.

- **Focus on "alternate" GHGs**: Saudi officials want more attention paid to GHGs, such as methane and nitrous oxides, which, although forming a smaller portion of overall emissions, carry much higher heat-trapping properties than $CO_2$. Nitrous oxides are largely used in fertilizers in the agricultural sector. Most methane pollution flows from leaks in upstream oil and gas infrastructure. Saudi Arabia has championed the Global Methane Initiative to reduce fugitive methane emissions (Global Methane Initiative n.d.).

The Saudi COP 22 brief also argues that fossil fuels should be retained in a future energy mix due to their synergies with renewables. The kingdom sees $CO_2$ emissions as a "harmful side effect" that can be mitigated with technological solutions (Ministry of Energy, Industry & Mineral Resources of Saudi Arabia 2017). In 2014, Saudi Aramco joined the Oil and Gas Climate Initiative, a group of 13 major oil companies each pledging $100 million for research into low-emissions fossil fuel technology.

### 6.1 *Renewables and Nuclear Power*

The Saudi climate approach also leans on aspirations for replacing oil-fired power generation with zero-carbon renewable and nuclear electricity generation within the kingdom. There are strong economic rationales for policies that push oil out of the Saudi power generation sector, given oil's international market value and the very low cost of solar power in a country with world-leading insolation and plenty of empty land. But, as of mid-2018, neither nuclear nor renewable generation had an appreciable presence in the kingdom. In 2017, Saudi Arabia produced just 0.04% of its electricity—135 gigawatt-hours of a total of nearly 376,000 GWhs—from solar means, the only non-hydrocarbon source of electricity currently online in the country (BP 2018). Instead, a rising share of the Saudi power market was being claimed by natural gas (Table 12.3).

**Table 12.3** Power generation by source in KSA

| *KSA power generation 2017* | |
|---|---|
| Natural gas | 59% |
| Oil-based fuels | 41% |
| Renewables (solar) | 0.04% |

Source: ECRA (2017) and BP (2018)

### 6.2 *Social License to Operate*

In coming years, the kingdom and Saudi Aramco appear likely to highlight these efforts, as well as the low carbon-intensity of its crude, lack of flaring and fugitive methane, and its investments in high-efficiency engines to claim credentials as an "environmentally responsible" supplier of necessary fossil fuels. The business associations it has supported around these issues may in future set minimum compliance standards and issue "green" certifications or endorsements for fuels that meet standards.

Oil's lack of substitutes in the transportation sector, particularly the aviation market, means that these sectors are unlikely to be decarbonized for decades. Oil-based fuels will be necessary far into a climate-constrained energy future. Attention will necessarily turn to the GHG footprints of the various grades of crude in the global oil supply. Suppliers that minimize $CO_2$ and GHG emissions from their upstream, midstream, and downstream supply chains will gain competitive advantage, allowing consumers, refiners, and importing states to differentiate among "clean" and "dirty" oil-based fuels.

Saudi Arabia is positioning itself to present a strong case as a "climate friendly" fuel supplier. As such, the social acceptability of Saudi crude oil is more than a "Join In" in climate action strategy. It also becomes a "Dig In" strategy, since it might also extend the longevity of Saudi supply in the global oil market—and the social stability in a kingdom where politics are structured around oil rents.

## 7 Strategy No. 3 "Throw In" and Accept Climate Damage

Finally, there is the "Throw In" strategy, a collective approach whereby producer governments lobby for a relaxation of the atmospheric GHG accumulation limit of 450 parts per million of $CO_2$ equivalent. This is the maximum GHG concentration associated with limiting the global average temperature increase to 2 °C, which forms the basis for the Paris agreement.

As an alternative, fossil fuel producers, scholars, and others have been developing a relaxed climate strategy they describe as a more "pragmatic" path toward decarbonizing global energy systems. The strategy amounts to a concession (i.e. "throwing in the towel") that costs of compliance with 2 °C emissions limits are unreasonably high, and that more climate damage would be preferable to the economic disruption implied by thoroughgoing and rapid decarbonization.[13]

Among the institutions developing the strategy are Saudi Arabia's Ministry of Energy, Industry and Mineral Resources; Japan's Ministry of Economy, Trade and Industry; as well as the U.S. Chamber of Commerce, the European Union and African Union, and academics from think tanks and universities. At the time of writing, no official document had been released.[14]

The "pragmatic" climate strategy stresses the deficiencies of fossil fuel substitutes and how these energy sources have failed to gain traction in the developing world, where much power demand growth continues to be met by fossil fuels, mainly coal. The narrative points out that climate policy must become cost-competitive for developing countries, and that market mechanisms are needed to motivate advances in technology. It repeats well-known arguments that renewables and electric vehicles are disadvantaged by price, energy density, and intermittency. Therefore, tradeoffs are needed among the three categories of climate spending:

- **Adaptation**: Investments and engineering techniques that reduce exposure to geographic and climatic changes
- **Mitigation**: Reducing emissions, including by reducing fossil fuel use
- **Damage**: Economic losses from climate change.

Most of the tradeoffs lie in the mitigation-versus-damage balance. If strict mitigation measures are imposed before substitute technology is mature, proponents of the strategy argue that costs of mitigation could outweigh the costs of damage and adaptation, resulting in a larger-than-optimal expense. In Fig. 12.5, Path 3 represents an "unbalanced" climate policy dominated by mitigation.

[13]These strategies are outlined in Samantha Gross and Yuhji Matsuo, "Towards More Pragmatic Global Climate Goals and Policies," scholarly paper (Riyadh: KAPSARC, October 2017), http://eneken.ieej.or.jp/data/7608.pdf. Also see slides 17–20 in: Yukari Yamashita, "Climate Change and Economic Growth in Asia: What Are Realistic Goals?" (Powerpoint presentation, March 16, 2017), https://www.bakerinstitute.org/media/files/files/0587096b/Pragmatic_Approach_Baker_Institute_yamashita.pdf.

[14]In the interest of full disclosure, the author declares his involvement in the process.

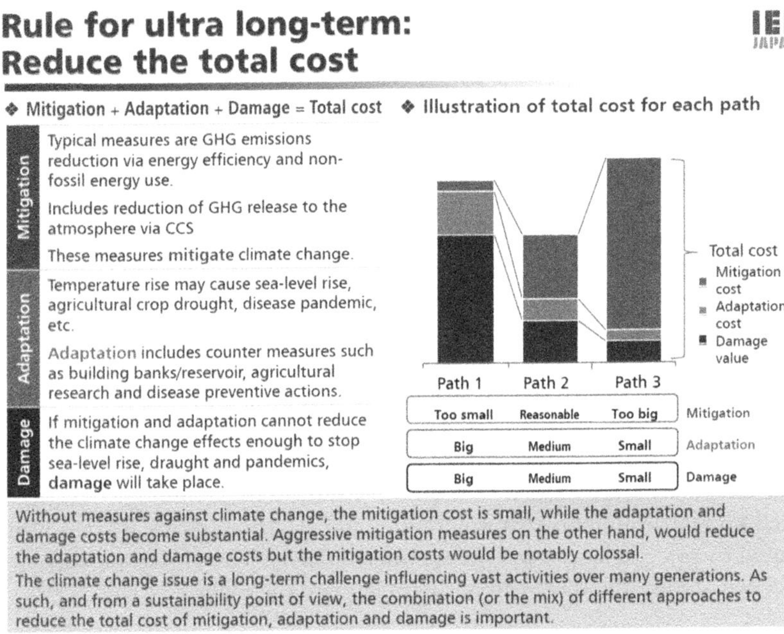


**Fig. 12.5** Comparison of cost paths with varying levels of spending on mitigation, adaptation, and damage. (Source: Institute for Energy Economics)

However, the pragmatists' "optimum" path (Path 2 in Fig. 12.5), which minimizes total cost, provides insufficient decarbonization to meet the 2 °C carbon target. Average warming might reach 3 °C and bring catastrophic climate damage. But the costs of widespread damage would—proponents argue—be more than offset by reductions in spending on mitigation. This finding also implies less harm to fossil fuel producers and to developing economies seeking low-cost energy services.

These raw estimates raise questions in regards to the accounting methods used, which are based on modeling carbon taxes required to bring about sufficient reductions in demand. A very rough comparison of *actual* climate mitigation and damage costs highlights potential weaknesses in the assumptions. For instance, between 1980 and 2018, the United States sustained some $1.5 trillion in damage from 219 weather and climate disasters (Smith 2018). The portion of those damages attributed to climate change is unknown. By contrast, the US government spent a tenth of that amount, less than $150 billion, between 1993 and 2014 (a shorter period) on climate research, technology, and assistance (US Government Accountability Office n.d.).

More specifically, in 2017, the United States experienced a record $306 billion in damages from weather and climate disasters (Fig. 12.6). That amount is five times the 2017 revenues of Saudi Aramco[15] and more than 70% of OPEC's 2016 oil export revenues.[16] If anthropogenic climate factors were responsible for 20% of the damage—due to intensified drought-induced wildfires and flooding from extreme rainfall—paying for that

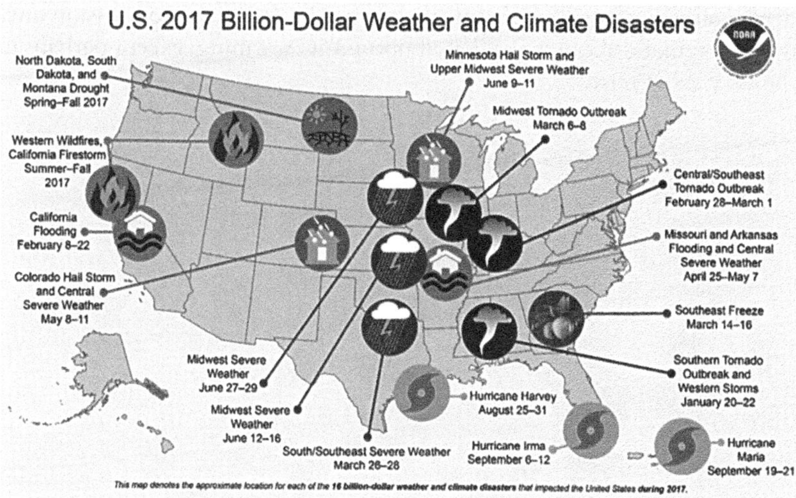


**Fig. 12.6** US 2017 billion-dollar weather and climate disasters. (Source: NOAA)

[15] The Saudi Aramco revenues estimate comes from a rough doubling of Saudi Aramco reported revenues of $33.8 billion in the first half of 2017. See: "The Aramco Accounts: Inside the World's Most Profitable Company," Bloomberg News. April 13, 2018 (https://www.bloomberg.com/news/articles/2018-04-13/the-aramco-accounts-inside-the-worlds-most-profitable-company). The US climate-related weather damages come from: Adam B. Smith "2017 U.S. billion-dollar weather and climate disasters: a historic year in context," Climate.gov; Jan. 8, 2018. (https://www.climate.gov/news-features/blogs/beyond-data/2017-us-billion-dollar-weather-and-climate-disasters-historic-year).

[16] The 13 OPEC member states earned a combined $433 billion in 2016. See: "OPEC Revenues Fact Sheet," US Energy Information Administration, May 15, 2017. https://www.eia.gov/beta/international/regions-topics.php?RegionTopicID=OPEC.

portion alone would require Saudi Aramco's *entire* 2017 revenues. Of course, 20% may be too large an estimate of the anthropogenic role, but the damages tallied are also incomplete and do not account for heat-related mortality, decreased crop yield, increased electricity demand, and other factors such as negative feedback loops from shrinking snow and ice cover, or methane releases from thawing permafrost (Fig. 12.7).[17]

Finally, the 2018 IPCC report estimates global economic damages from 2 °C warming by 2100 at the nearly unfathomable sum of $69 trillion (Intergovernmental Panel on Climate Change n.d.). Tallying the additional damage from a *further* degree of warming, including changes in coastlines and increased human migration and conflict, threatens to render an already tenuous concept inconceivable. It is hard to envision anything pragmatic about a goal based upon 3 °C warming, except perhaps in the very short term.

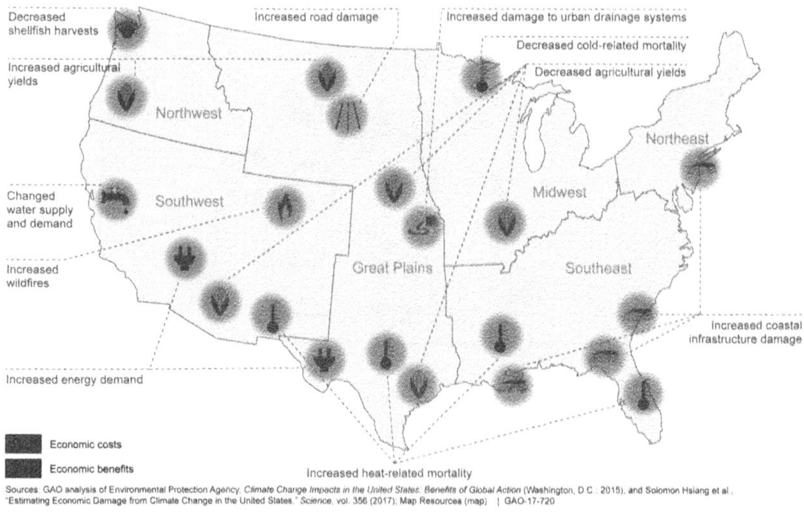


**Fig. 12.7** Examples of potential economic effects from climate change by 2100. (Source: TBD)

[17] It is worth adding that no amount of mitigation spending in 2017 would have affected climate change-related disasters in 2017. Results could only be expected in future years. Effective mitigation requires GHG reductions outside the United States as well. Further, the future "payoff" in tempering climate change could be difficult to measure given natural climate variability, making current spending justifications difficult.

In summary, the "throw in" strategy revolves around speculation that humanity will be better off by delaying strict mitigation because improved technology will emerge in the future and reduce GHG emissions without terminating the fossil fuel industry. Given that such technologies have neither been demonstrated nor deployed, the "throw in" strategy may be described as a multilateral version of Saudi Arabia's obstructionism approach.

## 8 Conclusion: A Fourth Strategy

The three strategy types above outline near-term practices producer states might adopt to coexist with climate action, seeking not just to survive, but to recast their businesses in ways that provide competitive advantages.

Despite the increasing climate consciousness in Saudi Arabia and its improved relations with the global climate regime, some of the strategies outlined here are best described as *alterations* in the manner in which fossil fuels harm the Earth's climate, geography, inhabitants, and their property. While improvements in upstream production techniques can assist at the margins, it is worth noting that the fossil fuel sector—oil, gas, and coal—remains responsible for two-thirds of the ongoing GHG emissions that continue to accumulate in the atmosphere. To the extent that strategies outlined here assist producers with marketing fuels that continue to be combusted in unabated fashion, they prolong damage to the Earth's climate, despite providing short-run economic benefits, particularly in developing countries.

A fourth strategy—briefly mentioned in the introduction—offers greater promise than the three above: Diversification beyond the oil and gas business. For companies, particularly shareholder-owned oil companies, diversification is part of the constant challenge of adjusting to evolving markets. When governments nationalized their oil concessions in the 1970s, the big Western oil companies created new opportunities elsewhere. Climate action is hastening the next major shift in the energy business.

For producer countries, nonoil diversification also makes sense. The more prudent of these states have actively begun creating new economic sectors that complement and eventually can replace those facing the risk of climate action. Ironically, a robust fossil fuel export sector is useful in funding investments aimed at diversifying away from fossil fuels. Some oil producers have taken steps in this direction. The United Arab

Emirates—and Dubai in particular—have built diversified economies that are already unwinding lopsided dependence on oil exports and prices. Saudi Arabia has announced a similar effort. Structural barriers in the GCC labor markets add complexity to diversification.

Perhaps the largest hurdle to diversification is, as mentioned, the continued availability of economic rents that have persisted in the oil business for more than a century. Inevitably, diversification will bring profits that appear disappointing in comparison to the lopsided earnings from low-cost oil.

Regardless, the oil market appears likely to grow more competitive as demand shifts into reverse in the OECD, and developing Asia becomes the main growth focus. How oil producing countries adapt in the long run remains in question.

A more chaotic oil market could be one result. Producers might seek a more bilateral route to marketing of oil, pairing with importers in an environment of enhanced competition. Hydrocarbon states which cannot compete may find themselves sidelined—either for geological (cost) reasons or above-ground political and institutional deficiencies, or even importer embargo.

Alternatively, producer countries could pursue a more cooperative path. One envisions an enhanced role for OPEC in allocating producers with equitable shares of a shrinking global oil market in the best interests of exporters and the global climate. If that happened, OPEC's organizing principle might shift. Rather than managing the price of oil, it might allot member countries a share of the market based not just on a member state's output capacity, but on the climate credentials of the oil it produced. The preparations outlined in Saudi Arabia, the world's low-cost, low-carbon oil producer, would place it in a strong position in either scenario.

## References

Boonen, Katrien, and Ive Vanderreydt. 2017. "Evolution of Environmental Impacts and Benefits of Motor Oil from a Lifecycle Perspective." Sept. 12, 2017. http://www.essenscia.be/en/Document/Download/16770.

BP. 2018. "Statistical Review of World Energy 2018." Accessed July 25, 2019. https://www.bp.com/en/global/corporate/energy-economics/statistical-review-of-world-energy/downloads.html.

Brelsford, Robert. 2018. "Aramco, SABIC let contract for crude oil-to-chemicals complex." Oil and Gas Journal, April 30, 2018. https://www.ogj.com/articles/2018/04/aramco-sabic-let-contract-for-crude-oil-to-chemicals-complex.html.

Depledge, Joanna. 2008. "Striving for No: Saudi Arabia in the Climate Change Regime." Global Environmental Politics 8, no. 4: 9–35. https://doi.org/10.1162/glep.2008.8.4.9.

El Gamal, Rania. 2018. "Saudi Aramco eyes partnerships as it expands refining, petrochems." Reuters, June 12, 2018. https://www.reuters.com/article/us-saudi-aramco-downstream-exclusive/exclusive-saudi-aramco-eyes-partnerships-as-it-expands-refining-petrochems-idUSKBN1J81M9.

Electricity Co-Generation and Regulatory Authority, "Annual Statistical Booklet for Electricity and Seawater Desalination Industries." Riyadh, ECRA 2017. https://www.ecra.gov.sa/en-us/MediaCenter/DocLib2/Lists/SubCategory_Library/ECRA-Statistical-Booklet-2017.pdf.

Emissions Database for Global Atmospheric Research. 2017. "Fossil $CO_2$ & GHG Emissions of All World Countries, 2017." European Commission science database. Accessed July 25, 2019. http://edgar.jrc.ec.europa.eu/overview.php?v=CO2andGHG1970-2016.

Emissions Database for Global Atmospheric Research. 2011. "Global Emissions EDGAR v4.2 (November 2011)," European Commission science database. Accessed July 25, 2019. http://edgar.jrc.ec.europa.eu/overview.php?v=42.

Ezekoye, Obi, Andjelka Milutinovic, and Theo Jan Simons. 2018. 2018, May. "Chemicals and capital markets: Back at the top." McKinsey. https://www.mckinsey.com/industries/chemicals/our-insights/chemicals-and-capital-markets-back-at-the-top.

Global Methane Initiative. n.d. "Partner Countries: Saudi Arabia." Accessed July 25, 2019. https://www.globalmethane.org/partners/country.aspx?country=saudiarabia.

Gramer, Robbie. 2017. "Saudi Arabia Now Controls the Largest Oil Refinery in North America." Foreign Policy, May 3, 2017. https://foreignpolicy.com/2017/05/03/saudi-arabia-now-controls-the-largest-oil-refinery-in-north-america-energy-middle-east-aramco/.

Intergovernmental Panel on Climate Change. n.d. "Impacts of 1.5 °C of Global Warming on Natural and Human Systems." p. 264. Accessed July 25, 2019. https://www.ipcc.ch/site/assets/uploads/sites/2/2019/02/SR15_Chapter3_Low_Res.pdf.

International Energy Agency. 2018. "The Future of Petrochemicals: Towards More Sustainable Plastics and Fertilisers." https://webstore.iea.org/download/direct/2310?fileName=The_Future_of_Petrochemicals.pdf.

Jackson, RB, C Le Quéré, RM Andrew, JG Canadell, JI Korsbakken, Z Liu, GP Peters, and B Zheng. 2018. "Global Energy Growth Is Outpacing Decarbonization." Environmental Research Letters 13, no. 12: 120401. https://iopscience.iop.org/article/10.1088/1748-9326/aaf303/meta.

Krane, Jim. 2014. "Stability versus Sustainability: Energy Policy in the Gulf Monarchies." PhD dissertation, University of Cambridge. https://doi.org/10.17863/CAM.5943.

Krane, Jim. 2015a. "A Refined Approach: Saudi Arabia Moves beyond Crude." Energy Policy 82: 99–104. https://doi.org/10.1016/j.enpol.2015.03.008.

Krane, Jim. 2015b. "Rationalizing Energy Demand through End-User Prices in the GCC." Oxford Energy Forum, no. 102 (November).

Lee, Heesu and Serene Cheong. 2018. "A Saudi-Backed Asia Refinery Is Going to Be a Fuel Juggernaut." Bloomberg, May 17, 2018. https://www.bloomberg.com/news/articles/2018-05-17/fuel-juggernaut-coming-as-saudi-backed-asia-refinery-giant-stirs.

Masnadi, Mohammad S, Hassan M El-Houjeiri, Dominik Schunack, Yunpo Li, Samori O Roberts, Steven Przesmitzki, Adam R Brandt, and Michael Wang. 2018. "Well-to-Refinery Emissions and Net-Energy Analysis of China's Crude-Oil Supply." Nature Energy 3, no. 3: 220. https://doi.org/10.1038/s41560-018-0090-7.

Ministry of Energy, Industry & Mineral Resources of Saudi Arabia. 2017. "Saudi Arabia Takes Its Place as Global Citizen at COP 22." https://docs.wixstatic.com/ugd/bc5fc9_455265c3bb8a41db987af3d3f72308cc.pdf.

Pal, Jeremy S., and Elfatih A. B. Eltahir. 2016. "Future Temperature in Southwest Asia Projected to Exceed a Threshold for Human Adaptability." Nature Climate Change 6, no. 2 (February): 197–200. https://www.nature.com/articles/nclimate2833?platform=oscar&draft=journal.

Plumer, Brad and Lisa Friedman. 2018. "Trump Team Pushes Fossil Fuels at Climate Talks. Protests Erupt, but Allies Emerge, Too." New York Times, Dec. 10, 2018. https://www.nytimes.com/2018/12/10/climate/katowice-climate-talks-cop24.html.

Saudi Aramco. 2018. "Saudi Aramco, Mazda Motor Corporation and AIST collaborate to develop the world's most advanced engine/fuel combination." Aug. 8, 2018. http://www.saudiaramco.com/en/home/news-media/news/Saudi-Aramco-Mazda-and-AIST-collaborate-on-advanced-engine.html.

Showa Shell Sekiyu K.K. 2016. "Corporate Report 2017." p. 101. Dec. 31, 2016. https://www.idemitsu.com/content/100866140.pdf.

Smith, Adam B. 2018. "2017 U.S. billion-dollar weather and climate disasters: a historic year in context." NOAA, Jan. 8, 2018. https://www.climate.gov/news-features/blogs/beyond-data/2017-us-billion-dollar-weather-and-climate-disasters-historic-year.

Smith, James B. 2010. "What Concerns Saudi Arabia about the Future of Energy and Climate Change." Diplomatic cable. Riyadh: US State Department via WikiLeaks, February 21, 2010.

S-Oil. n.d. "History." Accessed July 25, 2019. http://www.s-oil.com/en/company/History.aspx.

US Government Accountability Office. n.d. "Climate Change Funding and Management." Accessed July 25, 2019. https://www.gao.gov/key_issues/climate_change_funding_management/issue_summary.

Van de Graaf, Thijs, and Aviel Verbruggen. 2014. "Saving OPEC: How Oil Producers Can Counteract the Global Decline in Demand." Foreign Affairs, December 22, 2014. https://www.foreignaffairs.com/articles/persian-gulf/2014-12-22/saving-opec.

Van de Graaf, Thijs, and Aviel Verbruggen. 2015. "The Oil Endgame: Strategies of Oil Exporters in a Carbon-Constrained World," Environmental Science & Policy 54: 456–62. https://doi.org/10.1016/j.envsci.2015.08.004.

Van de Graaf, Thijs. 2018. "Battling for a Shrinking Market: Oil Producers, the Renewables Revolution, and the Risk of Stranded Assets." In The Geopolitics of Renewables, 97–121. Springer.

Vu, Khanh. 2018. "Vietnam's Second Refinery To Be Fully Operational By Early August." Reuters, June 6, 2018. https://www.rigzone.com/news/wire/vietnams_second_refinery_to_be_fully_operational_by_early_august-06-jun-2018-155842-article.

World Bank. n.d. "Global Gas Flaring Reduction Partnership (GGFR): Upstream Gas Flaring." Accessed July 25, 2019. http://www.worldbank.org/en/programs/gasflaringreduction#7.

CHAPTER 13

# The Sustainability of GCC Development Under the New Global Oil Order

*Ibrahim A. Elbadawi and Samir Makdisi*

## 1 Introduction

The six-member countries of the Gulf Cooperation Council (GCC) have been able to sustain fast growth for over 30 years, since the start of the oil price boom of the mid-1970s. Though growth has of course been volatile, following the various oil cycles, it continued to rise or remained stable at high rates throughout the period. And, some GCC countries, such as the UAE, and most notably the sub-economy of the Emirate of Dubai, have managed to achieve tremendous economic transformations towards more diversified and sophisticated economic structures. Moreover, the GCC monarchies have been able to avoid civil wars and maintain civil peace.

The authors would like to acknowledge the helpful comments of Layal Mansour on an earlier draft.

I. A. Elbadawi
Economic Research Forum, Cairo, Egypt

S. Makdisi (✉)
The American University of Beirut, Beirut, Lebanon



G. Luciani, T. Moerenhout (eds.), *When Can Oil Economies Be Deemed Sustainable?*, The Political Economy of the Middle East,
https://doi.org/10.1007/978-981-15-5728-6_13

They have also presided over a somewhat functioning social contract that helped them to fend off the fallout from the Arab Spring, which swept several major countries in their neighborhood.

At the other extreme, the populous oil Arab economies[1] have experienced the "resource curse" big time: a disappointing growth record, limited economic diversification, depleted physical and soft infrastructures, massive unemployment, especially among youth, and political instability, even civil wars.

And in terms of policies and economic institutions, the contrast of the two groups is equally stark (Elbadawi et al. 2017). The GCC countries have been able to maintain macroeconomic stability through a credible de facto institutional exchange rate peg regime, supported by massive foreign account surpluses in the form of reserves accumulation and sovereign wealth funds (SWFs). Moreover, the deep fiscal pockets of the GCC countries allow fiscal stabilization against the oil market shocks, most notably ensuring uninterrupted large fiscal outlays for financing infrastructure development and other social sector programs as well as expansion of public sector employment for nationals. Also, their open trade, capital accounts and labor market policies provided a good measure of macroeconomic competitiveness to the GCC economies, hence moderating the adverse consequences of the rigidity of the exchange rate regime. By contrast, the populous oil Arab economies have been plagued by intermittent bouts of inflationary spells, overvalued real exchange rates and fiscal dominance and less buoyant growth.

The received literature suggests that the "resource curse" is a long-term phenomenon but is conditional on poor political governance. Empirical evidence shows that in resource economies lacking political inclusiveness and effective "checks and balances" institutions, resource rents are likely to be a hindrance rather than a boon to growth. Moreover, resource rents tend to have corrosive effects on the quality of economic and political institutions (e.g. Collier and Goderis 2009; Elbadawi and Soto 2012a). Also, the evidence from the empirical strand of the civil war literature finds that, in the absence of political inclusiveness and accountability, the

[1] Though Saudi Arabia is relatively populous, in terms of rents per capita, it is more comparable to the high rents per capita GCC countries of Kuwait, Qatar and UAE than the other populous oil Arab economies of Algeria, Iraq and Yemen. As discussed in the literature, rents per capita above a certain threshold have been a major factor in explaining the superior performance of the GCC economies relative to the populous oil Arab economies (e.g. Ali and Elbadawi 2012; Elbadawi 2016).

presence of resource rents increases the risks of conflicts for both “loot” and “grievance” motives (e.g. Collier and Hoeffler 2004; Bodea and Elbadawi 2007).

However, while the literature seems to adequately account for the poor management experiences of the populous oil Arab economies and most other resource-abundant countries, it seems to provide little predictive power for the case of the GCC. In fact, Elbadawi (2016) has shown that there is a small group of 12 exorbitantly high oil rent per capita countries that are mostly non-democratic (hereafter HRPC), including the GCC, which seem to defy the fundamental tenets of the received literature that non-democratic governance affects growth negatively. He finds that there exists a resource rents threshold, beyond which countries might be able to achieve fast growth and civil peace as well as avoid revolt regardless of the prevailing standard in terms of accountable political institutions. This finding lends support to theory suggesting that high-enough resource rents per capita could lead to a developmental, if non-democratic, political equilibrium (Ali and Elbadawi 2012).

This chapter reviews the development experiences of the GCC countries in the context of three development goals: growth, civil peace and political stability; it undertakes a synthesis of the literature, largely drawing from the authors’ own work. However, the main contribution of the chapter relates to identifying the prospects for the sustainability of the GCC development model in the emerging new “global oil order”, expected to eventually lead to a low oil price “equilibrium” in the longer run.

Section 2 reviews the received literature on the causal links from rents per capita to growth, especially the evidence on the existence of non-monotonic growth effects. Section 3 reviews the literature on the scale-contingent role of resource rents as a determinant of conflicts and democratic transitions. In this context, it discusses the significance of threshold effects in explaining the apparent success of the GCC monarchies in avoiding civil wars as well as pre-empting potential revolts. Section 4, which is the hallmark of this paper, asks the pivotal question as to the sustainability of the GCC development model in the emerging new global oil order. Section 5 concludes.

## 2 Growth and Institutions in the GCC: Resource Rents Threshold Matters

As discussed, the received literature suggests that resource-abundant[2] economies tend to grow slowly and that such countries are likely to be impacted by the "resource curse", with only few exceptional cases of countries that happen to be endowed with robust political and economic institutions (e.g. Collier and Goderis 2009; Elbadawi and Soto 2012a). However, though this literature has been borne out by the experiences of most resource-abundant countries, the six GCC countries seem to have defied this empirical finding. They have been able to grow rapidly over the longer term, despite their entrenched non-democratic governance.

We argue that the level of resource abundance, as measured by resource rents per capita, is central to explaining the superior growth performance of the GCC, compared to other less endowed populous oil and mineral economies. The distribution of countries by this criterion reveals considerable heterogeneity. For the period 2000–2013, the median rents per capita for the HRPC group (comprising countries with rent per capita in excess of $2000) was $5322, more than eight times the median for the second most resource-abundant group of 15 countries, whose rents per capita are within the range of $400 to $2000 (Table 13.1). In view of their vast hydrocarbon resource base and relatively small populations, it is not surprising that the high resource-endowed GCC countries have dominated the list of the top ten countries in terms of rents per capita. It is, however, notable that they also dominated the top ten fastest growing countries during the last three decades (Figs. 13.1 and 13.2). By comparison, the populous oil Arab economies, among other resource economies, achieved much lower growth rates.

The comparative growth experiences of the GCC and other less resource-abundant countries suggest the existence of association, not necessarily causation, between unusually high rents per capita and long-term growth. Moreover, governance data also reveal that the median country from the HRPC group, including the GCC, receives the lowest scores in two global political governance indicators: Polity2, the widely accepted

[2] Throughout this paper, we use the concept of "resource abundance" (measured by resource rents per capita), as opposed to the alternative concept of "resource dependency", which is given by rents/GDP. The latter has been criticized in the empirical growth and conflict literature, due to reverse causation (Elbadawi and Soto 2012a; Ross 2004, 2009).

**Table 13.1** Typology of resource rents per capita (average: 2000–2013)

| *Resource rents per capita* | *Median* | *Min* | *Max* | *Number of countries* |
|---|---|---|---|---|
| Rents Pc < 20 USD | 9 | Iceland (0.06 USD/capita) | Cambodia (20 USD/capita) | 46 |
| 20 < Rents Pc < 200 | 47 | Switzerland (20.01 USD/capita) | Bhutan (189 USD/capita) | 88 |
| 200 < Rents Pc < 400 | 294 | Brazil (200 USD/capita) | United Kingdom (398.5 USD/capita) | 16 |
| 400 < Rents Pc < 2000 | 616 | United States (409 USD/capita) | Venezuela, RB (1424.3 USD/capita) | 15 |
| Rents Pc > 2000 USD | 5322 | Gabon (2326 USD/capita) | Qatar (70,537 USD/capita) | 12 |

Source: Elbadawi (2016)

Notes: (1) Total natural resources rents are the sum of oil rents, natural gas rents, coal rents (hard and soft), mineral rents and forest rents. (2) For the GCC, resource rents per capita were computed as a ratio of total rents to national population, where the latter was calculated using 49% in Bahrain, 68% for Saudi Arabia, 61% for Oman, 16% for the UAE, 32% for Kuwait and 16% for Qatar

measure of political inclusiveness; and Polcon, a new index of the effectiveness of political constraints on the executive and legislative branches of government[3] (Fig. 13.3).

At the same time, looking at the HRPC group, and specifically the GCC, their record of avoiding civil wars and maintaining stability is impressive. It stands in contrast to that of other resource-dependent

[3]The Polity2 is a global database developed by the "Polity IV: Regime Authority Characteristics and Transitions Datasets" project; and the Political Constraint Index (POLCON-V) is also a global index due to Henisz and Zelner (2010).

It is not surprising that most country groups, especially the HRPC, received low scores on both indices (Fig. 13.3). A more detailed country-specific analysis of the two indexes suggests that the Polity2 index (rescaled so that 1 stands for full democracy, and 0 stands for extreme autocracy) for the median political regime in the three country groups with rents less than $400 is likely to be "partial democracy". Instead, for the group with rents less than 2000 but higher than $400, which includes the populous oil economies of Algeria and Iran, it is likely to be an "autocracy"; and except for Norway and Trinidad and Tobago, the rest of the countries in the highest resource-abundant group are all judged to be extreme autocracies. Furthermore, the resource-rich countries tend to fare much worse according to the Polcon, with the median index below 0.4 for all groups, which suggests weak systems of political checks and balances. Again, except for the two democracies of Norway and Trinidad and Tobago, the Polcon index for the HRPC group is close to zero, suggesting that a system of "formal" checks and balances does not even exist in these countries (Elbadawi 2016).

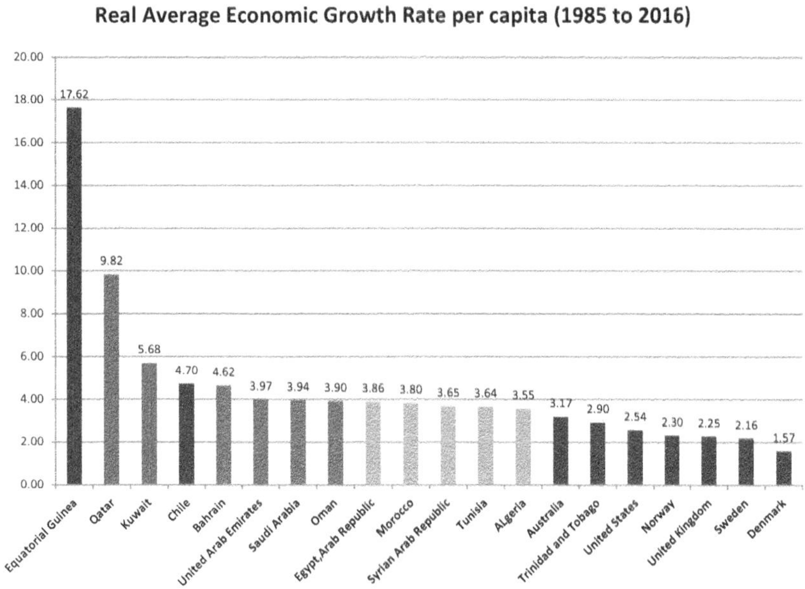


**Fig. 13.1** Average rent per capita (2000 to 2016). (Source: World Bank-International Financial Statistics (IFS) and authors' own elaboration. Notes: (1) Total natural resources rents are the sum of oil rents, natural gas rents, coal rents (hard and soft), mineral rents and forest rents. (2) National population for GCC countries are taken from Gulf Labor Markets, Migration and population (GLLM) programme (https://gulfmigration.org/glmm-database/) since national population represents a very small portion of the total population)

countries [populous?] which have been impacted by frequent and

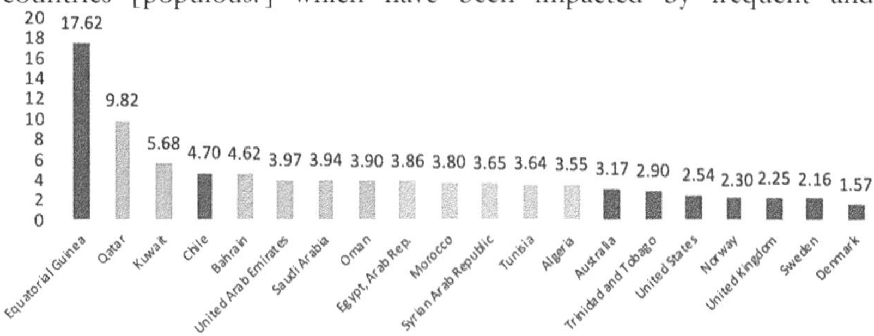


**Fig. 13.2** Real average economic growth rate per capita (1985 to 2016). (Source: World Bank-IFS Data and authors' elaboration)

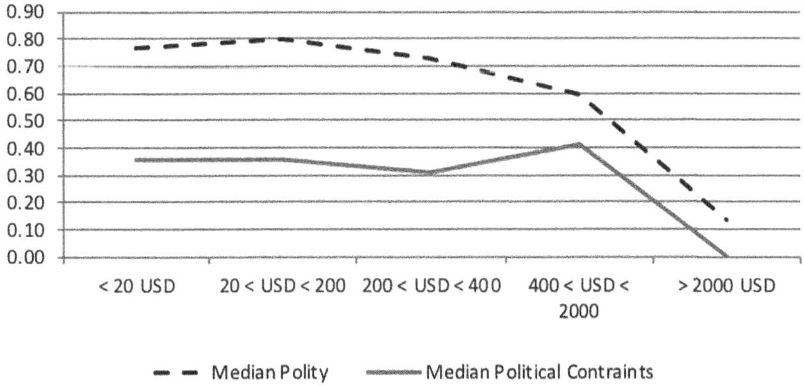


**Fig. 13.3** Median average polity and political constrains in resource-rich economies. (Source: Table 10.3 of Elbadawi 2016)

**Table 13.2** Armed civil conflicts, 1970–2010

| *Resource rents per capita* | *Number of countries* | *Longest war duration* |
|---|---|---|
| Rents Pc < 20 USD | 10 | 41 years |
| 20 < Rents Pc < 200 | 36 | 41 years |
| 200 < Rents Pc < 400 | 4 | 41 years |
| 400 < Rents Pc < 2000 | 7 | 26 years |
| Rents Pc > 2000 USD | 2 | 1 year |

Source: Table 10.2 of Elbadawi (2016)

long-duration conflicts (Table 13.2). Similarly, the long-reigning GCC monarchies have not only been able to maintain domestic political stability, but to avoid, so far at least, the contagion effects of the Arab Spring (Table 13.3). While it is not surprising for full democracies, such as Norway and Trinidad and Tobago, to enjoy political stability, this has been a challenging goal for authoritarian elites in relatively limited resource-abundant societies (Elbadawi 2016). The explanation, as discussed later, relates to the role played by relatively abundant resource rents.

The aforementioned evidence naturally motivates the question as to whether unusually high resource abundance (such as that of the GCC and other countries in the HRPC group) is likely to promote growth regardless of the quality of the underlying political institutions. This question was addressed by Elbadawi (2016), who, firstly, estimates a baseline

**Table 13.3** Political regimes types and years in power

| *Country* | *Duration and regime type* | | |
|---|---|---|---|
| Rents per capita > $2000 | | | |
| Kuwait | 1962 | Present | Monarchy |
| Saudi Arabia | 1928 | Present | Monarchy |
| Oman | 1742 | Present | Monarchy |
| UAE | 1972 | Present | Monarchy |
| Bahrain | 1971 | Present | Constitutional monarchy |
| Qatar | 1971 | Present | Monarchy |
| Brunei | 1984 | Present | Constitutional monarchy |
| Norway | 1886 | Present | Democracy |
| Trinidad and Tobago | 1962 | Present | Democracy |
| Gabon | 1961 | Present | Party-personal |
| Equatorial Guinea | 1968 | Present | Personal-military |
| Equatorial Guinea | 1979 | 1978 | Personal-military |
| Libya | 1952 | 1969 | Monarchy |
| Libya | 1970 | 2010 | Personal |

Source: Table 10.3 of Elbadawi (2016)

growth model that accounts for the level resource rents per capita and its squared value as well as other standard growth fundamentals other than political governance institutions. He finds that there exists a statistically significant but non-monotonic resource-abundance effect on growth, where resource abundance constitutes a hindrance to growth for countries with rents per capita lower than $1800, while those on the higher side of the tipping level should be able to *unconditionally* turn the resource rents into positive growth *regardless of the quality of the underlying institutions.* In view of such exorbitantly high threshold, only the HRPC group, especially the GCC, qualifies for such type of the growth process. In the extended model that also accounts for economic and political institutions, Elbadawi estimates a much lower tipping point at $148 when he controls for both Polity and Polcon. These results suggest that, conditional on good political governance, societies with moderate to high rents per capita (currently about 51) should be able to realize positive *conditional* growth from their resource endowments.

However, though the tipping point of the rent per capita was significantly reduced after controlling for the two indicators of political governance, the effect of rents remains. This suggests that whatever the GCC political construct, growth performance of the region has been anchored by their unusually high rents per capita, and that the ensuing growth-promoting political equilibrium has not been explicable in terms of the two global political indicators.

## 3 Maintaining Civil Peace and Political Stability: Resource Rents Threshold also Matters

Looking at the HRPC group, and specifically the GCC, noteworthy is their record of avoiding civil wars and maintaining stability. It stands in contrast to that of other resource-dependent countries [populous?] which have been impacted by frequent and long-duration conflicts (Table 13.2). Similarly, the long-reigning GCC monarchies have not only been able to maintain domestic political stability but to avoid, so far at least, the contagion effects of the Arab Spring (Table 13.3). While it is not surprising for full democracies, such as Norway and Trinidad and Tobago, to enjoy political stability, this has been a challenging goal for authoritarian elites in relatively limited resource-abundant societies (Elbadawi 2016). The explanation, as discussed later, relates to the role played by relatively abundant resource rents.

### *3.1 Resource Rents, Threshold Effects and Civil Wars*

Like growth collapse, civil wars have been prominently analyzed in the economic and political science literature as a major manifestation of the resource curse.[4] It has been argued that the existence of a "lootable" resource base is both a motive for rebellion and a facilitating factor. Because they need to purchase arms and recruit fighters, rebel organizations cannot survive without access to natural resource rents or other sources of easy financing, such as financial and military aid from sympathetic diaspora or friendly governments. For example, in Collier and Hoeffler (2004), the equilibrium level of the risk of conflict is given by equating the marginal product of rebel labor and government strength (measured by the economic strength and the size of the military) to their respective marginal costs for a given level of natural resources. This

[4] See, for example, Collier and Hoeffler (2004), Fearon (2005), Miguel et al. 2004), Sambanis (2004), Bodea and Elbadawi (2007), Caselli and Coleman (2013) and Elbadawi and Soto (2015), as representative examples on natural resource and conflicts.

framework solves for the risk of civil war onset as increasing with the natural resource endowment and decreasing with the opportunity cost of rebellion.

Again, though the linear effect of resource rents on the hazard of conflicts seems to account for the experiences of many, if not most resource-endowed countries, it does not, however, cohere with the evidence from the HRPC group, especially the GCC. As argued by Elbadawi (2016), "very high resource rents might in fact be a deterrent to conflict, either through the 'Hobbesian' effect of providing the state with substantial resources to crush potential insurgency or through strengthening the 'authoritarian bargain', which proved to be effective in preempting incipient conflicts in highly resource endowed societies". Subscribing to this argument, he estimates a model to test for the marginal effect of the resource per capita for HRPC group, which is given by an interactive term between rents per capita and the dummy for the HRPC group. In addition to the level effect of the rents per capita and the two political institutional variables, the model also accounts for the standard controls in this literature, including income per capita and population size.

As in the received literature, the overall linear rents effect remains positive and highly significant, but the threshold effect was negative and highly significant as well. This suggests that the HRPC group was able to substantially ameliorate the risk of civil war associated with resource rents. Moreover, marginally expanding the threshold to include the 15 countries in the following group (with rents per capita higher than $400 but lower than $2000) renders the threshold effect insignificant. As for the two political institutions, political checks and balances were found to be robustly and negatively associated with the hazard of civil war. Moreover, checks and balances appear to have weakened the linear rents effect, which was reduced in terms of order of magnitude and degree of significance when the former is included in the regression. Instead, and contrary to the model's prediction, democracy was not found to exert any significant influence.

### *3.2 Resource Abundance, Threshold Effects and Democratic Transitions*

Until recently, the Arab world has been immune to the various democracy waves that swept the developing countries since the end of WWII. This prompted some scholars to claim that Arab societies are culturally different, attributing lagging Arab democratic development to the so-called neo-patriarchy thesis. However, the popular uprising that swept the Arab world since 2010 suggests this is not what explains "Arab exceptionalism" to democracy. An alternative thesis for explaining the limited democratization in the Arab world is provided by Elbadawi and Makdisi (2011, 2017) and Elbadawi et al. (2011), who argue that, along with conflicts, the dominance of the hydrocarbon resource on Arab economies appears to be the main culprit, because it allowed incumbent autocrats to build a much more durable "authoritarian bargain" than has been possible in less resource-dependent regions.

However, again the resource rents as instrument for "greasing" the authoritarian bargain are not likely to have a linear effect. This is why several incumbent regimes in the oil-rich but populous Arab countries have experienced repeated revolts, coups and, more recently, mass protests, while the GCC monarchies remain fairly safe. This points to the limits to the effectiveness of the authoritarian bargain for the former; hence, suggesting that the role of resource rents as a hindrance to democratic transition might be subject to scale effects, akin to the cases of growth and civil peace.

Ali and Elbadawi (2012) develop a game-theoretic political economy model that accounts for two strategies that might be adopted by an incumbent autocrat to avoid or prevent a revolt by the citizenry (see also Desai et al. 2009; Esfahani and Gurakar 2014; Hodler 2006; Robinson et al. 2006). They show that there exists a threshold of rent per capita, above which the dominant strategy will be to spend on public employment and other public goods. Instead, for those less-endowed societies (i.e. below the threshold), elites will primarily rely on political repression as the main strategy for holding on to power. The predictions of this model are strongly corroborated by the evidence provided by Elbadawi (2016).[5] Indeed, the HRPC group seems to rely much less on political repression than does the median country in the rest of the resource countries. Furthermore, the former, especially the GCC countries, spend much more

[5] The evidence provided by Elbadawi (2016) is based on a direct measure of government repression that is available in the recently developed Cingranelli–Richards data set (2008). This measure, called *Physical Integrity Rights*, constructs an annual variable that ranges from 0 (worst repression) to 8 (repression free) and accounts for the incidence of torture, extrajudicial killing, political imprisonment and disappearances that are attributable to the government.

(on per capita basis) on education and health as well as building vastly superior infrastructure.

To assess the effectiveness of the two alternative strategies in preventing a revolt, Elbadawi and Makdisi (2013) estimate an extended version of the modernization democratic transition model that accounts for the potential role of resource rents per capita and political repression as hindrances to democratic transition. In addition to the standard modernization controls, such as income per capita and economic growth, their model also accounts for the effects of home wars, unemployment and the legacy of previous political instability, given by the number of times a country experienced a transition from democracy to autocracy.

They find that political repression, like unemployment, seems to fully explain the resource rents effect as a hindrance to democracy in low or moderate rents per capita group, such as the populous oil Arab economies. However, it does not fully account for the resource effect for the highly endowed group, most notably the GCC. Drawing on the implications of these findings, the two authors interpret their findings as corroborating the predictions of Ali and Elbadawi's model, that is, in high-resource but population-scarce countries, the elites are likely to rely more on expanding public employment and less on political repression. Instead, the opposite is likely to happen in moderately endowed but populous countries.

## 4 GCC Development Sustainability: Some Challenges Ahead

No doubt the GCC had a very strong growth record (Fig. 13.2). However, growth has been largely driven by factor accumulation in the form of massive investments and even more through large-scale utilization of unskilled and semi-skilled labor, mainly drawn from the Indian subcontinent. This growth model has obviously served the GCC well and allowed the building of state-of-the-art infrastructure at exceptionally low cost of labor. But one major drawback of the GCC growth strategy has been low and even declining total factor as well as labor productivity. This feature of the growth process raises serious doubts about the sustainability of growth, indeed of the entire GCC development model. Probing deeper into this issue, let us consider the case of the UAE, the second largest economy in the GCC and the Arab world. The economy of the UAE, most notably that of the Emirate of Dubai, has achieved the most significant transformation. In about 35 years (from 1975 to 2010), the country's output multiplied by five times (in real PPP prices), with an annual average growth rate for the period at 5.4%. However, though this growth was among the highest in the world, it has been associated with disappointing productivity growth rates.

The local economy of Dubai, the poster child of economic diversifying growth in the entire region, provides even more impressive economic development experience. The Emirate's economy has multiplied by 11 times during roughly the same period (1975–2008), growing by an exceptional nine percent per annum during the last 15 years of that period (Elbadawi and Soto 2012b). Nonetheless, despite its exceptionally commendable growth and much deeper economic transformation, Dubai has not been able to distinguish itself in terms of productivity growth. In fact, both average labor and total factor productivities remained stagnant for a long period of time (1987–2008[6]) following the substantial decline in the early 1980s. This contrasts sharply with the other city-states, where average labor productivity has grown steadily with time (Fig. 13.4).

It has been argued that the prevailing social contract in the GCC, which essentially underpins the political settlement for resource mobilization and distribution in these societies, is not capable of generating high productivity growth, even in the most successful growth model of the UAE, and especially Dubai. In the context of this social contract, nationals are provided assured public employment at high-enough wages reflecting their wage reservation levels as wealthy societies, while granting the private sector a privileged access to an unlimited supply of unskilled foreign labor. The public sector commitment to meet the demand for employment by nationals provides strong incentives, for entrepreneurs prefer to employ labor-intensive rather than more sophisticated and high productivity production techniques.

One frequently cited reason is the so-called kafala system, "which provides significant market power derived from the fact that an employee is forbidden to change occupations while the contract is in force and is required to leave the country upon expiration. This absence of horizontal mobility allows employers to extract economic rents: when choosing production technologies, therefore, employers would tend to focus on labor intensive techniques that, in addition to the normal profit obtained from

[6]This trend will likely to have continued following the post-2008 global economic recession.

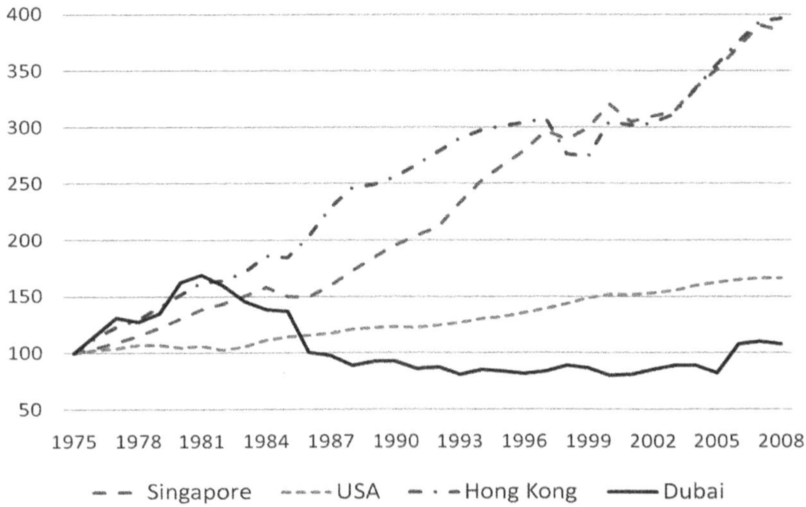


**Fig. 13.4** Real GDP per working-age person (2000 = 100). (Source: Elbadawi and Soto 2012b)

selling goods, would allow them to extract the highest rents from the worker. Naturally, this skews production towards labor-intensive technologies. It also biases employment toward low-skilled workers that have less ability to negotiate salaries in contrast with highly skilled workers are better educated and scarcer and, thus, better equipped to counterbalance the market power of employers. This aspect of the *kafala* would explain why labor productivity in the UAE is lower than in countries with similar GDP per capita" (Soto and Haouas 2012). However, even without the kafala system, the free access to the essentially unlimited supply of labor in the nearby southern Asian labor markets, coupled with large-scale public-sector contracts for entrepreneurs and developers in the construction and real estate sectors, would probably constitute a sufficient reason to steer their preferences toward the prevailing low-productivity model.

The 2008 global economic crisis and its aftermath have served a wakeup call to governments and societies at large about the need to reconsider the growth model and its underlying social contract. It is now a widely held view, by academic researchers and business experts alike, that eventually the price of oil will be converging to lower long-term trend, driven by the accelerating pace of technological advances in renewable energy as well as the shale oil and gas energy sources (Arezki and Matsumoto 2018). This realization has been borne out in the recent development strategies announced by GGC governments, all of which aimed at achieving transition to high-productivity, knowledge-based economies. Naturally, under the new oil order, such a transition is unavoidable, where these countries could no longer afford to use their deep fiscal pockets (e.g. reserves and SWFs) to finance expanding budgets or to maintain the credibility of their fixed exchange rate regimes. Moreover, guaranteed public sector employment of nationals is no longer possible in a few GCC countries, and major reforms to end the labor market segmentation will be required. However, the most difficult part of the new emerging GCC development agenda would be the realignment of political governance so as to allow for the required labor market and other public sector policy reforms.

At this juncture, perhaps the most compelling question to ask with regard to development sustainability of the GCC countries would be the following: In which direction might the political equilibrium in these societies shift as a consequence of the new global oil order?

To address this pivotal question, we consider the received institutional economics literature on the processes of formation and transition of economic and political systems in a society (e.g. North et al. 2009; Acemoglu 2008). According to this literature, a society is governed by a "limited access orders" (LAOs), when a relatively small but cohesive group of elites manage to exercise monopoly over the projection of organized violence and are able to restrict access to economic and political power. The elites, therefore, will tend to deploy the coercive power of the state in order to extract rents from the rest of the society. Furthermore, the restricted access to rents provides an incentive to the ruling elites to favor cooperation with each other and avoid or reduce violence against the population. An LAO might get trapped into a "fragile" state, when the ruling elites cannot agree on a formula for rents distribution due to the shifting balance of power among the various players. Instead, should the balance of power allow elite factions to establish a stable and predictable, if inter-personal, formula for distributing rents, the system becomes a "mature" limited access order (LAO). In this case, the elites have vested interest to invest and have the economy grow. However, even under mature LAOs, growth is not likely to be sustainable due to the marginalization of most of the population. Moreover, depending on external factors and internal developments alike, mature LAOs could very well collapse into a violent fragile state or, alternatively, graduate into an impersonal open access order (OAO).[7]

OAOs are built around impersonal rules that are broadly accepted by the society, and hence are self-reinforcing and not subject to ad hoc changes. These rules, or social contracts, must ensure open access as well as full accountability of those in positions of power to the general public. As articulated by Esfahani and Gurakar (2014), "The constraint on changes in the key rules under OAO is the cost of coordination and consensus building across large numbers of individuals and organizations. As a result, those who are elected to lead the state apparatus face constraints in getting other actors to help them change the rules in their favor, particularly because the other actors expect everyone else to hold them responsible if they deviate from the rules. Openness of access to political and economic leadership positions further means that those who are put in charge of political offices are replaceable by other contenders. So, they cannot expect to have much leverage over other members of the society to establish their own personal fiefdoms within the state. This in particular implies that the armed forces as state organizations cannot use the means at their disposal to violently gain direct or indirect control over the government. Rather, the military and security forces are placed under the consolidated control of civilians who represent the public" (p. 6).

Subscribing to the aforementioned conceptual framework, the prevailing GCC politico-economic institutional setup might be described as a relatively advanced form of "limited access order" (LAO), where projection of organized violence is firmly under the incumbent authorities and rents distribution is governed by inter-personal networks, but at the same time adjudication and settlement of disputes among the elites become increasingly formalized through courts and other formal arrangements. It follows that the prospects of significantly lower oil prices under the new global oil order, and the ensuing reductions in rents per capita, might very well push the GCC political equilibrium from an advanced LAO to a fragile state of LAO, akin to the one that prevails in some of the populous Arab oil countries. Such unfortunate transition would surely put an end to the GCC's relatively successful development model. On the other hand, it is possible that the path dependence of the political process in some at least of the GCC member countries might create some positive inertia,

[7] For example, Esfahani and Gurakar (2014) analyze the interplay of these factors in shaping the contrasting path from LAOs for the cases of Iran and Turkey.

leading them to develop into an impersonal "open access order" (OAO), whereby the society graduates to a stable institutional equilibrium.

## 5 Concluding Remarks

Thanks to relatively abundant resource rents, as measured by resource rents per capita above a specified threshold, the GCC countries have been able, thus far, not only to avoid the negative politico-economic consequences of the oil curse but also to generally maintain high rates of economic growth along with political stability. However, with the expected future persistence of a lower oil price equilibrium along with growing popular demands in the region for political change in the direction of democratic governance—whatever the outcome, so far, of the Arab uprisings of 2011—great challenges face the GCC countries.

Accordingly, we would argue, the stability of the prevailing "mature" LAOs that seem to characterize the economic and political order in the GCC will be seriously challenged by the prospects of lower oil prices in the future, along with other emerging regional and domestic developments, such as the youth bulge and the rising regional tensions and continuing civil wars and conflicts in the neighboring countries.

Thinking ahead, the fundamental question that would need to be addressed, in separate research, is whether or not the GCC countries are capable of initiating the required change in their political and economic model. And in either case, what would be the consequent politico-economic implications for the GCC in particular, but also for the wider region?

## References

Acemoglu, Daron (2008). "Oligarchic vs. Democratic Societies." Journal of the European Economic Association 6.1: 1–44.

Ali, Omer and Ibrahim Elbadawi (2012). "The Political Economy of Public Sector Employment in Resource Dependent Countries," ERF Working Paper # 673, the Economic Research Forum, Cairo, Egypt.

Arezki, Rabah and Akito Matsumoto (2018). Shifting Commodity Markets in a Globalized World. (Editors). International Monetary Fund, Washington DC, USA.

Bodea, C. and I. Elbadawi (2007). "Riots, coups and Civil War: Revisiting the Greed and Grievance Debate," World Bank Policy Research Working Paper 4397, Washington, DC.

Caselli, F. and W. J. Coleman II (2013). "On the Theory of Ethnic Conflict," Journal of the European Economic Association, 11(Supplement 1): 161–92.

Cingranelli, David L. and Richards, David L. (2008). Cingranelli-Richards (CIRI) Human Rights Dataset 2008 (Available from www.humanrightsdata.org).

Collier, P. and B. Goderis (2009). "Commodity prices, growth and the natural resource curse: reconciling a conundrum," CSAE Working Paper Series 2007–15, Centre for the Study of African Economies, University of Oxford.

Collier, P. and A. Hoeffler (2004). "Greed and Grievance in Civil War," Oxford Economic Papers 56: 563–95.

Desai, Raj M., Olofsgård, A. and Yousef, T. (2009). "The Logic of Authoritarian Bargains," Economics & Politics, 21:1, pp. 93–125.

Elbadawi, Ibrahim (2016). "Thresholds Matters: Resource Abundance, development and Democratic Transitions in the Arab World," in Ishac Diwan and Ahmed Galal (editors). The Middle East in Times of Transitions. Palgrave Macmillan.

Elbadawi, Ibrahim and Makdisi, Samir, ed. (2011). Democracy in the Arab World: Explaining the Deficit. Routledge.

Elbadawi, Ibrahim and Makdisi, Samir (2013). "Deconstructing Democratic Transitions in the Arab World," unpublished mimeo, Institute of Financial Economics, Department of Economics, The University of Beirut, Beirut, Lebanon.

Elbadawi, Ibrahim and Makdisi, Samir, ed. (2017). Democratic Transitions in the Arab World. Cambridge University Press.

Elbadawi, Ibrahim and Raimundo Soto (2012a). "Economic Growth During the Oil Cycle," ERF Working Paper # 678, the Economic Research Forum, Cairo, Egypt.

Elbadawi, Ibrahim and Raimundo Soto (2012b). "Sources of Economic Growth and Development Strategy in Dubai," in Ibrahim Elbadawi and Ali Alsadik (editors). The Global Economic Crisis and Implications for Dubai Strategy for Economic Development. Palgrave Macmillan.

Elbadawi, Ibrahim and Raimundo Soto (2015). "Resource Rents, Institutions and Violent Civil Conflicts," Defense and Peace Economics, 2015, vol. 26, issue 1: pp. 89–113.

Elbadawi, Ibrahim, Makdisi, Samir and Milante, Gary (2011). "Explaining the Arab Democracy Deficit," in Elbadawi and Makdisi (ed.), Democracy in the Arab World: Explaining the Deficit, Routledge, pp. 41–82.

Elbadawi, Ibrahim, Mohamed Goaid, Moez Ben Taher (2017). "Threshold Effects of FiscMonetary Interdependence and Exchange Rate Regimes in Oil-dependent Arab Economies," unpublished mimeo, the Economic Research Forum, Cairo, Egypt.

Esfahani, Hadi and Esra Ceviker Gurakar (2014). "Social Order, Rents, And Economic Development in Iran Since Early 20th Century," ERF Working Paper # 850, the Economic Research Forum, Cairo, Egypt.

Fearon, J. D. (2005). "Primary Commodity Exports and Civil War," Journal of Conflict Resolution 49: 483–507.

Henisz, W. and B. A. Zelner (2010). Measures of Political Risk Database, The McDonough School of Business, Georgetown University: http://www-management.wharton.upenn.edu/henisz/.

Hodler, R. (2006). "The curse of natural resources in fractionalized countries," European Economic Review 50: 1367–86.

Miguel, E., S. Satyanath, and E. Sergenti (2004). "Shocks and Civil Conflict: An Instrumental Variable Approach," Journal of Political Economy, 112(4), 725–53.

North, Douglass C., John J. Wallis, and Barry R. Weingast (2009). Violence and Social Orders: A Conceptual Framework for Interpreting Recorded Human History. Cambridge: Cambridge University Press.

Robinson, J. A., R. Torvik and T. Verdier (2006). "Political Foundations of the Resource Curse." Journal of Development Economics 79, 447–68.

Ross, M. (2009). "Oil and Democracy Revisited," unpublished mimeo, UCLA Department of Political Science Los Angeles, CA 90095, March.

Ross, M. (2004). "What Do We Know About Natural Resources and Civil War?," Journal of Peace Research 41: 337–56.

Sambanis, N. (2004). "What Is Civil War. Conceptual and Empirical Complexities of an Operational Definition," Journal of Conflict Resolution, 48(6): 814–58.

Soto, Raimundo and Ilham Haouas (2012). "Has the UAE Escaped the Oil Curse?," ERF Working Paper # 728, the Economic Research Forum, Cairo, Egypt.

CHAPTER 14

# When Can Oil Economies be Deemed Sustainable?

*Giacomo Luciani and Tom Moerenhout*

When can oil economies be deemed sustainable? We did not expect to find an unequivocal and final answer to this question when we set out collecting the chapters of this book, and in fact do not believe that an unequivocal and final answer exists. Our more modest purpose was to question a stereotype—that all Gulf economies are not sustainable—and start a critical discussion of what these economies and polities should do to guarantee themselves a relatively stable future.

Unsurprisingly, the different authors that contributed to this collection manifest different sensitivities and lay their emphasis on various aspects,

G. Luciani (✉)
Graduate Institute of International and Development Studies, Geneva, Switzerland

Paris School of International Affairs, Sciences Po, Paris, France
e-mail: giacomo.luciani@graduateinstitute.ch

T. Moerenhout
Columbia University, School of International and Public Affairs, New York, NY, USA
e-mail: tm2794@columbia.edu



G. Luciani, T. Moerenhout (eds.), *When Can Oil Economies Be Deemed Sustainable?*, The Political Economy of the Middle East, https://doi.org/10.1007/978-981-15-5728-6_14

but points of convergence are not missing, and they are important. In this last chapter, we shall try to pull together some of the strings and propose at least some overall conclusions, on which further debate and research may be encouraged.

## 1 The End of Oil

The most immediate way in which the issue of sustainability of the oil economies is commonly framed is to point to the time when oil runs out.

There really is no case of an oil-producing country having completely run out of oil, but having faced declines, yes. That said, globally, since the end of the first decade of the current century, we have ceased worrying about peak oil supply—a presumed physically determined fate—and now discuss peak oil demand and unburnable carbon, that is, the possibility that oil may lose its economic value because of the energy transition. The potential impact of declining oil rents on government finance is highlighted in almost all chapters in the book:

> *there is an inescapable imperative to prepare for a world in which hydrocarbons are no longer the main source of revenue, even if there may be alternative ways to monetize hydrocarbon resources that do not contribute to global emissions*—Al Saffar

However, there is also convergence on the view that the decline in oil rents will be a gradual phenomenon, and not a sudden one. On the one hand, global oil demand is likely to continue to be robust for a while, projected to increase at a compound annual average growth rate of 0.5% until 2040 (driven mainly by demand growth in India, China and the Middle East). On the other hand, oil-producing countries can also implement policies to resist or counter the process. On the first point, Fattouh and Sen write that a

> *key observation that can be drawn from the range of peak demand forecasts is that oil will continue to be an important part of the energy mix for the foreseeable future. The incumbent advantages of oil as an energy source, including its high energy density and an existing infrastructure ecosystem geared around it, imply that even if oil demand peaks, it is unlikely to 'fall off a cliff'.*

If we reason on the basis of a 20-year time horizon—certainly a very long-term one for fiscal, monetary, industrial and indeed almost all policies—we should accept that, while the oil rent is likely to decline over this period of time, it will not disappear altogether. The speed with which the oil rent may decline is a function of volumes sold and prices obtained. These two aspects are interlinked and very much depend on policies of major oil-importing countries, investment decisions of major national and international oil companies, producer subsidy allocations in hydrocarbon-rich countries and, more generally, how the economy will react to the combination of an energy transition and rising energy demand.

If the demand peak is driven by policies based on regulation or on imposing a high price on carbon, it is possible that supply may exceed demand and drive prices down. However, this is not the situation at the moment, and current concerns are rather that investment in the development of new oil sources may be insufficient to compensate for the natural decline of producing fields and unable to meet the still growing global demand. Therefore, the outlook may be one of growing, rather than declining, prices. The recent past has taught us that oil prices are on an unpredictable trajectory and that this uncertainty affects both investment in the sector and fiscal sustainability of many, but not all, producing countries. Limiting this volatility will be a key objective to secure a more stable oil rent (a point stressed by Manal Shehabi in Chap. 7).

Furthermore, a decline in global demand for hydrocarbons does not necessarily translate into a decline of demand for Gulf Cooperation Council (GCC) oil and gas exports. Hydrocarbons produced in the GCC are amongst the cheapest in the world, and there are strategies that the Gulf countries can adopt—and are adopting—to defend or increase their market share. On this point, several of our authors (notably Fattouh and Sen, Luciani and Krane) converge.

Fattouh and Sen believe that "As the world's lowest-cost oil producers, Arab oil-exporting countries will most likely be required to fill this gap", that is, the gap between investment required to meet expected future demand and what is actually being undertaken. Their main preoccupation is with the very significant investment effort required, as "any expansion in productive capacity will require massive investments running into billions of dollars". This in turn has political implications, because "Countries also require relatively stable political environments to make these investments. In the absence of such stability, it is possible that some countries with cheap reserves will be unable to develop these reserves".

On oil prices, they also stress the need for producing countries to have a proactive attitude towards coordination with major producers in other regions, beyond the temporary character of recent OPEC+ production decisions:

> *maintaining cooperation in a more competitive world is very challenging and while producers have the incentive to cooperate, the cooperation between producers has to take a different shape to what has existed in the past. For instance, producers should not only be concerned with low oil prices, but also be proactive when prices are too high, as high oil prices induce strong supply and demand responses and speed up the energy transition.*

Krane elaborates on three possible alternative strategies that the producing countries may follow in order to defend their hydrocarbon exports, which he dubs "Dig in", "Join in" and "Throw in". The first alternative includes developing non-fuel uses for hydrocarbons, decreasing the carbon emissions related to their production, improving the efficiency of internal combustion engines and integrating vertically into the markets of major importing countries, notably in Asia. The second alternative focuses on carbon capture and sequestration, reduction of flaring and emissions of greenhouse gases (GHG) different from $CO_2$ and promoting the uptake of non-carbon technologies, notably renewables and nuclear, to reduce the countries' own emissions. The last strategy envisages acceptance of global warming above endorsed goals in the Paris Agreement and is clearly not of the author's liking, but it pretty much corresponds to de facto behavior of all major emitting countries—whether we like it or not.

We find the same leading ideas expressed in other chapters of this book. Ali al Saffar writes about the need to "capture more value from hydrocarbons" and "using natural gas strategically in support of diversification goals", where he includes downstream integration and the development of petrochemicals. He also writes about "tapping the large, underutilized potential of renewables" and "phasing out subsidized use of energy" (the latter also the main focus of the chapter by Tom Moerenhout) to reduce domestic consumption of oil and redirect the same to exports. Finally, he advises "supporting the development of cleaner and more efficient energy technologies", where he includes Carbon capture, utilization and storage (CCUS). Similarly, Fattouh and Sen write: "Oil producers will need to be far more strategic in developing their energy sector, including the renewables sector, and strengthening forward and backward linkages to help diversify their economies".

Luciani supports essentially the same package of strategies, writing in particular about decarbonizing the production of oil, CCUS and integrating downstream toward non-fuel uses. He adds the perspective of the transition toward a hydrogen economy which would at least initially be based on producing hydrogen from hydrocarbons and capturing the resulting $CO_2$, rather than only looking at more expensive methods of hydrogen sourcing out of electrolysis or other biomass-based technologies.

Thus our discussion allows us to place the issue of sustainability in a perspective of gradualism where hydrocarbons remain an essential component of the GCC economies' competitiveness for several decades, which is as far into the future as one may meaningfully discuss. None of the authors believe that hydrocarbons alone can guarantee sustainability, but the view that the Gulf economies are not sustainable because oil and gas will soon and rapidly lose all economic value is not supported by current market trends as well as medium-term projections. Neither of course it is proposed that Gulf oil producers can be passive and keep enjoying the rent: on the contrary, much initiative and adaptation will be needed to defend the rent. But losing access to the oil rent is neither an inescapable fate nor likely a sudden development.

## 2 Need for Diversification?

Together with the belief that hydrocarbon rents may soon disappear, the discourse on sustainability is frequently associated with the need for economic diversification, that is, development of non-oil sectors, which is frequently interpreted as moving away from oil. In this view, all developments, either downstream or upstream from hydrocarbon production, are considered as doomed, since they depend on oil, which is expected to disappear. Unsurprisingly, when considering the trends and projections discussed earlier, our authors offer critical visions leading to a much more nuanced conclusion.

First of all, however, it is asserted that—contrary to the most widely accepted stereotype—diversification has already occurred. Luciani and Beutel both emphasize this point, based on different and complementary statistical evidence. Beutel, in particular, proposes a detailed analysis based on national accounts and input-output analysis, leading to some quite unexpected conclusions with respect to the degree of diversification of the Gulf economies, the pace of change and the international relativities—in particular with respect to Norway. Luciani rather relies on the Economic Complexity Index developed at Harvard University, and similarly points to some unexpected relativities. Skeptics will point to deficiencies in all of the statistical tools used in these two chapters: indeed, diversification is one of the most elusive characteristics of an economy, depending, as it does, on relative prices and peculiarities of classification. Nevertheless, the simultaneous use of several indicators and their convergence in showing that diversification has in fact taken place is a result to be taken into account. Whether such diversification has happened to a sufficient extent and whether it has occurred in sectors or stages of production that can contribute structurally to economic sustainability, remain open for discussion.

Does this mean that the level of diversification is not a problem? Certainly not, but one should keep in mind that "diversification" is not a panacea as not all diversification is necessarily positive. In Manal Shehabi's opinion, "setting diversification as the key recipe for economic sustainability in resource-rich states appears on face value inadequate and confusing". This is because diversification may take place with the development of activities, which may be commercially viable only contingent to existing conditions and policies, but not necessarily sustainable in the longer run. If such diversification is pursued at large scale, and subsequently locked in, then oil-producing states would find it difficult to be sufficiently adaptable, much like other non-oil producing states with the rise of global value chains.

From what was highlighted in the previous paragraph, it is clear that diversification may very well take place leveraging the comparative advantage in hydrocarbons and integrating upstream and downstream components of hydrocarbon production, pure and simple. In fact, many authors considered this as welcome, rather than optional. Progressively, this may contribute to the creation of a local market for more knowledge-intensive and higher value-added employment. This point is emphasized in particular by Martin Hvidt, who focuses on value chains as the directing principle for diversification: "adding steps of the value chain will imply a diversification of the product, and thus of markets".

> *this process of product diversification holds significant implications for a diversification of job content, and thus skills levels of the workers undertaking the jobs: The more diversification takes place within the value chain (adding more stages in the production chain), the greater varieties of jobs will be demanded. Generally, the more primary productive activities are pursued, the higher the demand for supporting activities, which implies jobs related to services. And*

*service related jobs generally have a higher knowledge content and provide more value added than jobs in 'pure' manufacturing or oil extraction.*

Similarly Shehabi writes: "economic specialization based on comparative advantage is not contradictory to that of economic diversification, and that both concepts are relevant to but insufficient for economic sustainability in resource-rich states". In a significant point of convergence, none of our contributing authors believe that diversification away from oil is a decisive factor of sustainability.

## 3 Fiscal Sustainability

In contrast to diversification away from oil, which is relativized in the discussion, the importance of fiscal sustainability is underlined by almost all our authors. The GCC governments remain excessively dependent on oil revenue for the financing of expenditure, and especially current expenditure. Individual authors offer somewhat different and interestingly complementary approaches to this crucially important aspect. Four main theses are proposed:

1. An important distinction must be made between higher- and lower-rent per capita countries within the GCC
2. Contrary to the view that the GCC implicit political pact is rigid, some significant fiscal adjustment has taken place since the 2014 collapse in oil prices
3. Reducing dependence on oil income is only possible if a sufficient domestic taxation base is developed, with very significant implications for the development strategy.
4. Less progress has been made in reducing public sector employment and nationalizing employment in the private sector, even though this is an imminent threat to the sustainability of GCC economies (discussed more later).

### *3.1 The Difference between Rentiers and Super Rentiers*

The importance of the level of per capita oil rent with respect to the question of sustainability is emphasized especially by Elbadawi and Makdisi, and Malik and Nagesh. Quoting previous research by Elbadawi, the first two authors assert that "there is a small group of 12 exorbitantly high oil rent per capita countries that are mostly non-democratic (hereafter HRPC), including the GCC, which seem to defy the fundamental tenets of the received literature that, non-democratic governance affects growth negatively. He finds that there exists a resource rents threshold, beyond which countries might be able to achieve fast growth and civil peace as well as avoid revolt regardless of the prevailing standard in terms of accountable political institutions". In other words, these super-rentiers, which currently in the GCC include Kuwait, Qatar and the United Arab Emirates (UAE) (certainly at least Abu Dhabi), have been able to escape the resource-curse and have sufficient financial resources to survive future social and political challenges.

Malik and Nagesh argue that "based on per-capita hydrocarbon endowments, the GCC countries can be broadly divided into two categories". On the one hand, there are countries with higher hydrocarbon reserves (and production) relative to their populations. They include Kuwait, Qatar and the UAE (Abu Dhabi led). On the other hand, there are hydrocarbon-poorer per capita countries with smaller hydrocarbon reserves (and production) relative to their populations. They include Bahrain, Oman and Saudi Arabia. Within this group, Oman and Bahrain are in relatively weaker positions versus Saudi.

The higher oil rent per capita countries have accumulated large foreign currency reserves, have maintained a budget surplus or small deficit even after the decline in oil prices and enjoy large sovereign funds whose revenue may in the future progressively substitute for a declining oil rent. It is therefore possible to envisage a scenario in which the state in these countries may continue to be financed from revenue directly accruing from abroad, rather than having to rely on domestic taxation. As a result, it can be expected that these countries will continue to implement pro-cyclical policies. These countries also display the highest share of expatriates in the resident population, meaning that the politically relevant share of the total population is small and relatively easy to be catered to. In contrast, in the three remaining members of the GCC, oil rents per capita are smaller either because oil revenue is less generous (Bahrain) or because the national population is much larger (Saudi Arabia) or a combination of both factors (Oman).

Thus, it is clear that the problem of sustainability is of a completely different nature in the two subsets of GCC member countries. For the higher-rent per capita countries, the essential problem is continuing to accumulate in their sovereign investment funds and making sure that returns on their international investment are sufficient to meet the fiscal requirements of the state. For the lower-rent countries, the issue of developing a domestic taxation base is inescapable and entails as a corollary, the need to develop competitive activities and sources of value added that the state may tap to raise the required income.

### 3.2 *Fiscal Adjustment*

A frequently repeated reason for doubting the sustainability of the Gulf economies is that the political order is based on an implicit pact between the rulers and their people that conditions the acceptance of the former on the part of the latter to the continued availability of historically established transfers in the form of employment in the public sector, free services and various consumption subsidies. The presumed rigidity of this implicit social contract is deemed to prevent adjustment and is therefore unsustainable.

The experience of fiscal policy reform since 2014 demonstrates that adjustment is on the contrary possible, at least partially. Adjustment has been sought through reduction of expenditure and, to a more limited extent, the imposition of new fees and taxes. "The initial response by governments was to substantially cut back government expenditure, especially on the capital front" (Malik & Nagesh). Abandoning or delaying investment projects has traditionally been the easiest step to reduce expenditure quickly, together with delaying payments to enterprises working for the government and implementing public projects—with the subsequent liquidity crisis, which inevitably ensues. But then steps have also been taken to reduce subsidies, especially on energy products.

Subsidies reform is the main focus of the chapter by Tom Moerenhout: "When the oil price plummeted in 2014, GCC revenues collapsed and all governments increased fuel and electricity prices. By doing so, they primarily avoided an even more menacing collapse of state budgets, which could have led to domestic instability". That is not to deny the fact that low fuel prices have justifications: they serve to contain increases in the cost of living, they enhance the competitiveness of energy-intensive industries and they contribute to political stability. But in the long run, adjustment is needed because consumption has grown out of control and endangers the export potential.

The pace of pricing reforms is a double-edged sword. On the one hand, there is urgency, not only for fiscal adjustment but also because of projected demand growth for oil and electricity in the Middle East and North Africa (MENA) region. This demand growth is substantial and carries large opportunity costs. Not for a small part, this demand growth will be caused by the need for energy-intensive end uses, such as running air-conditioning and salt water desalination as a result of climate change effects. On the other hand, however, Moerenhout warns that overly rapid price adjustments for industry might erode the comparative advantage of Gulf countries and complicate efforts to achieve within-sector diversification.

Since 2014, all GCC countries have enacted some reform of the subsidies system. Three countries—notably the UAE, which was the pioneer, Oman and Qatar—have been able to implement gradual but systematic reform, which created regular adjustment mechanisms linked to regional and international price movements. In contrast, Saudi Arabia, Kuwait and Bahrain have resorted to ad hoc adjustments that are not automatic—but the adjustment has nonetheless been very significant. Also important, the adjustments were not rolled back as soon as oil prices increased. Thus, we may conclude that the taboo of untouchable energy prices has been broken, and governments have learned that, under the appropriate conditions, they can impose adjustments to subsidies. Encouragingly, some countries have started with attempts to implement better targeted social welfare measures, such as cash transfers.

Similarly, new fees and taxes have been imposed by almost all of the GCC countries. True, there has been an attempt to increase the fees paid by expatriates rather than nationals, but this is mostly a question of appearances, because the cost of employing expatriates increases, and eventually this is felt in domestic prices paid by nationals. The imposition of fees on the presence of expatriates' family members has led to a wave of returns to their countries of origin, which has in turn translated into a slump in demand for real estate accommodation and a broad range of consumer goods. The net positive effect of such measures for enhancing government revenue is thus relatively limited.

The most important step undertaken in the GCC is the decision to implement a coordinated Value Added Tax (VAT), as detailed in Malik and Nagesh, although not all member countries may respect the engagement in the envisaged time window. The 5% rate at which the VAT is currently imposed is relatively small, but constitutes a very important development, especially from a political point of view. Certainly, VAT is a tax on consumption rather than income, and as such, it has less of a political impact in the relationship between each citizen and the state, but it is nevertheless a tax and may, in due course of time, prompt a request for greater fiscal transparency and accountability, especially if the rate were to increase significantly.

### *3.3 Developing a Domestic Taxation Base*

The difficulties experienced in imposing new taxes and diversifying the sources of state revenue point to the need for cultivating a domestic taxation base. This is the central message of Ishac Diwan in his chapter on fiscal sustainability and the labor market in Saudi Arabia. Although the chapter focuses specifically on the Saudi case, the crucial point about the need for developing a domestic tax base is valid for all lower-rent per capita countries; however, for higher-rent per capita states, much depends on whether they will be able to substitute the oil rent with income from international investment, thus dodging the need for diversification of income sources.

Diwan argues that "Expats do create a buffer for the economy, but are unlikely to be part of the tax base". This is because "expats are being paid their reservation wage, and while they can be fired at will, it is not possible to compress their net incomes by much and keep them employed". At the same time, the majority of nationals are government employees, and while they may earn extra income from private sector activities, the resources that may be squeezed out of household budgets by way of VAT or an income tax are perforce limited relative to government expenditure. Finally, corporate taxes are limited by total profits of private corporations, many among which are very small and presumably surviving on a precarious equilibrium.

Hence, the key conclusion and answer to "the question of interest is therefore what growth model can deliver a larger tax base?" is one that supports growth in the private sector and at the same time discourages employment of expatriates and encourages the employment of nationals and an increase in labor productivity, that is, value added per employee.

This crucial conclusion points to the fact that fiscal sustainability is not independent of the chosen growth strategy. Malik and Nagesh concur with this conclusion when they write, "Two factors will be central to boosting fiscal sustainability across the region, in our view: (i) the deepening of the tax base and (ii) lowering the wage composition of government spending". Of course, lowering the burden of wages on total government spending means reducing public sector employment and shifting some roles from the public to the private sector. Luciani writes about the need for developing taxation and notes "The need for nurturing a strong base for taxation is the main argument to conclude that economic sustainability eventually requires the consolidation of a competitive and independent private sector".

## 4 Sustainability and the Labor Market

Several of our authors underline the importance of reforming the GCC labor markets for sustainability. There are two opposing views of this issue, depending on whether we focus on high- or low-rent per capita countries. In the former group, dependence on expatriate labor is extreme, and provides an element of flexibility in the face of potential downturn of available rent, which may be said to considerably enhance sustainability. Shehabi writes "the expatriate labor exit offers another key stabilization mechanism of the Kuwaiti economy in the face of the volatility of petroleum revenue". The same can be said of Qatar and Abu Dhabi, although possibly not for the UAE as a whole. In other words, if push comes to shove and the revenues from oil exports and sovereign wealth funds collapse dramatically, these countries are able to send home a large share of their expatriate population. Radical and fast-paced changes would certainly allow for very significant reduction in government expenditure: it would also lead to a collapse of real estate and many businesses, which could also lead to the deterioration of effective standards of living for a part of the national population. That being said, a new equilibrium at a lower ratio of expatriates to nationals could be found without too much affecting the condition of most nationals.

The situation is different for low-rent per capita countries, because the flexibility offered by expatriate workers is proportionately lower (though still high compared to nearly any other producing country in other regions). Comparatively to high-rent GCC countries, the damage deriving from quickly drawing their numbers down would thus be higher. The fiscal equilibrium in these countries is also much more fragile, necessitating, as discussed here, the development of a domestic taxable base which expatriates cannot offer.

Elbadawi and Makdisi point to the importance of productivity:

*growth has been largely driven by factor accumulation in the form of massive investments and even more through large scale utilization of unskilled and semi-skilled labor, mainly drawn from the Indian sub-continent. This growth model has obviously served the GCC well and allowed the building of state-of-the-art infrastructure at exceptionally low cost of labor. But one major draw-back of the GCC growth strategy has been low and even declining total factor as well as labor productivity.*

The decline in productivity casts a shadow on the possibility of raising more revenue from domestic sources.

The open door to essentially unlimited importation of expatriate labor bids wages down and discourages nationals from active participation in the labor force.

*Huge gains could be made if they were instead encouraged to do so, because Saudi nationals are both grossly underemployed and increasingly well educated, thus increasing the opportunity cost of low participation. To give a sense of the potential gains if national labor was employed more effectively, a simple projection model suggests that with participation rates growing from 40 to 60 percent of the working-age population, and unemployment dropping to its natural rate, non-oil national income would more than double if the additional workers join the non-oil sector at current productivity levels. Improvements in labor productivity would add to this growth rate further. Altogether, it can be estimated that this addition to national wealth is comparable in magnitude to the kingdom's current oil wealth.* (Diwan)

Underemployment of nationals is a cause of unsustainability. For Diwan, the issue is closely connected to the prospects for raising domestic taxation:

*A path of labor intensive growth leveraged on expat labor can allow for larger taxation of corporate profits, but if it does not manage to increase employment among Saudis, it would require more social spending to preserve social peace, and thus, would not be conducive to correcting the macro-imbalances. On the other hand, growth based on the expansion of Saudi labor would lead to a broader tax base over time*

Already now, the GCC is confronted with high unemployment. In earlier work, Tom Moerenhout has pointed to increasing youth unemployment as a potential threat to national security, adding even more pressure when considering a poor female labor participation rate. Currently, most GCC countries have around 40% youth unemployment for 20–24-year-olds, declining to a still sizeable 15% for 30–34-year-olds. As Martin Hvidt highlights, approximately 500,000 nationals will enter working age each year during the next decade and a half in Gulf countries, exacerbating what is often labeled as a "ticking time bomb". Hence, political stability is directly challenged by the inability to mobilize the domestic workforce and increase its productivity level.

The issue is not simply one of substituting expatriates with nationals. Drawing down the number of expatriates has not so far succeeded in creating an equivalent number of jobs for nationals. Mobilization of the domestic work force can only be achieved through investment to change the capital per worker in the private sector:

> *The good news is that the education level among young Saudis has risen, which makes it possible for them to occupy jobs at relatively high productivity levels. The bad news, however, is that existing incentives have pushed firms in the private sector to create jobs that require either very low skills (especially in services) or very high skills (especially in the energy sector). To create jobs that are attractive for Saudis, who have mostly mid-level skills, there is therefore a need for structural change in the production methods used by the private sector, in response to changes in factor endowments and prices.* (Diwan)

## 5 Political Sustainability

> *At this juncture, perhaps the most compelling question to ask with regard to development sustainability of the GCC countries would be: in which direction might the political equilibrium in these societies shift as a consequence of the new global oil order?* (Elbadawi & Makdisi)

That is indeed an essential dimension of sustainability, because one cannot expect the needed economic policy adjustments to take place without parallel adjustments in political equilibria. This is especially true of the low-rent per capita countries, where greater reliance on a private sector less dependent on government expenditure and capable of walking on its own legs is a prerequisite of sustainability. In addition, the need to increasingly rely on taxation of domestic value added inevitably raises the issue of accountability and legitimacy of power structures. In other words, the low-rent per capita countries will need to gradually evolve from being pure rentier states to being states whose existence is based on the ability to extract a surplus from the domestic economy, which is a key determinant of demand for democratic participation, "No taxation without representation", also in the Persian Gulf. In earlier work, Luciani noted that in the late 1990s, when these states were fiscally challenged after 15 years of low oil prices and growing deficits, it seemed that a national bourgeoisie might emerge and become a significant political actor. Timid yet promising attempts were made in the direction of an enlargement of the political sphere, but following the Arab Spring, all hatches were closed down and authoritarianism has toughened.

Elbadawi and Makdisi analyze the situation using the concepts of limited versus open access order:

> *the prevailing GCC politico-economic institutional set up might be described as a relatively advanced form of 'limited access order' (LAO), where projection of organized violence is firmly under the incumbent authorities and rents distribution is governed by inter-personal networks, but at the same time adjudication and settlement of disputes among the elites become increasingly formalized through courts and other formal arrangements. It follows that the prospects of significantly lower oil prices under the new global oil order, and the ensuing reductions in rents per capita, might very well push the GCC political equilibrium from an advanced LAO to a fragile state of LAO, akin to the one that prevails in some of the populous Arab oil countries. Such unfortunate transition would surely put an end to the GCC's relatively successful development model. On the other hand, it is possible that the path dependence of the political process in some at least of the GCC member countries might create some positive inertia, leading them to develop into an impersonal 'open access order' (OAO), whereby the society graduates to a stable institutional equilibrium.*

Unfortunately, all evidence is in the direction of "fragilization" of LAO rather than movement toward OAO. This is all the more so since at the regional level, conflicts and tensions have been increasing systematically, and the GCC countries have implicated themselves deeper and deeper into such conflicts and proxy wars. With the row between Saudi Arabia and the UAE on one side and Qatar on the opposite side, the very significance of the GCC is also questioned. As Luciani has noted in an earlier work, regional disaggregation is a problem for sustainability, because it hinders the emergence of economic activities that are regional in nature and damages the development prospects of all countries in MENA.

As far as the social contract goes, Moerenhout has argued in an earlier work that the effects from the fragilization of LAO and the discontent with austerity measures can potentially be countered by the rapid development of more targeted social safety nets. As cash transfers can keep rentier state dynamics intact in the short term, they appear as the best option to compensate for loss of household welfare. However, they are not a panacea to resolve the structural deficiencies of GCC states. The success of cash transfers will ultimately depend on their design and implementation, as well as on the introduction of complementary short-term mitigation measures and longer-term adjustments in, among others, the labor market.

## 6 Can the GCC Economies Be Sustainable?

The discussion of sustainability that is presented in this book may not offer a definitive definition of the term, but nevertheless converges on some useful policy indications.

Sustainability does not necessarily mean diversification, and especially not diversification away from hydrocarbons. Rather, the comparative advantage in hydrocarbon production must be leveraged to integrate upstream and downstream in the value chain, while at the same time defending the acceptability of hydrocarbons through the decarbonization of their production, development of non-fuel uses and carbon capture and sequestration. At the same time, increasing reliance on renewable and other non-carbon energy sources must be pursued, as well as greater energy efficiency.

Diversification contributes to sustainability if it is accompanied by the emergence of a private sector less dependent on government expenditure and capable of offering better, more productive jobs to the national workforce. Reliance on expatriate labor must be reduced and taxes progressively introduced to pay for current public sector expenditure.

For the high-rent per capita countries, a scenario of growing revenue from international financial placements (sovereign wealth funds) complementing declining oil revenue cannot be excluded. The success of such a strategy, based on consolidation of, rather than transition from, the rentier nature of the state, cannot a priori be excluded. The outcome very much depends on the ability to invest wisely and attain a sufficiently high rate of return—not something that is guaranteed at this stage. It also very likely depends on a process of demographic retrenching, shedding a large portion of the expatriate population and raising the ratio of nationals to expatriates in the resident population. It is not inconceivable that these states may survive as patrimonial monarchies.

For low-rent per capita countries, including Bahrain, Oman and Saudi Arabia, a similar strategy is not possible. To the extent that the state will attempt to preserve its rentier nature through the accumulation of a large sovereign wealth fund (a possibility which in practice may be open to Saudi Arabia only), it will need to increasingly erode the benefits extended to the national population. In the absence of improved employment opportunities in the private sector, hidden or overt unemployment is already a growing problem, and discontent will inevitably ensue. Sustainability will be threatened not so much by immediate fiscal and/or trade imbalances, but by the inability to adapt institutions to the requirements of a sustainable economy. The status quo may be preserved through repression, possibly even for a long time, but the sustainability challenge will have been lost.

The manufacturer's authorised representative in the EU is Springer Nature Customer Service Centre GmbH, Europaplatz 3, 69115 Heidelberg, Germany. If you have any concerns regarding our products, please contact ProductSafety@springernature.com

Printed and bound by CPI Group (UK) Ltd, Croydon, CR0 4YY
30/09/2026
02225246-0004